SUNY Series in Science, Technology, and Society

Sal Restivo, Editor

# In Measure, Number, and Weight

## Studies in Mathematics and Culture

## JENS HØYRUP

STATE UNIVERSITY OF NEW YORK PRESS

Published by
State University of New York Press, Albany

© 1994 State University of New York

For information, address State University of New York Press,
State University Plaza, Albany, N.Y. 12246

Production by M. R. Mulholland
Marketing by Fran Keneston

**Library of Congress Cataloging-in-Publication Data**

Høyrup, Jens.
 In measure, number, and weight : studies in mathematics and
culture / Jens Høyrup.
    p.    cm. — (SUNY series in science, technology, and society)
 Includes bibliographical references and index.
  ISBN 0-7914-1821-9 (acid-free). — ISBN 0-7914-1822-7 (pbk. : acid
-free)
  1. Mathematics—Social aspects.  2. Mathematics—History.
 I. Title.  II. Series.
QA10.7.H69   1994                                              93-835
510'.9—dc20                                                        CIP

10 9 8 7 6 5 4 3 2 1

To Sonja
And in memory of
Hans and Ludovica

# CONTENTS

# PREFACE

"Mathematics and culture" is a phrase that may be interpreted in numerous ways. A bit more precision is achieved when the concept of "culture" that is involved is specified to be the one which is current among anthropologists.

In order to characterize my approach I have therefore often spoken of the "anthropology of mathematics." Even this phrase does not correspond to any discipline or generally known field of interest, nor will it probably ever do so. Then what do I mean by it?

Twelve years ago, when using the term for the first time, I explained (Høyrup 1980: 9) that I did so

> because of dissatisfaction with the alternatives. History and social history of mathematics both tend as ideal types to concentrate on the historically particular, and to take one or the other view (or an eclectic combination) in the internal-external debate when questions of historical causality turn up. "Historical sociology" would point to the same neglect of cognitive substance as present in the sociology of science. "Sociology of mathematical knowledge" would suggest both neglect of the historically particular and a relativistic approach to the nature of mathematical knowledge, which may be stimulating as a provocation but which I find simplistic and erroneous as it stands.[1]
>
> What I looked for was a term which suggested neither crushing of the socially and historically particular nor the oblivion of the search for possible more general structures: a term which neither implied that the history of mathematics was nothing but the gradual but unilinear discovery of ever-existing Platonic truths nor (which should perhaps be more emphasized in view of prevailing tendencies) a random walk between an infinity of possible systems of belief. A term, finally, which involved the importance of cross-cultural comparison.
>
> The latter term suggested social anthropology, a discipline whose cognitive structure also seemed to fulfil the other requirements mentioned [ . . . ].

This approach is one which makes me feel like a sociologist among historians, a cross-breed between a philosopher and a historian among sociologists, and a bastard historian among philosophers. It makes the actual content of mathematics, in particular the changing mode of mathematical thought, stand out as crucially important, both for the functions mathematics can fulfill, and for the way the pursuit and development of mathematics is conditioned by the wider social and cultural context. Reduced to essentials, (my brand of) the anthropology of mathematics is thus an approach to the history of mathematics that, first, rejects the distinction between "internalism" and "externalism"; second, even when investigating the contributions of individuals, sees these as members of a particular culture, or rather as members of one or perhaps several intersecting subgroups within a general cultural matrix; third, tends to use the evidence which can be found in the production of individuals as anthropologists use the testimonies furnished by their informants.

The question of externalism versus internalism earns further discussion. This, however, presupposes some preliminary considerations on the concept of causation.

General quasi-philosophical lore distinguishes two concepts: The "Humean" cause and the "Aristotelian" cause, of which the Humean cause is said to correspond to the "efficient" Aristotelian cause and to make the remaining Aristotelian causes superfluous.

This is wrong already for the reason that Hume (*Enquiries Concerning Human Understanding* VII,ii:59) considers causation to be simply *an expectation on the part of the observer* produced by habit. Leaving this finesse aside, and identifying Hume with the "Hume" of quasi-philosophical folklore, being hit by billiard ball $A$ is the Humean reason that ball $B$ starts to roll.

To this an Aristotelian will object that there are many answers to the question why $B$ moves as it does. Being hit is evidently one; but if the balls had consisted of soft clay the outcome would have been different; so it would if $A$ and $B$ had not been spherical, or if $B$ had been located at the very edge of a table not provided with a cushion. A complete answer to the question *why* will thus involve efficient causes (the hitting); material causes (ivory, not clay; the surface of the cloth); and formal causes (the laws of semielastic impact and of sliding/rolling, as well as the geometrical forms involved). It will also have to mention that somebody plays billiards and wants $B$ to move (perhaps as it does, perhaps otherwise), ultimately wanting to win the game and to gain the stake; both of these are final causes.

Thus causations are manifold, as Aristotle remarked, even though they can be grouped in classes according to their character (*Physica* 195$^a$28). Only the folklore and the Aristotelian textbook tradition speak in the singular about *the* efficient, *the* material, *the* formal, and *the* final cause.

Evidently, the rigid framework of precisely four causes is of scant value if we want to explain historical processes. Even more irrelevant, at least when we discuss the history of culture, ideas, or the sciences, is the single "Humean" cause. It is the search for such single, instantaneous, and thus efficient causes which produces stories like Newton's falling apple. Even if the anecdote had been true and the observation in the garden had indeed been the efficient cause of—i.e., the occasion for—Newton's formulation of the law of gravitation, this answer would only be of interest for the psychology of scientific creativity (and then only within the framework of Newton's total mental make-up); different questions are asked in the history of ideas and of the sciences—questions that have much more to do with the remaining Aristotelian headings.

This brings us back to the problem of internalism and externalism. From time to time we are told that internalist explanations are real explanations because they alone are ("Humean") causal; the socio-cultural context in which scientific development takes place is of course as necessary as the soil is necessary for the forest (cf. Whitehead 1926: 23), yet context and soil are "incidents," not *causes*. This justification, however, is untenable; the results of earlier science and the questions raised by these results are just as much background to the actual events in scientists' lives as the sociocultural context. The (still irrelevant) efficient causes of their doings are just as likely to be found outside the framework of scientific ideas and results (say, in their job situation) as inside (the reading of a particular book or a discussion with a colleague at a particular moment).

A better justification is provided by a "local separation of variables." Looking at what goes on during a particular epoch one may quite legitimately take the actual institutional and cultural framework within which scientists move for granted and as relatively constant, and look at how scientists react to and continue the scientific tradition that they encounter—or, just as legitimately, one may take the level and character of the science of the time as a given and look at how it is able to respond to social needs and how it is stimulated or hampered by sociocultural circumstances, pursuing thus the externalist road.

Valid as this justification is, it also shows the restricted validity of both the internalist and the externalist approach as only first approximations. As soon as the development over longer periods or the comparison between different cultures or epochs are undertaken, the character and substance of scientific thinking and the aims pursued by the sciences, as well as institutions, ideologies, and general social needs, change. This will not prevent the historian from making (valid) internally or externally oriented *descriptions* of events, or some eclectic mixture—but it deprives such descriptions of explanatory capability, as long as no dialectical synthesis takes place.

The practical necessities of exposition prevent most of us from honoring such ideal claims in full. In the following essays, accordingly, the content side of mathematics is mostly dealt with in rather general terms (concentrating on cognitive organization and mode of thought), and emphasis is on its interaction with the sociocultural setting. Studies where I investigate *the mathematics* of various cultures have been omitted from the present collection and only appear in the footnotes. The collection as it presents itself thus verges more toward externalism than it should ideally do.

The collection contains in total eight essays, the individual publishing histories of which are told in the corresponding introductions. Each essay starts from a definite perspective or a set of specific, acknowledged questions. It is, indeed, impossible to ask about everything at one time, and the claim that one makes (for example) history *simpliciter* is at best naive. However, the attempt to answer questions asked from one perspective will by necessity raise other questions and thereby introduce new perspectives: thus also here. The perspectives of essays written at a later date may therefore be complements, at times perhaps correctives, to others written before. In this way, I hope, a more comprehensive picture of the "anthropology" of mathematics will emerge from the totality than the one that presents itself in its single constituent parts.

The ordering of the essays is thematic, and does not correspond to the order in which they were written, nor to the dates of publication. A first cluster consists of essays that are primarily sociological or anthropological in orientation (Chapters 1–4). Chapter 1 compares features of Sumero-Babylonian, ancient Greek, and Latin medieval mathematics, whereas Chapter 2 is an attempt to trace the specific character and the history of pre-Modern practitioners' mathematics, which I characterize by the term *subscientific*. Chapter 3 investigates the interplay between state formation processes, scribal culture, and mathematics in ancient Mesopotamia, and Chapter 4 explores the specific character of Islamic mathematics and the sociocultural roots of its particular accomplishments: an unprecedented synthesis between mathematical theory and practice. Together, the four essays may be read collectively as an attempt to delineate some of the main facets of early "Western" mathematics.[2]

Chapters 5–7 probe the impact of ideologies, ideas, and philosophy on mathematics from the Latin High Middle Ages through the late Renaissance. Chapter 5 does so broadly, asking in particular to what extent formal philosophies and quasi-philosophical attitudes influenced the aims and ideals pursued by "mathematicians" from the early twelfth through the late sixteenth century. Chapter 6 studies the thirteenth-century mathematical author Jordanus de Nemore through his works (nothing is known directly about the person), tracing how he oriented himself with regard to the contradictory currents and attitudes that surrounded him. Chapter 7 scrutinizes the received persuasion that Platonism was a decisive motive force in the development of

Renaissance mathematics, suggesting the alternative thesis that the dominant ideology of humanist mathematics can be characterized as "Archimedism" and trying to trace the changing form and the impact of this ideology.

All these essays deal with the anthropology of pre-Modern mathematics. Chapter 8, written in collaboration with my friend and colleague Bernhelm Booß, focuses on the character and setting of Modern and contemporary mathematics. It does so through the perspective defined by the impact of militarization and warfare on mathematics. It may justly be argued that this perspective is not only limited but also narrowing and even distorting. We have exerted ourselves, however, not to use the distorting mirror for the purpose of caricature but in order to make visible and understandable features of contemporary mathematics that tend to be neglected; and, also important, to see which features are not deformed even by this disfiguring strain, and why.

Across these divisions, four recurrent themes (beyond the rejection of the internalist/externalist dichotomy) run through the book. I shall list them without arguing here for their pertinence—it is sufficiently done in what follows (so I hope), inasmuch as at all necessary.

One theme can be characterized as *desacralization without denigration.* The alternative to seeing mathematics as an ever-existing Platonic truth toward which mathematicians of all epochs strive when not hampered by obstructive forces need not be total relativism.

Another theme is that *actors participate in institutions,* not only being shaped by these but also shaping them. Thus, institutions mediate the influence of general sociocultural forces on actors—but actors, to the extent that institutions are not totally rigid, also contribute to the shaping of these in interaction with the general sociocultural forces to which they are also submitted through other channels.

The third theme is the dialectic between tradition and actual situation. In mathematics, as in other branches of culture, what is done in one generation presupposes what has been done before; but it presupposes it *through the form in which it is actually known and understood.* The tradition is always understood—which by necessity means *mis*understood—through concepts and mental habits formed through an actual practice; but coming to grips with the tradition is in itself an essential part of the practice of (*in casu*) the mathematician, thus contributing to the formation of concepts and habits.

The fourth theme explains the title of the book: *The relation between "high" and "low" knowledge. Measuring, counting* and *weighing* are indeed the (most practical and hence "low") starting points for mathematics. The phrase *in measure, number, and weight,* however, is borrowed from Wisd. 11:21, where it describes the principle of the Lord's Creation (quoted by almost every Christian author between Augustine and Pascal writing about the importance of mathematics). This transformation of "low" into "high" knowledge is a constant characteristic of pre-Modern mathematics, whereas the reverse process, after modest beginnings in ancient Alexandria and rise to equal prominence during the Islamic Middle Ages, became the cardinal ideology of utilitarian mathematicians from the late sixteenth century onward while remaining in actual reality only one facet of a twin movement.

With one exception, the single essays carry dedications, which are those of the original publications. Some of them are of private but most of public character. Insofar as I have considered them to be of public interest yet not self-explanatory, the dedications are explained in the introductions of the individual parts.

The book as a whole I dedicate to my mother, and to the memory of my father. This I could do for many reasons—but the one I will mention on the actual occasion is that rich stimulation of my intellectual curiosity which I received from them.

I also dedicate it to the memory of my beloved wife Ludovica, who was so eager to have this book published, and without whose enthusiasm and tender support I might never have completed it.

# ACKNOWLEDGMENTS

The individual essays were originally published or circulated as follows:

1. "Varieties of Mathematical Discourse in Pre-Modern Sociocultural Contexts: Mesopotamia, Greece, and the Latin Middle Ages." *Science & Society* 49 (1985), 4–41. © 1985 *Science & Society*.
2. "Sub-Scientific Mathematics. Observations on a Pre-Modern Phenomenon." *History of Science* 28 (1990), 63–86. © 1990 Science History Publications.
3. "Mathematics and Early State Formation, or, The Janus Face of Early Mesopotamian Mathematics: Bureaucratic Tool and Expression of Scribal Professional Autonomy." Revised contribution to the symposium "Mathematics and the State," 18th International Congress of History of Science, Hamburg/Munich, 1–9 August 1989. *Filosofi og videnskabsteori på Roskilde Universitetscenter*. 3. Række: *Preprints og Reprints* 1991 nr. 2.
4. "The Formation of "Islamic Mathematics": Sources and Conditions." *Science in Context* 1 (1987), 281–329. © 1987 Cambridge University Press.
5. "Philosophy: Accident, Epiphenomenon, or Contributory Cause of the Changing Trends of Mathematics. A Sketch of the Development from the Twelfth Through the Sixteenth Century." *Filosofi og Videnskabsteori på Roskilde Universitetscenter*. 1. Række: *Enkeltpublikationer* 1987 Nr. 1. A Croatian translation has appeared as "Filozofija: Slučaj, epifenomen ili sinergijski uzrok promjene trendova u matematici. Obriz razvitka od dvanaestoga do šesnaestoga stoljeća." *Godišnjam za povijest filozofije* 5 (Zabreb 1987), 210–74.
6. "Jordanus de Nemore: A Case Study on 13th-Century Mathematical Innovation and Failure in Cultural Context." *Philosophica* 42 (Ghent, 1988), 43–77. © 1988 *Philosophica*.
7. A somewhat different version has appeared as "Archimedism, not Platonism: On a Malleable Ideology of Renaissance Mathematicians (1400 to 1600), and on Its Role in the Formation of Seventeenth-Century Philosophies of Science," in C. Dollo (ed.), *Archimede: Mito Tradizione Scienza*. (Biblioteca di Nuncius. Studi e testi, IV). Firenze: Olschki, 1992. The present text was prepared for the symposium "Renaissance-

Philosophie und neuzeitliches Denken III: Die klassischen Interpretationen im Lichte der modernen Forschung,'' Dubrovnik, 1990.

8.  An earlier version appeared in German as: Bernhelm Booß & Jens Høyrup, *Von Mathematik und Krieg. Über die Bedeutung von Rüstung und militärischen Anforderungen für die Entwicklung der Mathematik in Geschichte und Gegenwart.* (Schriftenreihe Wissenschaft und Frieden, Nr. 1). Marburg: Bund demokratischer Wissenschaftler, 1984. A draft of the present text was prepared by Jens Høyrup for the Sixth International Congress on Mathematical Education, Fifth Special Day on Mathematics, Education, and Society, Budapest, July 31, 1988.

It is a pleasure to express my gratitude for the permission to republish the papers in the present context.

The form in which the essays appear herein is in *grosso modo* unchanged (with the reservation concerning 7 and 8, which was already stated). The reference system, however, has been harmonized and unified, and in part updated (in particular concerning autoreferences to preprint versions that have been published in the meantime), and cross-references between the essays have been inserted. Some linguistic filtration has also taken place. The subject matter, on the other hand, has not been tinkered with.

References are mostly made according to the author-date system. A few standard abbreviations are used, and articles from encyclopedias (in particular *DSB—Dictionary of Scientific Biography*) are referred to by author and title; further identification is given in the bibliography. Many primary sources are also referred to by author (if known) and title; in such cases, the bibliography contains a cross-reference to the edition I have used. Other primary sources are directly referred to by editor/translator-date.

All translations from original languages where nothing else is indicated are mine. An indication such as ''Russian trans. Krasnova 1966'' means that I have translated from the Russian translation in question, whereas ''trans. Colebrooke 1817'' indicates that I quote Colebrooke's English translation.

# Varieties of Mathematical Discourse in Pre-Modern Sociocultural Contexts: Mesopotamia, Greece, and the Latin Middle Ages

*An homage to Dirk Struik on the beginning of his tenth decade*

## Introduction

*This first essay is an elaborated and revised version of a paper presented to a seminar, "Mathematics as a Science and as a School Subject," arranged by the Greek Mathematical Society, Branch of Thessaloniki, in April 1983. The main body of participants were Greek secondary school mathematics teachers, and, as will be seen, this audience has determined both the rhetorical surface structure and the kind of conclusions—viz., on the obligations and responsibility of mathematics education—that are drawn in Section 4.*

*The article as it stands is close to the version that was published by* Science & Society *in 1985. There as here, I preferred to stick to the form of the original oral exposition, letting the text present the main argument and illustrations, and hiding away documentation, conceptual clarifications, and qualifying remarks in the notes. A few points in the text have been modified, mostly for stylistic reasons or in order to avoid misunderstandings, and a few references have been added. Only in one place (note 36, the final paragraph) has genuinely new information been inserted.*

*The primary aim of the essay is to investigate how the character of mathematical thinking depends on the institutional situation in which mathematics is practiced as knowledge—perhaps as theory, perhaps as techniques one should know in order to apply them—in interplay with the wider cultural settings and societal determinants of institutions. The method is cross-culturally and cross-historically comparative, but no effort is made to find the same parameters in all cases, apart from the choice of teaching as a critical factor and institution and from the general focus on mathematics. Nor do I, indeed, believe that a schematization aiming at finding a rigid common grid of explanatory factors makes much sense in cultures as widely divergent as those dealt with here, in general character and hence also in institutional make-up. As will ap-*

*pear, even the two key notions—*teaching *and* mathematics—*correspond to widely divergent phenomena, not only if we compare Mesopotamia, ancient Greece, and the Latin Middle Ages, but also when the historical development within one of these contexts is considered.*

*Cross-cultural comparison within limited space by necessity entails a rather coarse treatment of the single cultures. This also holds for the present article, as well as for the following one. The present essay, in addition, was written before the other items of the collection, and in some sense epitomizes the research program within which these others items belong. More fine-grained pictures will thus be found in Chapters 3 to 7.*

*A secondary aim of the essay is to draw explicit attention to possible consequences of the analysis that are pertinent to present social practice. Even though present concerns have been a recurrent motive for the choice of theme and approach for all articles in the collection, I have mostly left this kind of conclusion to the reader.*

*The dedication of the essay to Dirk Struik is due, it hardly needs to be said, to his role as a pioneer precisely in this integration of historical sociology of mathematics with contemporary political engagement and open-minded Marxist thought.*

As Hellenes, you will know the Platonic view that mathematical truths *exist,* and that they are eternal, unchanging, and divine.

As mathematicians you will, however, also know that mathematics is not adequately described as a collection of unconnected "truths"; the whole point in the activity of the mathematician, of the mathematics teacher, and of the applied mathematician, for that matter, is the possibility to establish *connections* inside the realm of mathematics—between one theorem and another; between problems and procedures; between theorems, procedures, and sets of axioms; between one set of axioms and another set, etc.—connections that in some sense (which I am not going to discuss here) map real connections of the material or the human world.[1]

You will also know that such connections are established by means of proofs, demonstrations, *arguments.* Mathematics is a *reasoned discourse.*[2] Further, you will probably agree that the eagerness of the ancients, not least Plato and Aristotle, to distinguish scientific argument from arguments concerned with mere opinion—the arguments of the rhetor and the sophist—undermines its own purpose: What creates the need for such eager distinctions, if not the close similarity between the two sorts of arguments? On the other hand, you will also concede that *mathematical* discourse is often organized in agreement with the Aristotelian description of scientific argumentation, as argument from indisputable premises.

As teachers, finally, you will know that an argument—be it a mathematical argument—is no transcendental entity existing from before the beginnings of time. It is a human creation, building on presuppositions that in the particular historical (or pedagogical) context are taken for granted, but which on the other hand cannot be taken over unexamined from one historical (or pedagogical) situation to another. What was a good argument in the scientific environment of Euclid was no longer so to Hilbert; and what was nothing but heuristics to Archimedes became good and sufficient reasoning in the mathematics of infinitesimals of the seventeenth and eighteenth centuries—only to be relegated again to the status of heuristics in the mid-nineteenth century (cf. Grabiner 1974).

So, one aspect of mathematics as an activity—other aspects I shall take for granted—is to be a reasoned discourse. The corresponding aspect of mathematics as an organized body of knowledge is to be the *product of communication by argument,* i.e., communication where a "sender" convinces by means of arguments a "receiver" that some statement or set of statements is true; in many cases the "receiver" is of course only a *hypothetical* average, professional interlocutor of the mathematician, defined through the sort of arguments that are thought adequate (the "model reader" of semiotic literary theory; cf. Eco 1979: 50–66). Normally, either the truths communicated, or at least some broad base for the communication, is supposed to be fixed in advance: The discourse is not fully open. In principle this sort of communication can be described as *reasoned teaching,* the concept taken as a general philosophical category. So, teaching is not only the *vehicle* by which mathematical knowledge and skill is transmitted from one generation to the next; it belongs to the *essential characteristics* of mathematics to be *constituted through teaching* in this broad philosophical sense.

However, teaching, and even teaching regarded philosophically, belongs no more in the eternal Platonic heavens than do mathematical truths and arguments. Quite the contrary. Teaching is an eminently social activity, depending on context, personal and group characteristics of the persons involved, social norms, purpose of the teaching, material, cultural and linguistic conditions and means, and the like. And so, if mathematics is constituted through being taught, we must expect it to be very much molded by the *particular* teaching through which it comes about.

It will be my aim in the following to trace that molding in the restricted sense of *institutionalized teaching.* That is, I shall eschew the airy philosophical definition of reasoned teaching where the sender convinces the receiver by means of arguments organized inside a closed or semi-closed discourse; instead, I shall concentrate on teaching as something involving a teacher and his or her students, ordered according to some fixed social and societal pattern, and normally taking place in a more or less formalized school. I shall stick to the pre-Renaissance (and so, pre-Gutenberg) era, and through three extended key episodes I shall try to trace the relations between the development of mathematics, the character of mathematical discourse, and the institutional setting of mathematical teaching (mainly the teaching of adults).

## 1. Mesopotamia: Scribal Computation, and Scribal School Mathematics

The first of my extended episodes is the development of Sumerian and Babylonian mathematics, in a process of several phases covering some 1500 years. Parallel with the rise of the early proto-Sumerian city-states in the late fourth millennium B.C., a first unification of a variety of proto-mathematical techniques (for primitive accounting, practical geometry, and measurement) into a single coherent system (*mathematics,* in our terminology), united by the common application of numeration and arithmetic, appears to have taken place. The social environment of this unification was that of the Temple corporation—indeed, the institution that molded society into a *state* was, according to all available evidence, the Temple.[3] It seems, however, that *mathematics* did not grow out of the mere environment of administrators of temple property and taxation: various types of evidence suggest that the unification and coherence did not really correspond to practical administrative needs, which could have been provided for by isolated extensions of the existing separate techniques. Instead, like the development of more genuine writing, mathematical unification and coherence seem to be products of the school where future officials were trained, and where the techniques that they were going to apply were also developed.

The sources for this are few and scattered, and what I have just told is a reconstruction arising from the combination of many isolated pieces of evidence.[4] As times go on, however, the picture becomes clearer. Toward the end of the third millennium B.C. southern Iraq had been united into a single, centralized royal state (the ''neo-Sumerian Empire,'' or Ur III), where the Palace directed large parts of the total economy through a vast bureaucracy. This bureaucracy was carried by a body of *scribes* who, at least since the mid-third millennium, had emerged as a specialized profession, and who had long since been taught in specialized schools. By the end of the third millennium, when sources describing the curriculum of the scribe school turn up, it is clearly dominated by applied mathematics (see Sjöberg 1976: 173). At the same time, the professional ideology of the scribal craft (as inculcated in the scribal school) becomes visible in the sources: The scribe is, or is at least expected to be, proud of his service to the royal state, which is presumed to serve general affluence and justice.

The centralized neo-Sumerian state had a short life. After only a hundred years it crumbled, not least under the weight of its own bureaucratic structure. Still, it created two very important innovations in mathematics, of which one has served since then.

The first innovation is the introduction of *systematic accounting,* on occasion with built-in controls.[5] Such accounting systems were created or at least very suddenly spread in the administration of the whole empire[6] in a way that has rarely been equaled in history (the spread of double-entry book-keeping, an analogous process in the Renaissance, took around 300 years). The second innovation (the one that was to survive) was the introduction of the *place-value system* for both integers and fractions—not with base 10 as in our "Hindu" system, but with base 60 (as we still use it in the subdivisions of the hour and the angular degree taken over via Hellenistic astronomy). Even this system appears to have spread quickly. It was not used in official documents but for the intermediate calculations of the scribes.

Both innovations built on foundations that had been laid during centuries of scribal school activity. Regarded from that point of view, they may be said to represent mere continuity. Still, mere continuity does not guarantee theoretical progress, and so both the occasions on which progress occurs and the precise character of the progress occurring should be noticed—in this case, the ability of the scribal school to respond to the demand created by a royal administrative reform, introducing well-functioning and sophisticated tools of applied mathematics. Another fact to be noticed is the seemingly total absence of mathematical extrapolation beyond the range of applications, in striking contrast to the tendencies of the following period. The scribal school and the scribal profession appear to have been so identified with their service function to the state that no interest in mathematics *as such* grew out of a curriculum of applied mathematics, in spite of clearly demonstrated mathematical abilities. The objective requirements and regularities of mathematical structures had created a tendency toward coherence (perhaps because of the way they manifested themselves when mathematics had to be made comprehensible in teaching?); but mathematical discourse and practice seem to have remained a non-autonomous, integrated part of administrative discourse and practice until the very end of the third millennium.

This was to change during the following, so-called "Old Babylonian," period, c. 1900 B.C. to c. 1600 B.C. As already mentioned, the centralized neo-Sumerian state crumbled under the weight of its top-heavy bureaucracy, and it was brushed aside by barbarian invasions. When organized states and civilized life had stabilized themselves once more, a new economic, social, and ideological structure had appeared. Large-scale latifundia had been replaced by small-plot agriculture, mostly held by tenants; the royal traders had become independent merchants; and the royal workshops had been replaced by private handicraft.[7]

This socioeconomic change was mirrored by social habits and ideology. In the neo-Sumerian Empire, only official letter-writing had existed; now, private and even personal letters appeared. The seal, once mainly a prerogative of the royal official, turned up as the private mark of the citizen. The gods, with whom one had once communicated through the temples of the royal state, now took on a supplementary role as private tutelary gods. And, of course, the street scribe required to write personal letters on dictation appeared, together with the freelance priest performing private religious rites. All in all, the human being was no longer bound to the role of subject in the state (which he once took on when loosing his roots in the primitive community); he was *also a private man.* Most strikingly, perhaps, this is seen in the case of the king. He is still the State in person, but *at the same time* he is a private person, with his private tutelary god, who may differ from the tutelary god of the State.[8]

Through the scribal profession and the scribal school, all this is reflected in the development of mathematics—such is at least my interpretation of the great changes in mathematical discourse and practice. The scribes remained scribes, i.e., they continued to fill their old managerial and engineering roles; the school went on to teach them the accounting, surveying, and engineering techniques needed for that; those activities and the school still imbued them with pride of serving as scribes for the royal State and the king, and so, supposedly, for general affluence and justice.[9] But the professional ideology of the scribes, as revealed in school texts, is not satisfied by mere usefulness. Even the scribal role has become a *private* identity. The scribe is proud of his scribal *ability* rather than of the functions that his ability permits him to carry out; and he is proud of a *virtuosity* going far beyond such abilities as would be functionally useful. A *real* scribe is not one who is able to read and write the current Akkadian tongue; these are presuppositions not worth mentioning. No, a scribe's cunning is demonstrated only when he is able to read, write, and even speak the dead Sumerian language; when he is familiar with the argot of various crafts; and when besides the normal meanings of the cuneiform signs (a task in itself, since many signs carry both one or several phonetic meanings and one or several ideographic interpretations) he knows their occult significations. All these qualities, this virtuosity, had its own name: The scribes spoke of their special *humanity.* So, a human being *par excellence,* i.e., a *real scribe,* is one who goes beyond vulgar service, one whose virtuosity defies normal understanding. He is, however, still one whose virtuosity lies within the field defined by the scribal functions; the scribe can only be proud of *scribal* virtuosity.[10]

You may already have thought that mathematics must be an excellent tool for anybody needing to display special abilities beyond common understanding. And indeed, the so-called "examination texts," which permit us to decipher the scribal ideology, align mathematical techniques with occult writing and craftsmen's argot.[11]

There is, however, more than this to the connection between the rise of private man, scribal "humanity," and mathematics. Indeed, an interest in mathematics beyond the purely useful develops in the Old Babylonian period,[12] leading to large-scale development of techniques at best described in modern terms as "second- and higher-degree algebra."[13] This algebra becomes a dominating feature of Old Babylonian mathematics; second-degree problems are found by the hundreds in the cuneiform texts of the period. Many of them look like real-world problems at first; but as soon as you analyze the structure of known versus unknown quantities, the complete artificiality of the problems is revealed—for example, in the case where an unfinished siege ramp is presented, of which you are supposed to know the total amount of earth required for its construction, together with the length and height of the portion already built, but not the total length and height to be attained.[14] These "algebraic" problems can be understood—in my opinion, can *only* be understood—as being on a par with correct Sumerian pronunciation and familiarity with occult significations of cuneiform signs: scribal ability, true enough, but transposed into a region of abstract ability with no direct practical purpose. As scribal discourse in general, mathematical discourse had been disconnected from immediate practice; it had achieved a certain autonomy.

Thus, Old Babylonian "pure mathematics" must be understood as the product of a teaching institution that no longer restricted itself to teaching privileged subjects what they needed to perform their future function as officials—i.e., it can be understood as the product of an institution where teachers and students (especially perhaps the teachers[15]) were also persons with a private identity as scribes.

Another aspect of Old Babylonian "pure mathematics" can, however, only be understood if we see it as a product of a *scribal* school. This aspect is the fundamental difference between Old Babylonian and Greek "pure mathematics." Old Babylonian mathematics grew out of its *methods,* whereas Greek mathematics grew out of *problems,* to state things very briefly.

This may sound strange. Indeed, Babylonian mathematics is known only from problems; it does not contain a single theorem and hardly a description of a method.[16] That, however, follows from the training role of the Babylonian school, which did not aim at the theoretical understanding of methods but at the *training of methods*—first, of course, the training of methods to be used in practice, but next also of methods that would permit the solution of useless second-degree problems. Such training could only be obtained through drill. Indeed, the problems that occur appear to be meant exactly for drill; at least, many of them appear to have been chosen not because of any inherent interest but just because they *could be solved* by the methods at hand.[17]

Greek mathematics, on the other hand, for all the theorems it contains, grew out of problems for the solution of which methods often had to be created anew. We may think of the squaring of the circle, the trisection of the angle, and the Delic problem (doubling the cube). But we should not forget

the Eudoxean theory of proportions or Book X of the *Elements,* on the classification of irrationals and the algebraic relations between the resulting
classes.[18] They, too, are investigations of problems—problems arising from
and only given meaning through the development of mathematical theory.

The Greek effort to solve fundamental problems is clearly related to the
whole effort of Greek philosophy to create *theoretical understanding.* That
could never be the aim of a scribal school. There, skillful handling of methods was the central end. Problems were the necessary means to that end, and
in the Old Babylonian school they became the necessary pretext for the display of virtuosity.[19] Paganini, not Mozart, plays the violin.

*Mutatis mutandis,* we can speak of a Babylonian parallel to the effects
of the "publish-or-perish" pressure on contemporary publication patterns.
When submitted to such pressure, the easiest way out is to choose your problems according to their accessibility, given your ability and the methods with
which you are familiar; current complaints that the scientific literature is
drowned in publications treating problems of no other merit than that of treat-
ability indicate, even if great exaggeration is allowed for, that some researchers *have* found this way out.[20]

Apart from its determination by methods rather than problems, another
characteristic of Old Babylonian "pure mathematics" seems to derive from
its particular background: *viz.,* that you have to analyze the structure of its
problems in order to decide whether they are practical or artificial, i.e.,
"pure" (cf. the siege-ramp problem mentioned previously; superficially regarded, it looks like a typical piece of militarily applied mathematics). In
other words, even when Old Babylonian mathematics is "pure" *in substance,* it remains *applied in form.* In contradistinction to this, the prototype
of Greek mathematics is pure in form as well as in substance, to such an extent that the applications of geometry to astronomy are formulated abstractly
as dealing with spheres in general—so in Autolycos's *Moving Sphere* and
Thedosios's *Spherics.*[21] Greek mathematics had become pure in form, i.e.,
fully abstract, even when it was applied in substance.

This "applied form" of Old Babylonian mathematics could hardly
have been different; at least, it is in full harmony with other aspects of scribal
"humanism." The virtuosity to which a scribe could aspire had to look like
*scribal* virtuosity. The scribe, however, was an *applier of mathematics,* a *calculator;* there were no social sources, and no earlier traditions, from which a
concept of mathematics as a concept *per se* could spring, and there was thus
no possibility that a scribe could come to think of himself as a *virtuoso mathematician.* Only the option to become a virtuoso calculator was open; so,
Babylonian "pure mathematics" was in fact calculation pursued as *art pour
l'art,* mathematics applied in its form but disengaged from real application.
Expressed in other terms, Old Babylonian mathematical discourse had

achieved autonomy for its actual working; it remained, however, *defined* through scribal professional practice.

So much for Sumerian and Babylonian mathematics. The argument could be supported by telling the story of Babylonian mathematics after the disappearance of the individualized Old Babylonian social structure and of the scribal school as an institution—"pure" mathematics vanishes from the sources for more than a thousand years—or by comparing Egyptian and Babylonian developments. This, however, I shall bypass,[22] and I shall close my first episode by emphasizing that the overall social characteristics of the institution in which Mesopotamian mathematics was taught and developed influenced primarily the overall formal characteristics of Mesopotamian mathematics as a discourse and as a developing system, and normally *not* its factual contents. The Babylonian and the Greek would calculate the diagonal of a rectangle with sides 5 and 12 in the same numerical steps; so far, of course, mathematics *does* consist of socially and historically transcendental truths. But whereas the Greek mathematician would argue from a strict concept of the rectangle built upon a concept of quantifiable angles, the Old Babylonian scribe would only distinguish quadrangles with "right" angles from those with "wrong" angles.[23]

## 2. Greek Mathematics: From Open, Reasoned Discourse to Closed Axiomatics

My next episode will be the development of Greek mathematics. Whereas the role of argumentation in the Mesopotamian development follows mainly from indirect evidence supplemented by a few clay tablets where something like a didactical explanation occurs,[24] Greek mathematics is argument through and through.

Greek mathematics is thus a product of "reasoned teaching" in the general philosophical sense that I gave to this expression. It is more dubious whether it is a product of institutionalized teaching; it may be that the overwhelmingly argumentative character of Greek mathematics should be sought in its initial *lack* of school institutionalization.

At first, however, a very different hypothesis suggests itself. Much of Greek mathematics has been seen as a search for harmony and completeness—and do these ideals not look as if they were taken over from the *paideía eleúthera,* the liberal education of the free citizen as a harmonious and complete human being?[25] Furthermore, once institutionalized (from the fourth century B.C. onward) the *paideía* came to contain a fair measure of mathematics. Finally, Proclos tells us that Pythagoras gave mathematics the "character of a liberal education," *schêma paideías eleuthérou.*[26]

On closer investigation, however, there seems to be no causal connection leading from the *paideía* to mathematics. Apparently, we are confronted with two analogous but distinct structures, partly with a causal chain leading the opposite way, partly perhaps with a neohumanist optical illusion. Therefore, instead of analogies, I shall try to build my exposition on the actual evidence,[27] incomplete as it is.

I shall try to distinguish three periods, of which those of most critical interest for the investigation are unhappily the most difficult to document:

- The rise of reasoned mathematics, in the sixth and fifth centuries B.C.— "pre-Socratic mathematics."
- The creation of deductive and axiomatic mathematics, in the fourth and early third century—from Plato and Eudoxos to Euclid.
- Finally, the mature period from Euclid onward, in which the style and character of Greek mathematics was already fixed, and in which every Greek mathematical text known to us, a few fragments aside, was created.[28]

The first period is that of the Pythagorean order, the philosophical schools, and the sophists. Of these, the sophists, who were creators of a theoretical concept of education, and whose *paideía* aimed exactly at the creation of the complete human being or citizen, have the least to do with mathematics. Truly, Hippias, Bryson, and Antiphon interested themselves in the trisection and circle-squaring problems. But Hippias's curve for trisection[29] is nothing but a smart trick, and Bryson's and Antiphon's treatments of the squaring shows them (according to Aristotle's polemics) to lie outside the mainstream of Greek mathematical thought.[30] One is tempted to assume that the sophists' treatment of these problems reflects the necessity for those professional teachers to deal with mathematical subjects *à la mode* in order to satisfy their clientele. If such is the case, sophist education underwent the influence of mathematics, not the reverse.

In the case of the Pythagorean order,[31] it is difficult to distinguish legend from history. It seems sure, however, that the Pythagorean movement was the place where *mathémata* changed its meaning from "doctrines," i.e., matters being learned, to "knowledge of number and magnitude," i.e., to "mathematics,"[32] no later than the late fifth century B.C. It is well established that an essential part of the teaching of the order, be it secret or not, was by then dedicated to theoretical arithmetic, to geometry, to harmonics, and to astronomy—the four *mathémata* listed by Archytas. Finally, the active role of the Pythagoreans in the development of arithmetic, harmonics, and astronomical speculation during the fifth century can be trusted with confidence.[33]

According to tradition,[34] the order consisted of an inner and an outer circle, *mathematikoí* and *akousmatikoí*, "mathematicians" and "listeners"—if the two groups were not, as seems more likely, the result of a split in the order. The latter are supposed to have been literal followers of the tradition, whereas the former are supposed to have been taught rationally, and perhaps to have made rational inquiry (a supposition which is independent of the question whether their group went back to Pythagoras himself, or was a later fabrication). If this tradition is reliable—as it appears to be—the teachings of the *akousmatikoí* will at most have contained semi-mystical numerology, more or less shared with folk traditions. Some Pythagoreans may, on the other hand, have started rational inquiry from this basis; and their investigations of theoretical arithmetic, of harmonics, and of the problems of the irrational, may have become part of a cumulative research tradition *because of* integration with a stable tradition for reasoned teaching of the *mathematikoí* (the tradition giving rise to the very name of the group). More than this can hardly be said, given the lack of adequate sources.[35]

Still, I find it doubtful whether the rational and abstract character of Greek mathematics can have originated inside the circle of *mathematikoí*. It would seem to be in better harmony with the picture of the Pythagorean order as a mystical, religious, and ethico-political movement, if available philosophical knowledge, including mathematics, was borrowed from outside in cases where it could serve the overall world view and aim of the movement (cf., for example, Schuhl 1949: 242–57)—possibly in a process of several steps, where the most elementary abstract arithmetic was taken over (as numerology and for use in musical and cosmological speculation) already in the late sixth or early fifth century. In fact, various sorts of evidence points in that direction.[36] It thus appears that the mathematical activity of the Pythagoreans consisted of work in agreement with the rational tradition when it was already established, and refinements of the same tradition. Therefore, I shall direct attention to the open philosophical "schools."

Here, a word of caution may be appropriate. The philosophical schools were probably not schools in an institutional sense. Even when we go to the mathematical "schools" of the fourth century, what translators designate, e.g., "the school of Menaechmos," is spoken of by Proclos as "the mathematicians around Menaechmos."[37] No doubt the philosophers had disciples; but they are distinguished from the sophists by not being *determined* as professional teachers.[38] The philosophers made rational inquiry, some of them in mathematics; and they taught. Both activities must be understood as implying rational discourse. But nothing indicates that the philosophers' inquiry was determined in style and structure by their teaching.

Instead, the reasoned and abstract structure of fifth-century mathematics and its orientation toward *principles* must be sought elsewhere. That such a tendency was there is obvious even from the scant source material at our disposal, be it Hippocrates' investigation of the lunes[39]; his writing of the presumed first set of *Elements* (Proclos, *In primum Euclidis* 66[7-8]);

Oenopides' presumed first theoretically founded *construction* of dropping and ascending perpendiculars and his singling out of ruler and compass in that connection[40]; the description of the solar movement in the ecliptic as an inclined great circle (equally ascribed to Oenopides[41]); or a number of Platonic passages, from the references to investigations of incommensurability (*Theaetetus* 147d–148b; *Laws* 819d–820c) to the slave boy guided by suggestive questions to the doubling of a square (*Meno* 82b–85b).

It is also evident that all these pieces of evidence point toward various locations inside the "philosophical movement." The immediate background to the rise of reasoned mathematics is thus the condition of mathematical discourse and inquiry as part of general philosophical discourse.

This observation is also in harmony with chronology. Even though a number of mathematical discoveries are ascribed to Thales, and such ascriptions can neither be proved nor disproved (not least because it is not clear what, precisely, Thales is supposed to have discovered), the philosophical transformation of myth and cosmogony into philosophical cosmology seems to precede the rise of genuine reasoned mathematics. Indeed, even the Eleatic critique of natural philosophy appears to precede, if not the first steps toward a reasoned approach to arithmetic, then at least the techniques of proof that came to characterize Greek mathematics as we know it (cf. note 36).

So, the rationality of fifth-century Greek mathematics appears to build on general philosophical rationality, and on an open, non-hierarchical type of discourse: Not the one-way master-student relationship of institutionalized teaching, but a discourse of mutual disagreement, conflict, and common search.[42] Whereas the former may be more effective for the assimilative expansion of a knowledge system, the latter, open discourse, may be the presupposition for fundamental change.[43] So, if I am right in my interpretations, the *specific* formation of Greek mathematics may have originated in the *lack* of didactical institutionalization of the soil from which it grew during the first phase, as claimed already.

The open-type discourse of the philosophical environment may have had wider, social backgrounds: perhaps not so much in fixed social institutions as in the break-up of institutions. Indeed, the Solon reforms, which averted social conflict by instituting reasoned constitutional change, are contemporary with the earliest Milesian philosophy—and they are not the first attempt of their kind (cf. G. Smith 1956). About a century before Solon and Thales, Hesiod presents us with an instance of conceptual analysis by dichotomy,[44] in a way that reminds one very much of Plato's dialogues, but which in historical context shows that the germs of logical analysis are older than philosophy in Greece. Still earlier, at the dawn of Greek literature, the rhetoric of the Homeric heroes contains clear dialectical, syllogistic figures (used even to persuade the gods).[45] Ultimately, the discourse of early Greek rational philosophy may go back to the open discussion of the popular assembly and the *agora*[46]—whereas, as we saw, the discourse of Mesopotamian mathematics, *explaining* procedures and *training* scribes rather than *investigating problems* or questioning, goes back to the more closed discourse of organized school teaching.[47]

The second phase of the development of Greek mathematics, going from Archytas, Plato, and Theaetetos, not only brings a marked quantitative growth of mathematical knowledge, explained by Proclos/Eudemos as an increase in the number of theorems (*In primum Euclidis*, 66[16f]), but also a fundamental qualitative change, "a more scientific arrangement."[48] Indeed, what happens is a continuation and accentuation of a process inaugurated by Hippocrates of Chios when he wrote the presumably first set of *Elements*. Gradually, mathematics comes to consist of larger, theoretically coherent structures; no longer just reasoned, it becomes deductive and, in the end, axiomatic. The ideals for the organization of mathematical knowledge are clearly delineated by Aristotle in the *Analytica posteriora*, where these ideals are even put forward as paradigmatic for all "scientific" knowledge (cf. note 48).

No doubt the search for coherent structures gave extra impetus to the quantitative growth of knowledge; and probably the quantitative growth called for better organization. But this internal dynamic of late fifth and early fourth century mathematics was only made possible because mathematics had become something possessing a *social identity*. Mathematical activity had become institutionalized; the very successes of mathematics toward the late fifth century had made it a field of learning of its own.

Mathematics was institutionalized on several levels. It was introduced into the *paideía* of adolescents; but that was in all probability without effect on the dynamics of mathematical knowledge (apart from the recruitment thereby procured), the mathematics taught to adolescents being quite elementary. But mathematics also became something that one might study as a philosopher taught by a teacher—e.g., as one of the "mathematicians around Menaechmos." The increasing systematization of reasoned and argumentative mathematics performed as an autonomous activity automatically led toward deductivity and axiomatization. Systematization relentlessly revealed flaws and circularities in argumentation; no mathematician gathering a circle around himself could then avoid trying to get rid of such defects.

In the phase of merely reasoned mathematics, it would have been possible to prove that the sum of the angles in a triangle equals two right angles by drawing the parallel without questioning its existence[49]; in other connections, it would be possible to argue for the existence of a parallel in a way that ultimately involved knowledge of the sum of the angles of a triangle. In fact,

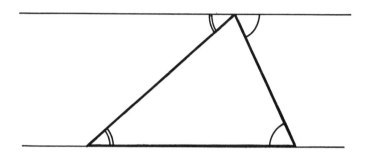

such attempts to prove everything are discussed by Aristotle in *Analytica posteriora* (72$^b$33–73$^a$20). When geometry became an integrated system, the circularities arising from the combination of such piecemeal demonstrations would become evident. Even this is borne out by Aristotle when he speaks of "those persons who do think that they are drawing parallel lines; for they do not realize that they are making assumptions which cannot be proved unless parallel lines exist" (*Analytica priora* 65$^a$5–7). The recognition would then force itself upon the mathematicians that some things had to be presupposed, in agreement with the initial sentence of the *Analytica posteriora,* that "all teaching and learning that involves the use of reason proceeds from pre-existent knowledge" (71$^a$1–2). In the end, Euclid found a way out of the complex problem by a combination of his fifth postulate (which implies that at most one parallel exists) and a tacit assumption about the figure used to prove prop. I.16—an assumption that holds true only in geometries where parallels exist (and hence not on a sphere).

Trying to describe the character of mathematical discourse of this second phase, we can say that it becomes closed into itself, i.e., autonomous: Mathematics builds up its own system of scientific and epistemological norms; when Protagoras argues against the geometricians that "the circle touches the ruler not at a point [but along a stretch of finite length],"[50] he just disqualifies himself (in the eyes of fourth-century geometricians) by being unaware of this closing of mathematical discourse—dealing, as Aristotle argues, not with sensible and perishable lines (etc.) but with the ideal *line in itself*—to non-mathematical reason.[51] In the elementary *paideîa,* the mathematical discourse also becomes closed in the sense of being one-way, dependent on authority and open to no alternative thinking. The same process is on its way at the philosophical level, but here only as a goal pursued: Mathematics is understood as concerned with eternal, immovable truth, and thus it cannot admit of alternatives and discussion of its foundations; so, mathematics must by necessity *aim* at the closure of its own discourse.[52]

By the end of our second period, this process was carried to its end. Already in the outgoing fourth century, "a common fund of theorems existed, in plane as well as spherical geometry, which had already taken on their definitive form, and the formulation of which was to be perpetuated for centuries with no change whatever" (Aujac 1984: 10). In the early third century B.C. (or, if recent proposals to make Euclid a contemporary of Archimedes are correct, around 250 B.C.—see Schneider 1979: 61f.), Greek mathematical thought has been shaped in the Euclidean *Elements* as a specialized, hard, sharp, and immensely effective tool for the production of new knowledge. In the third period, it produced astonishing quantitative accretions, e.g., in the works of Archimedes and Apollonios. But its whole style and formal character was fixed. It was deductive, axiomatic, abstract, and formally "pure"; and it was totally "Euclidized." Commentators such as Pappos, Proclos, and Theon of Alexandria explain and extend (for good or for bad); *grosso modo* they raise no doubts. Truly, Archimedes extended the range of aims of mathematics by his numerical measurement of the circle, in a way that (in spite of its "pure" form) was noticed as a deviation by some commentators (see Vogel 1936: 362). But even if his results were adopted, they inspired no further renewal (apart from the Heronian tradition, the connections of which to Archimedes are clear, but which misses on the other hand the high level of mainstream Greek mathematics). Greek mathematics had, in the *Elements,* got a paradigm in Thomas Kuhn's original sense of that word: a book "that all practitioners of a given field knew intimately and admired, achievements on which they modeled their own research and which provided them with a measure of their own accomplishment" (1963: 352). Mathematical discourse now became really closed, in agreement with the "Platonic" intentions; the closing was, however, not effected by a teaching institution, but through a book—or, better, it was effected by a teaching institution whose main institutional aspect was the use of that book.

This institutionalization did not give Greek mathematics of the mature phase any *new* character. But it was the precondition for the perpetuation of a character that Greek mathematics had once acquired through a series of settings which had now disappeared: institutionalized and non-institutionalized, discursively open and discursively closed.

The social carriers of mathematical development in the mature phase had less to do with teaching, institutionalized or not, than their predecessors of any earlier phase. Only a few, and not the greatest, were connected to those institutions of higher learning that, from Plato's Academy onward, had succeeded the earlier philosophical circles or informal schools. Most great mathematicians are best described as "professional full- or part-time amateurs" (strange as this mixing may sound to modern sociological ears), who after a juvenile stay, for example, at the centers of higher learning in Alexandria remained in mutual contact through letters, and whose impregnation with the professional ideals of the discipline had been so strong that their lonely work and their letters could maintain them as members of one stable scientific community.

## 3. The Latin Middle Ages: A Discourse of Relics

In order to prevent any simplistic—or just simple—picture from emerging, I shall discuss one more episode. I shall bypass the very interesting relations between socioeconomic and cultural background, traditions, and institutions, types of mathematical discourse, and the development of mathematics in ancient and medieval India and in the medieval Islamic world.[53] Instead, I shall concentrate on the Latin Middle Ages of Western Europe.[54]

The Roman part of the ancient world had never shared the Greek interest in theoretical mathematics; as Cicero remarks, the Romans restricted their mathematical interests to surveying and computation.[55] Truly, the education of a Roman gentleman was built on the seven Liberal Arts of the Greek *paideía:* grammar, rhetoric, dialectics (i.e. logic), arithmetic, geometry, harmonics ("music"), and astronomy. But what was taught in the latter four mathematical arts was utterly restricted. Only a few of the Greek mathematical works were ever translated into Latin (only one of which, part of the *Elements* translated by Boethius around A.D. 500, was of scientific merit), and only superficial popularizations were ever written in Latin.

The Christian takeover of education in late antiquity did nothing to repair this: on the contrary. Still more unquestionably than the gentleman, a good Christian should definitely not be secularly learned, even though good manners required him to be culturally polished. The breakdown of the Western Empire and the rise of loose barbarian states deprived him even of the possibility to be polished.

Still, if not saving much of the ancient learned legacy, the Christian Church saved at least the faint memory that *something had been lost.* At every occasion of a cultural revival, be it Visigothic Spain in the early seventh century, Anglo-Saxon England in the early eighth, the Frankish Kingdom of the early ninth, or the Ottonian empire of the late tenth, the recurrent characteristic is an attempt to reconquer the cultural ground that had been lost. This is the reason why every revival looks like a renaissance and has been labeled so by its modern historians.

Until the end of the first millennium, the reconquests at the mathematical front were restricted to arithmetical Easter calculation (*computus*); ancient presentations of theoretical arithmetic and harmonics for non-mathematicians; and some surveying manuals that had to play the role of geometry. Now, by the onset of the High Middle Ages, things were going to change. But in the actual moment, mathematics was (like every remnant of ancient culture) as much a sacred relic as something to be learned and understood or as a type of discourse; furthermore, even as a subject to be learned or as a discourse, mathematics was profoundly marked by being a relic in a culture given to worshipping relics.[56]

An economic and demographic leap forward in the eleventh and twelfth centuries was the occasion for a revival of trade and monetary economy and for the rise of towns achieving a certain degree of autonomy (ideological autonomy, often officially recognized juridical autonomy, and, in economically advanced regions, even *de facto* political autonomy). Men participated in the social and political life of these towns as members of more or less institutionalized groups, inside which they acted as equals; presumably it was on this background that an interest in open reasoned discourse grew in the eleventh century urban environment.[57]

The economic revival was also the occasion for the growth of *cathedral schools*, the students of which were taught the seven Liberal Arts to the extent that competent teachers and the necessary text materials were at hand. From the point of view of the Church, the schools were designed to provide future priests and other ecclesiastical functionaries with the knowledge necessary in a new social context where the priest had to be more than the main actor of rituals and the magician of relics.[58] At the same time, the growth of the schools can be seen as yet another spontaneous expression of the recurrent tendency to translate cultural blossomings into "renaissances," revitalizations of ancient learning. Finally, the clerks trained at the cathedral schools would often come to serve in and outside the Church in cancellarian and secretarial functions, the non-engineering aspect of the old scribal function. As long, however, as the ideological interest in free discourse and the cultural need for a renaissance prevailed, the schools did not take on the character of scribal training schools. The disciplines of autonomous thought, which had not been known to the Babylonians, were now at hand, and they were fundamental to the scholars' view of their world and of their own identity.

For mathematics, the eleventh century school meant little directly, apart from a firmer possession of the insecure conquests of the late first millennium. An awakening interest in astrology, nurtured by a few translations of Arabic treatises on the subject in the tenth and eleventh centuries (see J. W. Thompson 1929 and van der Vyver 1936), is probably best understood as an expression of the search for natural explanation distancing direct divine intervention,[59] an essential search in a society on the way to rationalize its world picture. Even though partly carried by cathedral school masters, it was no product of the school institution as such, whose whole heritage both from ancient learning and from the Fathers of the Church would rather have made it separate liberal-arts astronomy from astrology.[60]

Indirectly, though, the eleventh century school had great importance for the future of mathematics. Taken together, the rational, discursive spirit of the times and the training and opportunities provided by the schools gave rise to great changes in every corner of Latin learning. On mainly native ground, figures such as Anselm of Canterbury, Abelard, Hugh of Saint-

Victor, the "twelfth century Platonists," and Gratian recast philosophy, theology, and canon law in the late eleventh through the mid-twelfth century. More important for mathematics, the background provided by the schools made possible the translation of Muslim learning and of Arabic versions of ancient Greek works (and, initially to a lesser extent, direct translations of Greek works) by creating the scholarly competence and, not least, the enormous enthusiasm of the translators; furthermore, the schools provided a public that could receive the translations and have them spread. So, during the twelfth century, most main works of ancient and Judeo-Muslim astrology (including Ptolemy's *Almagest*), the *Elements,* al-Khwārizmī's *Algebra,* and several expositions of "Hindu reckoning" (the decimal place value system for integers and its algorithms) were, together with many other mathematical and non-mathematical works, translated and spread.[61] Even the larger part of Aristotle's *Organon,* his *Metaphysica,* and part of his natural philosophy were first transplanted in the mid- to late twelfth century.

In the late twelfth century, learning at those cathedral schools which were to develop into universities was in a situation of suspense. Learning was not seen as something being created in an active process. Learning already existed in the form of great scholarly works, "authorities;"[62] the enthusiasm for knowledge still had something of the enthusiasm for relics.[63] And yet, the "new learning" was really something *new;* as a body of relatively *coherent* knowledge it was in fact something that was being actively created. In spite of its concentration on already existing authority, the discourse of the new learning was anything but closed: There were still tight connections between the open "political" discourse of the corporations ("universities") of masters and students and the discourse of learning.[64] The "political" discourse of the scholars, for its part, was of the same genre as that which had manifested itself in the turbulent urban environment already in the later eleventh century, reinforced in a synergetic process by the increasingly "dialectical" organization of the teaching institution.

The openness and the semi-political character of the learned discourse of the late twelfth and the early thirteenth century was not mistaken at the time. As Bernard of Clairvaux had attacked the rationalizing theology of Abelard and the "Platonist" philosophers, so many smaller theological minds attacked the new learning, complaining, for example, that the Christians (and even monks and canons) not only wasted their time but also endangered their salvation studying the

> philosophical opinions, the [grammatical] rules of Priscian, the Laws of Justinian, the doctrine of Galen, the speeches of the rhetors, the ambiguities of Aristotle, the theorems of Euclid, and the conjectures of Ptolemy. Indeed, the so-called Liberal Arts are valuable for sharpening

the genius and for understanding the Scriptures; but together with the Philosopher they are to be saluted only from the doorstep. [ . . . ] Therefore, the reading of the letters of the Pagans does not illuminate the mind, but obscures it.[65]

It will be observed that even interest in the *Elements* is presented as a danger to theological order (and, beyond that, to social order in the ecclesiastical universe).

That, however, was only for a brief period. Euclidean mathematics was not fit to serve the construction of a coherent counterdiscourse to the discourse of the conservative theologians. Interest in mathematics dropped back from the front in thirteenth century learning. The great conflicts, the prohibition of dangerous works, and the executions of heretical scholars were all concerned with Aristotelian philosophy and its derivations (including pseudo-Aristotelian occult science).[66] Except for a few active researchers, the scholars of the thirteenth century looked at mathematics as a venerated part of the cultural heritage. In the context of the thirteenth century university world, mathematics was better fit as a modest part of the peaceful synthesis than as a provider of revolutionary counterdiscourse.

Such synthesis did take place in the mid-thirteenth century. It took place at the social and political level, where the liberation movement of the towns attained a state of equilibrium with the prevailing princely power, and where even princely, papal, and feudal power learned to live together; and it was seen at the level of religious organization, where the specific urban spirituality got an authorized expression through the orders of friars, but where it lost its *autonomous* expression through lay pauper movements. Closer to our subject, synthesis forced its way through in the matter of Aristotelian philosophy. By the 1230s, the position of the conservative theologians had become untenable: Aristotelian metaphysics had penetrated their own argumentation to the bones. But for all the fluidity and turbulence of the university environment, general social and ideological conditions were not ripe to overthrow the power of the ecclesiastical institution. This scholarly stalemate was solved through the great Christian-Aristotelian synthesis due to Saint Thomas and Albert the Great. Thanks to their work, the "new learning" was fitted into the world conception of Latin Christianity, as that cornerstone which had been missing (and missed) for so long. One is tempted to paraphrase St. Thomas's famous dictum, *Gratia non tollit naturam sed perficit,* to the effect that "natural philosophy did not abolish the world view of Divine Grace; it made it complete."

By making it complete, however, Latin Aristotelianism was given an orthodox interpretation that deprived it of its character of open discourse. Only for a short time (roughly speaking, the fourteenth century) was it able to develop answers to new questions, and to procure a world-view that could really pretend to being a view of a world in change; after that, Aristotelian learning (and the whole of medieval university learning, which had come under its sway) was upheld only by institutional inertia and by the external social forces guaranteeing its survival. Gradually, the universities were to develop into training schools for priests, lawyers, physicians, and officials, of scribal-school character. Their much-beloved dialectical method, once the reflection of an open, critical discourse, could be derided as a display of empty virtuosity by the satirical authors of later ages, from Thomas More to the eighteenth century.

Mathematics was no main pillar in the synthesis. But on a modest level, it had prepared the way. Neither the mathematics of the "new learning," nor mathematics at all or its single constituent disciplines, were ever practiced on a larger scale as something autonomous. The traditional mathematical disciplines belonged to the total scheme of Liberal Arts. The commentaries written into mathematical treatises, be it the eleventh-century low-level presentations of theoretical arithmetic (see Evans 1978) or the twelfth- and thirteenth-century *Elements* (see, for instance, Murdoch 1968), demonstrate their attachment to a teaching tradition where the establishment of connections to the non-mathematical disciplines of the curriculum were as important as mathematics itself. These traditional disciplines remained—the short-lived tendencies of the twelfth century and a few mathematicians by inclination[67] disregarded—integrated parts of a larger cultural whole, and parts of a heritage. Mathematical discourse was never, as it had been in antiquity from the fourth century B.C. onward, autonomously closed on itself; its main epistemological responsibility was not inwardly but outwardly directed.[68] But it was closed on past performances, closed to fundamental renewal, closed to alternatives.

The non-traditional disciplines were necessarily less closed: Algebra, optics, the "science of weights" (i.e., mathematical statics) did not fit into the traditional curriculum and could only be merged with it through a creative process. But only a handful of people engaged in these disciplines, and of the works of that handful of scholarly eccentrics, only those that were congruous with traditional thought gained general acceptance.[69]

The prevailing scholarly synthesis was a real synthesis of all the important interests present: Only a few of those concerned did not feel good inside its frame. Those who did not (and their number increased as the fourteenth century approached and especially during its progress) would rather leave the universe of closed rational discourse altogether and fall into mysticism and skepticism, than try to open it to alternative rationalities—be they mathematical or philosophical.

The late thirteenth and fourteenth century offer what might appear to be a pair of exceptions to this generalization: A violent increase in the semi-rational, semi-mystical field of astrology; and an attempt at mathematization of Aristotelian natural philosophy (and other fields that had taken over the Aristotelian conceptual structure), through a quantification of "movement" and "qualities."[70] Neither, however, led to the establishment of lasting mathematical disciplines.[71] Instead, the mathematical renewal of the Renaissance came from elsewhere: first, from the "abacus schools" of Northern Italy, training schools for merchant youth, spiritually (and perhaps also by vague descent, by way of schools and teaching traditions of the Islamic world) related to the Old Babylonian training schools for scribes;[72] and second, from alternative approaches to the mathematical disciplines of antiquity inspired by the humanist movement and by the problems of perspective painting and architecture.[73] The real explosion took place in the confrontation of these two sources for renewal, for example, in the famous episode[74] when Cardano published Tartaglia's solution to the third-degree equation (giving "due credit"), thereby bringing the secret method that had made Tartaglia a virtuoso in the genre of a traditional abacus schoolmaster or an Old Babylonian scribal teacher[75] into the light of open scientific discourse.

## 4. Perspectives

The third episode ends when the merging of traditions and the reorganization of all learned and literate discourse made possible by printing opened the way to modern mathematics. Much has changed since then, even concerning the contents, structure, obligations, and social organization of mathematics. If any key science can by pointed out in the ongoing scientific-technological revolution, mathematics is the key. Indeed, a recurrent condition for the integration of the theoretical sciences with the management of practical affairs is an analytic reformulation of theories that either amounts to mathematization or opens the way to mathematization—when necessary, through the development of new mathematical disciplines which serve that mathematization.[76] So, mathematics and mathematicians must be held co-responsible, both for the fulfillment of the promises of that revolution and for its dangers and perversions.

If this is so, what, then, is the use of a discussion of the historical development of mathematics as a system of organized knowledge and as a discourse stopping around A.D. 1500—be it inspired from historico-materialist points of view? Does it have indications beyond itself (500 years beyond itself!), or is it merely a piece of historical research practiced as pure, immaculate science?

I think the story has important implications for the present, politically as well as pedagogically. Politically, it suggests that the concepts of "open" and "closed" discourse are central,[77] but that it makes no historical (and hence, when applied to the present, no political) sense to posit a mechanistic and abstract dichotomy, in the manner of a Karl Popper, between the Open Society and the Open Discourse, on the one hand, and the Closed Society and the Closed Discourse on the other. Things are more complex than that, and therefore the character and mutual relations of openness and closedness must be understood in their social concreteness, not accepted as abstract postulates. This is as true today as in the pre-Renaissance era.

The pedagogical implications of the story concern the obligations of mathematics education. The contemporary world is confronted with immense problems: problems of survival and problems of welfare. Disregarding the dangers of nuclear war, how shall we avoid climatic and ecological catastrophes and the wasting of resources necessary for the future of civilization; how shall we provide everybody with food, housing, and leisure? These are not only mathematical and technological problems, and our present capitalist order may well be as unable as it is unwilling to solve them at all. But they are *also* technological and mathematical—*no* social order will solve them without using adequate technologies—*and many of the necessary technologies have not yet been developed, nor has the mathematics required for their development been shaped.* Therefore, mathematics must be allowed to develop, and mathematical development furthered at our best; according to historical experience mathematics must hence be allowed autonomous existence.

On the other hand, many of the problems of our contemporary world are due to unconscious, restricted technological and scientific (sub-)rationality (behind which stand, of course, political interests and economic structures favoring the development and perpetuation of unconscious and unscrupulous pseudo-rationality). The acceptance of this situation, the belief that such rationality is the only possible and perhaps the only genuine breed of rationality is supported by the appearances of technology, science, *and mathematics* as they present themselves in their current closed discourses. If the transmission of mathematics to the next generation is to contribute to making a better world, these closed and self-sufficient discourses must be combated from the inside; mathematics must be made known also under the aspect of an open discourse, as a contribution to the critical opening up of technological discourse, and as an integrated part of human knowledge.

It must be the duty of the teacher, from primary school to university, to present a double picture of mathematics to students: as a field of human knowledge with its own *integrity*, requiring its own autonomous further development; but a field which at the same time must achieve *integration* as part of human knowledge and human life in general.

# 2

# SUBSCIENTIFIC MATHEMATICS: OBSERVATIONS ON A PRE-MODERN PHENOMENON

In memoriam
N. I. Bukharin

## *Introduction*

*In 1931, a delegation from the Soviet Union suddenly turned up in London at the Second International Congress of the History of Science and Technology. The impact of the event was great, first on the circle that Werskey (1979) has baptized the "Visible College": Joseph Needham, John Desmond Bernal, Lancelot Hogben, J. B. S. Haldane, and others, and through these on the study of the history of science and on the "social studies of science"; the latter discipline, indeed, owes its very conception to this event and to members of this group.*

*This story has been told often enough, among others by Joseph Needham (1971), who was there. I was not, and I shall not repeat what I know only at second or third hand. What makes me refer to the event is Bukharin's contribution (1931), which I had recently reread when I formulated the following essay in the fall of 1988. Here he emphasizes the complexity of the relation between theoretical and what the Greeks would call "productive" knowledge, not only going beyond the idea that productive knowledge is a simple derivation from theory but also transcending the idea that theoretical knowledge is brought forth as a direct reflection of the needs of social practice (the thesis that critics read into Boris Hessen's famous contribution to the same Congress[1]—not to speak of the reduction of theory to* nothing but *practical knowledge.*

*This is also the primary reason that my analysis was made* in memoriam N. I. Bukharin, *rather than the chronological coincidence that Bukharin was born in 1888, executed in 1938, and rehabilitated in the very year 1988. My aim was indeed to systematize insights about how, in the pre-Modern epoch, the mode of mathematical thought depended on how mathematical activity was located with regard to practical use—in particular, to show how practitioners' mathematics possessed a more complex epistemological structure than hitherto recognized, and to characterize the manner in which such "sub-scientific" specialists' knowledge was adopted into "scientific" mathematics.*

*Even though they have relatively little impact upon the presentation of the final argument, my investigation was inspired by insights borrowed from the study of folktales and other aspects of oral culture. Besides, the reader will notice that the distinction between* determination through problems *and* determination through methods, *which was introduced in Chapter 1 in order to characterize the difference between Greek and Babylonian pure mathematics, rises to greater prominence in the present study.*

*Apart from the usual normalization of the reference system and the introduction of cross-references, the text is practically identical with the one that was published in* History of Science *in 1990.*

## 1. The Concepts

In the post-Renaissance world, mathematics used for technical or administrative purposes utilizes results and techniques derived from some level of "scientific" mathematics, though often transformed or simplified in order to be adequate for practitioners. Furthermore, modern practitioners of mathematics have been taught their mathematics by teachers who have been taught by teachers (who . . . , etc.) who have been taught by mathematicians. In both senses, practitioners' mathematics can then be regarded as "applied mathematics."

In the pre-Modern world, the situation was different. A quotation from Aristotle's *Metaphysics* ($981^b14$-$982^a1$—trans. Ross 1928)—dealing not with mathematics but with knowledge in general—will help us see how:

At first he who invented any art whatever that went beyond the common perceptions of man was naturally admired by men, not only because there was something useful in the inventions, but because he was thought wise and superior to the rest. But as more arts were invented, and some were directed to the necessities of life, others to recreation, the inventors of the latter were naturally always regarded as wiser than the inventors of the former, because their branches of knowledge did not aim at utility. Hence when all such inventions were already established, the sciences which do not aim at giving pleasure or at the necessities of life were discovered, [ . . . ]

So [ . . . ], the theoretical kinds of knowledge [are thought] to be more of the nature of Wisdom than the productive.

First of all, of course, this passage establishes the distinction between "theoretical" and "productive" knowledge, between that knowledge which aims at nothing beyond itself (or possibly at moral improvement of the knowing person) and that knowledge which aims at application. We also observe, however, that Aristotle assumes the two kinds of knowledge to be carried by *different persons*.

This is no new observation if we consider the history of Greek science and technology. In the main, the two kinds of knowledge were indeed carried by *separate social groups and traditions* (the exceptions do not concern us for the moment). Among other things, because of the source situation, the history of ancient science and especially that of ancient mathematics has normally focused upon that tradition which cared for theoretical knowledge. It has done so to the extent that the specific character (or even the separate existence) of the other traditions has often gone unnoticed. This might be relatively unimportant as long as nothing but ancient mathematics itself was concerned. When we turn to those mathematical cultures inspiring or inspired by ancient Greek mathematics, however, neglecting the existence and distinctive character of the "productive" traditions makes us blind to many facets of history; this is what I shall try to demonstrate in the following by pointing to new observations on mostly well-known material, which are made possible by the explicit distinction between "scientific" and "subscientific" knowledge and traditions.

The distinction concerns the *orientation* of knowledge, the purpose intended in the acquisition, the conservation and the transmission of knowledge. *Scientific knowledge* is knowledge that is pursued *systematically* and *for its own sake* (or at least without any intentions of application) *beyond the level of everyday knowledge*. "Scientific knowledge" is thus the same as "theoretical knowledge" in the Greek sense. The term *scientific* is chosen because this is the kind of knowledge upon which descriptions of the history of science will normally concentrate (again, the exceptions are immaterial for the present argument). In order to avoid confusion with the vaguer everyday meaning of the word, the term will be kept in quotes throughout the article.

*Subscientific knowledge,* on the other hand, is *specialists' knowledge* that (at least as a corpus) is acquired and transmitted *in view of its applicability.* Even subscientific knowledge is thus knowledge beyond the level of common understanding, and it may well be much more refined than "scientific" knowledge—as is evident if one compares the level of the "scientific" knowledge presented in Nicomachos' *Introduction to Arithmetic* with that of Old Babylonian "algebra," the subscientific character of which shall be argued.[2]

All this might sound like nothing but new words for the familiar division of knowledge into "pure" and "applied." This division, however, is bound up with the specific Modern understanding of the connection between the two levels. Roughly speaking, fundamental knowledge is assumed to be

found by "pure scientists" and then to be worked upon, recast, and synthesized in new ways by "applied scientists" or technologists in consideration
of the problems and possibilities of current practice. These implications of
the Modern terms constitute one reason to avoid them in descriptions of the
non-Modern world, where things were different, where practitioners' mathematics was not a separate *level* but an autonomous *type* of mathematics.

Another reason is the existence of a "pure" outgrowth of the subscientific corpus of knowledge—i.e., that subscientific knowledge *itself* consists of (at least) two levels, of which one is intentionally *non-applicable*. I
shall return to this question in Section 2; first, however, I shall point briefly
to the two issues that have led me to formulate the distinction. Both concern
*mathematical knowledge,* which is also going to be my sole subject in the
following, although the distinction itself is of more general validity.

One issue is the contrast between Greek and Old Babylonian mathematics.[3] The naive first-order explanation of the difference between the two
says that Babylonian mathematics aimed at practical application, whereas
Greek mathematics considered abstract entities and aimed at Aristotle's
"Wisdom."[4] This view derives from the formulation of Babylonian problem
texts, which formally deal with fields, siege ramps, and the like. Many problems, it is true, do train skills in practical computation; other problems, however, and indeed whole branches of Babylonian mathematics, are "formally
applied" but "substantially pure"—i.e., the entities dealt with are those encountered in practical scribal (surveyor's, military engineer's, accountant's)
life, but the problems about these entities are not only unrealistic as far as
numerical magnitudes are concerned but also with regard to structure.[5] The
second-order explanation is, then, that the Babylonians had also discovered
the joys of pure mathematics.[6] However, if these joys were the same to the
Babylonians and, for example, to the Greeks, how shall we explain the fundamental difference between Greek and Babylonian "pure" mathematics—
Greek mathematics being, roughly speaking, determined by *problems,* for the
solution of which new methods would have to be developed (cf. below),
whereas Babylonian non-applicable mathematics was determined by the
*methods already at hand,* for the display of which new problems were continually constructed (cf. Chapter 1, and Section 2).

These difficulties are cleared away if we look into the function of Babylonian mathematics for the scribes and in the scribal school. The social basis
of Babylonian mathematics is "subscientific" rather than "scientific" in
character, and the peculiar character of Babylonian "pure" mathematics is
close to that of non-applicable subscientific mathematics, as I shall discuss in
the following section.

The other issue is the question of the sources of Islamic mathematics.
When the tenth century Baghdad court librarian al-Nadīm wrote his *Cata-*

*logue* (*Fihrist*), he mentioned 21 Greek mathematical authors, but not a single pre-Islamic non-Greek mathematician.[7] This corresponds to the current picture, where Greek mathematics is still considered the all-important source. Still, India is also taken into account nowadays, and is in fact unavoidable if we trace, for instance, Islamic trigonometry; at times, even Syrian, Pahlavi and Khorezmian learning are mentioned at least as undocumented or indirectly documented possibilities. It is rarely observed, however, that Indian mathematics cannot have been taken over directly from Indian high-level mathematicians like Āryabhaṭa and Brahmagupta[8]; nor is the question discussed whether (for example) Khorezmian learning will have been comparable to Greek science in structure.

There are, in fact, good and even compelling reasons to take non-Greek sources into account. But why, then, were they not known to the learned al-Nadīm, who was in a position to know better than anybody else? Once again, the distinction between "scientific" and "subscientific" may be called in: the sources unknown to al-Nadīm were subscientific in character—practitioners' traditions that were transmitted not in books but orally, and often "on the job."[9]

## 2. The Character and Structure of Subscientific Mathematics

I shall address the character of Babylonian mathematics as well as the subscientific sources of Islamic mathematics in somewhat more detail. First, however, I shall consider the general structure of subscientific mathematics.

Subscientific mathematics was carried by specialists' "professions,"[10] and the chief aim for building it up and transmitting it was its practical applicability as a tool for the profession in question. Evidently, then, its main contents were methods and techniques for practical numerical and geometrical computation (and, in certain professional contexts, practical geometrical *construction*). Still, practical computation has mostly been lost from the sources,[11] and often we only know the global results—accounts have been made, heritages have been distributed, taxes have been levied, armies have been supplied with food and pay, goods have been bought and sold. Paradoxically, then, we know the basis and *raison d'être* of subscientific mathematics less well than its "pure" outgrowth: "recreational mathematics."

Recreational mathematics was once described by Hermelink (1978: 44) as "problems and riddles which use the language of everyday but do not much care for the circumstances of reality." "Lack of care" is an understatement: A funny, striking, or even absurd deviation from the circumstances of reality is *an essential feature* of any recreational problem. It is this deviation from the habitual that causes amazement, and which thus imparts upon the problem its recreational value.

One function of recreational mathematics is that of teaching. Thus the following

> example of reduction of residues: A traveller, engaged in a pilgrimage, gave half his money at *Prayága;* two-ninths of the remainder at *Cási;* a quarter of the residue in payment of taxes on the road; six-tenths of what was left at *Gáyá;* there remained sixty-three *nishcas;* with which he returned home. Tell me the amount of his original stock of money, if you have learned the method of reduction of fractions of residues,[12]

which obviously serves to train a method of practical use, *viz.,* that of "reduction of fractions of residues." The fancy application of a staple method may serve in part as an appetizer, in part also to suggest abstract or general validity: Look, this method may be used for *any* problem of a similar structure, not just for trite commercial calculation.

This end of the spectrum of recreational mathematics passes imperceptibly into general school mathematics, which in the Bronze Age as now would often be unrealistic in the precision and magnitude of numbers without being fun in any way. Whether funny or not, such problems would be determined *from the methods to be trained.* This organization around an existing stock of methods is also a valid description of the other end of the spectrum, the purpose of which was described in one example by Christoph Rudolph in 1540 as the way "to find out, *not without exceptional amazement of the ignorant,* how many penning, creutzer, groschen or other coin somebody possesses."[13] Recreational problems *stricto sensu* are *riddles;* like an *argot,* they belong to the cultural superstructure of the profession. They set aside the members of the craft as particular, and particularly clever, people (whether in the opinion of others or in their professional self-esteem)—and set aside those who are able to solve the problems as especially clever members of the craft. Since the problems occurring in everyday practice will soon become trivial, these are not fit for kindling anybody's vanity. Instead, more complex problems are constructed, which still look as if they belonged in the professional domain, and which are still solvable by current professional techniques—but only on the condition that you *are* fairly clever.

As a typical example illustrating the matter we may mention the "purchase of a horse," for example in Leonardo Fibonacci's first version:

> Two men in possession of money found a horse which they wanted to buy; and the first said to the second that he wanted to buy it. If you give me ⅓ of your money, I shall have the price of the horse. The second asked the first for ¼ of his money, and then he would equally have the price. The price of the horse and the money of each of the two is asked for.[14]

All elements of the problem are familiar to the commercial calculator or merchant; but the situation as a whole is certainly not, both because of the funny coincidence of numbers and because the problem is indeterminate. In which market would the price of a horse be determined only as an arbitrary multiple of 11?

A more extreme example can be found in the Carolingian *Propositiones ad acuendos juvenes:*

> A paterfamilias had a distance from one house of his to another of 30 leagues, and a camel which was to carry from one of the houses to the other 90 measures of grain in three turns. For each league, the camel would always eat 1 measure. Tell me, whoever is worth anything, how many measures were left. [15]

Again, the elements of the problem will have been familiar in the Near or Middle Eastern environment where its origin is to be sought. The solution given, however, is based on an unexpected trick (an intermediate stop after 20 leagues), as it is indeed characteristic of riddles (and on the tacit assumption that the camel eats nothing while returning), and not on mathematical reasoning; if a bit of elementary mathematical reasoning is applied to the trick, furthermore, the solution is seen not to be optimal even on its own premises.

The latter case is extreme, but still illustrative of an important aspect of recreational mathematics. Actually, the problem-type is still alive in the contemporary "global village," where one may find it dealing, for example, with jeeps and petrol in Sahara. Once introduced, a specious trick will, in fact, often be adopted into the stock of current techniques (even though it is of no use in professional practice), and new problems will be constructed where it can be used (often but not necessarily amounting to nothing but new fancy dressings of the same mathematical structure).

Over the whole range from school mathematics to mathematical riddles, the methods or techniques are thus the basic determinants of development, and problems are constructed that permit one to bring the methods at hand into play. This not only contrasts with that foundation of applications out of which the superstructure had grown and to which it referred, but also with the structure of "scientific" mathematics as embodied, for example, in Greek mathematics.

That problems are primary and the methods used to solve them derivative is a matter of course (and almost of definition) when practical applications are concerned: the Eiffel Tower, built in order to demonstrate the possibilities of modern iron constructions, remains an exception. That the same holds for Greek pure mathematics follows from historical scrutiny.

For one thing, we know the importance of the three "classical problems" as foci of interest: doubling the cube, trisecting the angle, squaring the circle. When these were formulated as geometrical problems,[16] no theoretically acceptable methods were known that allowed them to be solved; their whole history throughout antiquity is the story of recurrent attempts to solve them by means of methods more satisfactory than those found by earlier workers.[17] But we may also look at the theory of irrationals. The first discovery of irrationals led to the problems *how to construct* according to a general scheme lines that are not commensurate with a given line (or whose squares are not commensurate with a given square); *how to classify* magnitudes with regard to commensurability; and *which are the relations* between different classes of irrationals. The first problem is the one that was addressed by Theodoros according to Plato's *Theatetus* 147D; further on in the same passage, Theaetetos makes a seemingly first attempt at the second problem; *Elements* X, finally, is a partial answer to all three problems.

This role of *the problem* is no exclusive distinction of Greek mathematics. It is a global characteristic of all those later traditions which can be characterized as "scientific."[18] As suggested already, it does distinguish, on the other hand, Greek mathematics from the non-utilitarian level of Babylonian mathematics. To see how, we may look at an Old Babylonian text.[19]

A trapezoidal field. I cut off a reed and used it as a measuring reed. While it was unbroken I went 1 three-score steps along the length. Its 6th part broke off for me, I let follow 1,12 steps on the length. Again, ⅓ of the reed and ⅓ cubit broke off for me; in 3 three-score steps I went through the upper width.

I extended the reed with that which [in the second instance] broke off for me, and I made the lower width in 36 steps.

The surface is 1 bur [ = 30,0 nindan²). What is the original length of the reed?

You, by your making: Pose the reed which you do not know as 1.

Break off its 6th part, then 0;50 remain for you

Detach its igi, raise [the resulting] 1,12 to 1 three-score

Append [the resulting] 1,12 to 1,12; it gives 2,24, the false length.

Pose the reed which you do not know as 1.

Break off its ⅓, raise [the remaining] 0;40 to 3 three-score, the upper width; it gives 2,0.

Accumulate 2,0 and 36, the lower width.

Raise [the resulting] 2,36 to 2,24, the false width; 6,14,24 is the false surface.

Repeat the [true] surface until twice, that is 1,0,0; raise this to 6,14,24; it gives 6,14,24,0,0, and raise ⅓ cubit, which you broke off, to 3 three-score.

Raise [the resulting] 5 to 2,24, the false length; it is 12,0.
Break ½ of 12,0, to two, make it confront itself.
Append [the resulting] 36,0,0 to 6,14,24,0,0; it gives 6,15,0,0,0.
6,15,0,0,0 makes 2,30,0 equilateral.
Append that 6,0 which you have left back to 2,30,0; it gives 2,36,0.
The igi of 6,14,24, the false surface, cannot be detached. What shall I
    pose to 6,14,24 which gives me 2,36,0?
Pose 0;25.
Because the 6th part broke off, write 6, let 1 go away; you leave 5.
⟨The igi of 5 is 0;12; raise 0;12 to 0;25: It gives 0;5⟩
Append 0;5 to 0;25: It gives you [0;30, i.e.] ½ nindan as original reed.

The problem pretends to deal with surveying practice, i.e., with an essential constituent of Babylonian scribal practice. But it is obviously no real-life problem, but a puzzle. So far it belongs to the genre of recreational mathematics. Mathematically, however, it is above the level of normal recreational problems: First, it leads to a mixed, non-normalized second-degree equation[20]; second, this equation is itself only established by means of fairly complex first-degree operations. The first-degree operations involve repeated use of the "single false position," which was a staple method of practical computation. Second-degree problems, on the other hand, would never occur in *real* scribal practice. Their solution built on a special trick (the quadratic completion), which appears to have been designated "the Akkadian method."[21] In one sense, this corresponds to the trick by which the paterfamilias avoided having all his grain eaten by the camel. But even though not relevant for daily mensuration practice the completion is a *mathematical* trick; it is, moreover, the basis of a whole mathematical discipline (second-degree "algebra") which was explored extensively and systematically in the search of problems permitting the use of the trick.

The problem of the broken reed is thus a nice illustration of the general character of Babylonian "pure" mathematics: It is similar to the recreational genre, but it is much more technical. It is, like recreational mathematics, governed by the stock of available techniques and methods, and its purpose is to display and/or to train these; and, like recreational mathematics, it goes beyond the range of practically relevant problems and makes use of techniques of no practical avail. But being taught in a formal and highly organized school system, it becomes systematic, building up quasidisciplines according to the possibilities of the fund of methods. In this respect it is very different from medieval recreational mathematics, which was practiced by reckoners who (when exposed to the problem of repeated doublings of unity)

strain themselves in memorizing [a procedure] and reproduce it without
knowledge or scheme, [and by others who] strain themselves by a
scheme in which they hesitate, make mistakes, or fall in doubt,

as it was formulated by al-Uqlīdisī in Damascus in 952 A.D.[22] Though basically subscientific in character, Babylonian mathematics demonstrates to what extent subscientific mathematics could mimic "scientific" mathematics under appropriate cultural and institutional conditions (and to what extent it could not).[23] Babylonian mathematics could be said to represent *scholasticized* subscientific mathematics. Al-Uqlīdisī's reckoners, on the other hand, represent a *lay,* and presumably *orally transmitted,* type of subscientific mathematics.

Even Egyptian mathematics belongs to the scholasticized type: it differs from Babylonian mathematics regarding mathematical substance (and differs fundamentally); but it shares the overall subscientific orientation coupled with the rigorizing effects of a systematic school system. Quite different, however, is another apparent intermediate form between the subscientific and the "scientific" orientation: Diophantos' *Arithmetic.* This work, too, shares many of the features of subscientific mathematics, from single problems to the guiding role of methods. Still, the work is written on the background of Greek "scientific" mathematics, and this background had provided Diophantos with his perspective: What he does is to *adopt a corpus of subscientific knowledge* into the domain of "scientific" mathematics (and expand it immensely). Diophantine arithmetic is therefore not to be regarded along with the scholasticized, quasi-"scientific" variants of subscientific mathematics but under the heading of subscientific *traditions* and their role as *sources* for "scientific" mathematics.

Before we leave the discussion of the structure of subscientific mathematics and take up the question of traditions we should take note that the subdivision into "scholasticized" and "lay" types is not the only relevant subdivision. First, one can differentiate roughly between "computational" and "geometrical" orientation, as we shall do in the following sections. Secondly, the concept of "practitioners" is of course vague; the computations of a caravan merchant and those of, say, a Sassanian royal astrologer are distinct not only according to subject matter but also, and more decisively, when we ask for the level of mathematical sophistication. These distinctions are not only of internal relevance but also important if one wants to investigate the impact of various subscientific traditions upon "scientific" mathematics. They can only be made, however, if we have a reasonably detailed knowledge of the subscientific tradition in question, which is often not the case. I shall therefore not pursue them in any detail.

Another highly interesting question regarding the possible relation between "scientific" and subscientific mathematics I shall leave as an open question, this time not because adequate source material does not exist but because I am not sufficiently familiar with it: How are we to describe the relation between the two types in India, in China, and in Japan? It is my preliminary impression from the secondary literature that at least Japanese *wasan* exemplifies a process where the "pure" level of subscientific mathematics gives rise to a direct and smooth creation of "scientific" mathematics. Is this really so? If it is, can a similar process be traced in India and China? And can this provide us with new insights into the possibly distinctive nature of Indian, Chinese, and Japanese "scientific" mathematics?

## 3. Traditions and Index Fossils

As stated initially, the term *subscientific* describes an orientation of knowledge. In principle, such an orientation could be individual and idiosyncratic, as are contemporary deviations from the "scientific" orientation (cf. note 18). But since it belongs with specialists' professions, which themselves are continuous over time, subscientific mathematics can be described historically as being carried by traditions.

Fundamental arithmetical and geometrical rules are interculturally true—these are the "2 · 2 = 4-truths" that Karl Mannheim (1965: 71f., 251f., and *passim*) excluded from the concern of the sociology of knowledge. Similar global orientations and interest in specific problems may arise not because of diffusion, but on similar sociological and technical backgrounds; and even common errors may be explained as parallel simplifications or as random incidents. Much of the subject matter and many of the techniques of subscientific mathematics, however, are so specific that they are indubitably witnesses of the duration and intertwinement of traditions. Often, such shared elements are even the only indications of otherwise purely hypothetical connections; we may regard them as the index fossils of cultural history.

One such index fossil is the usage of "parts of parts" and its extension into a system of ascending continued fractions."[24] Parts of parts are expressions such as "⅖ of ⅓" for ²⁄₁₅; ascending continued fractions can be exemplified by the expression "⅓, and ⅖ of ⅓" (= ⁷⁄₁₅). Both expressions are found in medieval Islamic mathematics, and they are mostly discussed with reference only to this area and to the post-Islamic tradition in medieval and Renaissance Italy. Though rare, they can, however, be found in Middle Kingdom Egyptian and Old Babylonian sources, often (in Egypt exclusively) in connections that suggest popular usage or in problems of definite riddle character. In late Greek antiquity, they turn up in the arithmetical epigrams of the *Anthologia graeca*—but only in problems concerned with trade routes, with the partition of heritages, or with the hours of the day (whether working hours, astrological timekeeping, or questions to the *gnomon*-maker with no further explicit purpose); the only exception among those problems dealing with other subjects concerns a banquet taking place in Hellenistic Syria. In medieval Islam, the usage goes together with the so-called finger-reckoning tradition, which in Arabic was called *ḥisāb al-Rūm waʾ l-ʿArab* (computation of the Byzantines and the Arabs). Even in the Carolingian *Propositiones ad acuendos juvenes* can it be seen—but only in a few problems of the same mathematical type as one in which it occurs in the Rhind Mathematical Papyrus.

The evidence does not allow us to determine whether the Babylonian and Egyptian usages were borrowed from some common contact (e.g., a commercial intermediary), or whether shared computational tools or techniques or common Hamito-Semitic linguistic structures called forth parallel developments. But there is little doubt that the Greek use of composite fractions was borrowed from the Near East, together with the techniques for timekeeping, astrology, and notarial and mercantile computation. The idiom of composite fractions turns out to be an important index fossil, demonstrating that these techniques belonged in a common cluster, which was inspired or borrowed from the East as a relatively connected whole. The usage may have been in general use among Semitic speakers in the Near East, as the later Arabic sources suggest; but in the Greco-Roman world it never spread beyond the circle of practitioners of the techniques together with which it had been borrowed. It went inseparably with a specific subscientific tradition.

The usage (and hence probably the living tradition) may even have remained restricted to the Greek orbit. This is actually intimated by the *Propositiones.* Several problems from this collection point to the Eastern trading connections of the Roman world; but none of them contains composite fractions. As observed, those simple ascending continued fractions which occur suggest an Egyptian connection—which may be indirect but which appears in any case *not* to be mediated by the channel reflected in the *Anthologia graeca.*

The latter conclusion already involved another sort of index fossil: shared problems. Mathematical problems dealing with real applications may, of course, be shared already because features of social or technological organization are common (exceptions to this rule in the sphere of practical geometry will be discussed in Section 4); recreational problems, on the other hand, will often be so specious (regarding dressing, mathematical structure, and/or numerical constants) that chance identity can safely be ruled out. Shared recreational problems of this category therefore constitute firm evidence of cultural connections.

Much work has been done, especially by German historians of mathematics, on the distribution of specific recreational problems. The basic results on each type are conveniently summarized in the new edition of Tropfke's *Geschichte der Elementarmathematik* (Tropfke/Vogel 1980: 573–660), and there is no need to repeat the details. One very important category of problems is found in medieval India and Islam, and then again in late medieval Western Europe. It includes the "purchase of a horse," the "hundred fowls," and many others. Some of these problems are also found in China, mostly in a different dressing but recognizable because of a peculiar mathematical structure; the same stock was also drawn upon by Diophantos, who, of course, stripped the problems of their concrete dressing.[25] Some of the problems, finally, turn up in the *Propositiones*. Disregarding perhaps the latter work, the distribution of this group of problems coincides with the trading network bound together by the Silk Route and its medieval descendants (when the *Propositiones* acquired their final form, Francia was no longer an active part of the network; instead, the very eclectic collection reflects the various influences to which the Roman world had been exposed in antiquity). At least one problem belonging with the Silk Route group, however, can be traced to much earlier times: the successive doublings of unity treated so confusedly by al-Uqlīdisī's reckoners. In the same passage, he tells that these problems always deal with 30 or 64 successive doublings. The oldest version of the problem known to date is found in a mathematical tablet from Old Babylonian Mari.[26] This text has direct affinities both to the (Islamo-Indian) chessboard problem and to the version in the *Propositiones*, and probably also to a Chinese version that I have not seen.[27] Other problems may have originated elsewhere, but the Mari problem indicates that one of the centers from which the caravan-route merchants' culture grew was located in the Middle East (from where also the composite fractions appear to have diffused, without meeting with great success outside the Semitic-speaking area).

The "Silk Route group" encompasses many but far from all popular recreational problem types. The Egyptian flavor of certain problems from the *Propositiones* was pointed out previously. Similarly, the "non-Near-Eastern" problems of the *Anthologia graeca* may have been borrowed from the same location as the unit fraction system of which they make use—*viz.*, from Egypt. True, they are not specifically akin to anything known from Middle Kingdom Egyptian sources (the shared interest in inhomogeneous first-degree problems is too unspecific to allow any conclusion); but a large group appears to reflect precisely that kind of elementary mathematics teaching which Plato ascribes to the Egyptians in *The Laws* 819A–C.

The riddle character of recreational mathematics was referred to repeatedly; like other riddles, recreational mathematics belongs to the domain of *oral literature*.[28] Recreational problems can thus be compared to folktales. The distribution of the "Silk Route group" of problems is also fairly similar to the distribution of the "Eurasian folktale," which extends "from Ireland to India" (see Stith Thompson 1946: 13ff.). However, for several reasons (not least because the outer limits of the geographical range do not coincide) we should not make too much of this parallel. Recreational problems belong to a *specific subculture*—the subculture of those people who are able to grasp them. The most mobile members of this group were, of course, the merchants, who moved relatively freely or had contacts even where communication was otherwise scarce (mathematical problems appear to have diffused into China well before Buddhism).

The parallel between recreational (and most subscientific) mathematics and oral literature is more illuminating in another respect. Around 900 A.D., Abū Kāmil described the recreational problem of the "hundred fowls" as

> a particular type of calculation, circulating among high-ranking and lowly people, among scholars and among the uneducated, at which they rejoice, and which they find new and beautiful; one asks the other, and he is then given an approximate and only assumed answer, they know neither principle nor rule in the matter.[29]

This is not far from the scornful attitude of literate writers to folktales up to the Romantic era. Their disdain would not prevent many of them from using folktale material; but Apuleius and Boccaccio (to name only two of them) would then rework the material and "make it agree with good taste." This is also exactly what Abū Kāmil does with the problem of the hundred fowls: Instead of presenting without "principle nor rule" a particular haphazard solution to the indeterminate problem, he shows how to find the complete set of solutions. He was not the only "scientific" mathematician reworking subscientific mathematics in this way. Diophantos's treatment of the "purchase of a horse" reflects the same attitude and intention; others, from al-Khwārizmī and Leonardo Fibonacci to Cardano, Stifel, and Clavius would follow similar programs. Subscientific "sources" for "scientific" mathematics were thus accepted and used in a way that differed fundamentally from that in which "scientific" sources were used. Pointedly and paradigmatically: When medieval mathematical cultures got hold of Euclid he supplied them both with material and with norms for mathematical "good taste." With these norms in their luggage, mathematicians would look around in their own world and discover there a vast supply—anonymous and ubiquitous—of mathematical problems and techniques beyond the range of what was already available in acceptable form; they would pilfer it (at times appropriate it wholesale), either without recognizing their debts (as Diophantos, Apuleius, and Boccaccio) or while criticizing those poor fools who thought themselves competent (like Abū Kāmil and eighteenth-century *literati*). No wonder that sources dealt with like this were not reflected in al-Nadīm's *Catalogue*.[30]

## 4. Practical Geometry

Both composite fractions (at least when used outside the Semitic-speaking area) and the cluster of "Silk Route problems" belonged with the calculators' craft. They should not make us forget other traditions of subscientific mathematics engaged not in commercial transactions and accounting but in practical geometry. Their practitioners were surveyors, architects, master builders, and the like. In some cultures, all or some of these preoccupations were taken care of, if not by the same persons, then at least by the same groups as those engaged in accounting and such. So, surveying, accounting, and the allocation of rations were the responsibility of scribes with a common schooling both in Old Babylonia and in Middle Kingdom Egypt[31] (architects and master builders, on the other hand, may have belonged outside the scribal craft). In other cultures, however, the tasks were socially separated (Roman gentlemen might write on surveying, but not on accounting). It is therefore reasonable to deal with geometrical practice as a distinctive topic, the subscientific character of which need not be argued.

A reason why this topic should be discussed at all and not merely be dismissed with a *ditto* is the character of its index fossils. Recreational problems reflecting geometrical practice are relatively rare, and therefore not very informative (some exceptions will be mentioned); the methods used in real life, however, are more significant than in the case of computation, because they often make use of one of several possible approximations or techniques.

One familiar exemplification of this is the treatment of the circle. The Old Babylonians would find its area as $\frac{1}{12}$ of the square of the circumference (corresponding to the value $\pi = 3$) and the diameter as $\frac{1}{3}$ times the circumference.[32] The Egyptians would find the area as the square on $\frac{8}{9}$ of the diameter.[33] These methods are obviously unconnected. The ratio 3 between the diameter and the circumference is used in the Bible,[34] which fits well with that dependency on Babylonia which follows from borrowed metrologies (but which would of course prove nothing in itself). Ptolemaic Egyptian texts (written well after the conquest of Egypt by Assyrian and Persian armies and shortly after the establishment of Greek rule) determine the circular area as $\frac{3}{4}$ of the square on the diameter,[35] which looks like a grafting of the Babylonian numerical assumption upon the traditional Egyptian habit to determine the area from the diameter (and not from the circumference, as the Babylonians had done). Combined with other evidence this fossil can be taken to imply that the originally separate traditions had amalgamated in the Assyro-Persian melting pot. The adoption of a mathematically poorer way to calculate the circular area in Egypt shows that the amalgamation did not change the subscientific character of the domain. Increased mathematical precision ("truth") was apparently no decisive criterion for preferences; the new way may have been accepted because it offered greater calculational ease—or simply because it was used by the engineers and tax collectors of the conquerors.

Later practical geometries, from the Roman agrimensors to the Hebrew *Mishnat ha Middot,* began using the Archimedean approximation ($\pi = \frac{22}{7}$).[36] This then confronts us with a new phenomenon. The ratio $3\frac{1}{7}$ can hardly have had any practical advantages over $3\frac{1}{8}$—except perhaps in the hodometer, where Vitruvius appears nonetheless to have considered the latter value satisfactory.[37] For easy computation in current metrologies, $3\frac{1}{8}$ was certainly to be preferred. The only reason to embrace $3\frac{1}{7}$ is that it *derived from "scientific" mathematics.* The adoption is thus a portent of later tendencies to transform the autonomous subscientific traditions into *applied science.*[38] For a long time, however, it remained an isolated portent. The new value occurs as indubitable (or at least undoubted) *truth,* and often in eclectic connection with other techniques with no "scientific" merit, just as it is characteristic of the subscientific traditions. Not only were societal conditions (in particular teaching systems) not ripe for the autonomous subscientific practitioners' traditions to be absorbed; "scientific" mathematics itself was hardly developed to a point and in a direction where purposeful simplification and restructuration would make it solve the specific problems of geometrical practice. Even in late antiquity and the early Middle Ages, the subscientific category remains important if one wants to understand the specific character of geometrical practice and its relation to the "scientific" knowledge of the day.

## 5. Algebra

This also comes true if we consider the distinctive mathematical innovation of the early Middle Ages: algebra.[39] Conventionally, the rise of this discipline is traced to al-Khwārizmī's book on the subject.[40] It is, however, obvious already from al-Khwārizmī's preface that not only the subject but also the name was already established before his times. A few decades after al-Khwārizmi, Thābit ibn Qurrah wrote a treatise "On the Rectification of the Cases of *al-jabr*" (ed., trans. Luckey 1941), where the discipline was ascribed to a group of "*al-jabr* people." On close investigation the sources leave no doubt that these were the carriers of a subscientific tradition, the doctrines and techniques of which had been systematized, excerpted, and exposed by al-Khwārizmī from a "scientific" perspective and then put on a Euclidean basis by Thābit. The practitioners of the field appear to have been "reckoners," people engaged in accounting and juridical and commercial computation.[41]

The roots of this subscientific *al-jabr* tradition are not easily extricated, and may be diverse. Indian "scientific" algebra is out of the game, its organization being clearly different from (and on a higher level than) that of the tradition known to al-Khwārizmī. But some connection to India must be present, since the (far from self-evident) term used metaphorically for the first power of the unknown in a second-degree problem (*jidhr*, "root, stem, stub," etc.) is shared with Indian sources from the first century B.C.[42] But the term for the second power (*māl*, "property, possessions, fortune, assets," etc.; translated *census* in medieval Latin texts) coincides first with the term used for the unknown in a whole class of first-degree problems, and second with the *thēsaurós* used in analogous Greco-Egyptian first-degree problems in the same function.[43] As far as I know no equivalent term is found in the same standardized role in Indian sources (cf. also Datta and Singh 1962: II, 9f.).

It is thus not very plausible that the subscientific *al-jabr* tradition was borrowed directly from India; since both early writers of the subject are of Turkestanian descent, the best guess is perhaps that its home was somewhere in Central Asia (Khorezm? Iran?), and that it had developed in this cultural meeting place *par excellence*. Still, this can be nothing more than a guess. Nor can it be known for sure whether the tradition had roots back to Babylonian "algebra"; *if* it had, it had been much transformed before reaching the form in which we know it.

This form was that of a *rhetorical algebra*. What is meant by this concept can be shown on an example borrowed from al-Khwārizmī (I follow Rosen's translation (1831: 41f.), which is adequate for the present purpose):

If a person puts a question to you as: "I have divided ten into two parts, and multiplying one of these by the other, the result was twenty-one"; then you know that one of the two parts is thing, and the other ten minus thing. Multiply, therefore, thing by ten minus thing; then you have ten things minus a square [i.e., *māl*], which is equal to twenty-one. Separate the square from the ten things, and add it to the twenty-one. Then you have ten things, which are equal to twenty-one dirhems and a square. Take away the moiety of the [number of] roots, and multiply the remaining five by itself; it is twenty-five. Subtract from this the twenty-one which are connected with the square; the remainder is four. Extract its root, it is two. Subtract this from the moiety of the roots, namely, five; there remain three, which is one of the two parts. Or, if you please, you may add the root of four to the moiety of the roots; the sum is seven, which is likewise one of the parts.

In the first half, everything goes by "rhetorical," i.e., verbal argument. If we replace the term *thing* with *x*, and express the arithmetical operations by symbols instead of words, the reduction of the problem follows exactly the same path as we would follow today. The second half, then, solves the reduced problem ($10x = 21 + x^2$) according to a standard algorithm, without giving any arguments at all. This is indeed characteristic of the tradition.

True, al-Khwārizmī introduces geometrical justifications of these algorithms (justifications of the same character as the procedure used to solve the "broken reed" problem). These are, however, grafted upon the contribution of the subscientific tradition, in agreement with al-Khwārizmī's "scientific" perspective. This does not mean that he invented them himself,[44] nor that they were borrowed from Greek geometry, from which they differ in style. Instead, they were taken over from *another* subscientific tradition.

This tradition is reflected in another book, a treatise *On mensuration* written by some unidentified Abū Bakr and only known in a Latin translation from the twelfth century, the *Liber mensurationum* (ed. Busard 1968). Its first half has little to do with practical mensuration; instead, it contains a series of problems that we would today describe as "algebraic": A square plus its side equals 110; the 4 sides of a square plus its area equals 140; in a rectangle of area 48, the sum of length and width is 14; in a rectangle, where the sum of the area and the four sides is 76, the length exceeds the width by 2; etc. In most cases, two ways to solve the problem are indicated. The first, fundamental method is based on "naive" geometric arguments in Babylonian style, and formulated in a language that, down to grammatical details, repeats the phrasing of Babylonian texts.[45] The second method, which is stated to be "according to *aliabra*" and which is not given in all cases, is identical with al-Khwārizmian *al-jabr*.

Detailed analysis leaves no reasonable doubt that the trunk of the first part of the treatise reflects a subscientific tradition going directly back to Old Babylonian "algebra" (the time span might inspire doubts, but it coincides precisely with the span from the Mari doublings to the earliest versions of the chessboard problem, and from the Old Babylonian ascending continued fractions to those known from medieval Islam). This will have been the source from which al-Khwārizmī drew his geometrical justifications of the standard algorithms of *al-jabr*.

Who were the carriers of this tradition? The inclusion in a "treatise on mensuration" suggests that they were surveyors and possibly master builders. This guess is bolstered by a source from the late tenth century, Abū'l-Wafā''s *Book on What is Necessary from Geometric Construction for the Artisan*. One passage of this work[46] refers to discussions with artisans from this category, where these stick to ideals and methods corresponding precisely to those displayed in the *Liber mensurationum*. Algebra, as explained to posterity by al-Khwārizmī, is thus a merger of calculators' and practical geometers' subscientific traditions.[47]

This is, of course, a nice point. But does it imply that Old Babylonian scribal mathematics, which in its own times had combined calculation and practical geometry, was transmitted by practitioners of the latter field alone?

Not necessarily. Even though the *Liber mensurationum* looks like a direct continuation of Old Babylonian scribal traditions, its basis may indeed be different: A practical geometers' tradition that had also once inspired the Old Babylonian scribe-school teachers and was systematized by them.

For the moment, this is a hypothesis—but still a hypothesis derived from the sources. The main source is an Old Babylonian text dealing very systematically with problems concerning squares and sides.[48] It starts from the elementary beginning: Given the sum of area and side, to find the latter; then (in symbolic mistranslation) $x^2 - x = c$ (problem 2); $ax^2 + bx = c$ and variations on that pattern (problems 3, 4, 5, 6, 16); $x^2 + y^2 = a$, $x \pm y = b$ (problems 8–9); $x^2 + y^2 = a$, $y = bx$ (problems 10–11, 13); $x^2 + y^2 = a$, $x\,y = b$ (problem 12); $x^2 + y^2 = a$, $y = bx + c$ (problem 14); $x^2 + y^2 + z^2 + u^2 = a$, $x = bu$, $y = cu$, $z = du$ (problem 15); etc.— all in technical and standardized language. Then, suddenly, in problem 23, we find the following (my translation and restitution of damaged passages):

> In a surface, the four fronts and the surface I have accumulated,
> 0;41,40.
> 4, the four fronts, you inscribe. The igi of 4 is 0;15.
> 0;15 to 0;41,40 you raise: 0;10,25 you inscribe.
> 1, the projection, you append: 1;10,25 makes 1;5 equilateral.
> 1, the projection, which you have appended, you tear out: 0;5 to two
> you repeat: 0;10 nindan confronts itself.

Abstractly seen, this is once more the simple problem $x^2 + ax = b$. But the wording is different from that used in the beginning, and so is the procedure,[49] which locates "the four fronts" as four rectangles of width 1 (the "projection") along the edges ("fronts") of the square.

The same figure is used by al-Khwārizmī in the first of two alternative proofs of the algorithm solving the case $x^2 + ax = b$.[50] The problem of a square or rectangle to which is added the sum of all four sides will also be remembered from the *Liber mensurationum*. In contrast to the series of technicalized problems in the beginning of our Babylonian tablet this looks like a typical recreational problem created and transmitted in a surveyors' environment. At the same time, school systematization of the techniques involved in the solution of the recreational problem would automatically lead to something like the initial technical series. It is thus at least a reasonable assumption that the *square/rectangle plus four sides* was originally invented in a surveyors' environment as a recreational problem and then taken over by the scribal school and used to sharpen the wits and kindle the self-esteem even of future accountants and other calculators.

Another typical recreational problem for surveyors will have been the bisection of a trapezium by a parallel transversal. This was also popular in the Old Babylonian school; but it goes back at least to the twenty-third century B.C. (Jöran Friberg 1990, section 5.4.k). Old Babylonian texts presenting its solution share the characteristic vocabulary of second-degree ''algebra.'' ''Neo-Sumerian'' texts (dating from the twenty-first century B.C.), on the other hand, seem to ignore these subjects. It is therefore my guess that the surveyors' environment that created the quasi-algebraic recreational problems was Akkadian (in agreement also with the name ''Akkadian method'' of the completion trick, (see note 21) and non-scribal (whence illiterate). As the Akkadian tongue (in Babylonian dialect) became a literary language in the Old Babylonian period, adoption and ensuing systematization into the scribal school took place. At the same time, however, the original surveyors' tradition survived, and it is this tradition (I guess), rather than a direct descendant of the Old Babylonian scribal school, that surfaces for the last time with its pet recreational problems in the *Liber mensurationum* and in the early ninth century justifications of the algorithms of *al-jabr*.

## 6. The End of Subscientific Mathematics

Already Hero had attempted to improve the practitioners' tool kit by teaching them Archimedean formulae, and thus to bring the autonomous subscientific traditions under the sway of theoretical knowledge—we may regard his work as an early attempt to make *applied science*. He and other Alexandrians had some success, insofar as they had some practitioners' groups accept the Archimedean value for $\pi$ and the formula for the triangular area. As we saw, however, the subscientific character of practitioners' mathematics was not changed. Nor was it affected when Diophantos and others borrowed subscientific material and reworked it from a ''scientific'' perspective.

Changes set in, however, in the medieval Islamic world. Islamic scholars were more prone to accept the *specific* problematic of practitioners as legitimate without refusing for that reason the legitimacy of the ''scientific'' perspective; they were thus able to combine the ''Heronian'' and the ''Diophantine'' approaches, transforming material from both sources in a way that was relevant for practitioners. ''Applied science'' began making its way.[51] Thābit ibn Qurrah would still know the *al-jabr* people as a subscientific group in the mid-ninth century. Already one or two generations later, however, *al-jabr* would be known to Abū Kāmil only as al-Khwārizmī's discipline, and it would be listed by al-Fārābī exclusively under the heading of "ingenuities" with reference to *Elements* X, and not with subscientific, "practical" arithmetic or geometry (cf. note 30).

The process was never brought to completion in the Islamic world. Nor was it during the European Renaissance, where similar developments would take place from Jean de Murs to Adam Riese. In the mid-sixteenth century, however, from Stiefel's and Riese's times onward, the impetus of the process had become irresistible, and from then on specialists' mathematical practice was no longer semi-autonomous but dependent upon "scientific mathematics." In early Modern Europe, then, the concept of "subscientific knowledge" loses definitively its heuristic value, at least as far as mathematics is concerned (in most other fields, the change took place later); instead, we encounter the problem, to which extent (and in which sense!) "applied science" can be described as *an application* of "science." This, however, is a different problem, which I shall leave to historians and philosophers of Modern technology.

Recreational mathematics did not die with subscientific mathematics. But it stopped being the exclusive property of the minority of specialists who were able to grasp it. Like mathematical literacy, recreational problems became a possession of the majority—eventually, of virtually everybody.

# 3

# Mathematics and Early State Formation, or The Janus Face of Early Mesopotamian Mathematics: Bureaucratic Tool and Expression of Scribal Professional Autonomy

In memoriam
Kenneth O. May

## Introduction

*The following essay grew out of an invitation to present "something Babylonian" at the symposium "Mathematics and the State" at the XVIIIth International Congress of History of Science, Hamburg/Munich, August 1989. I took advantage of the opportunity to undertake a critical examination of the traditional, simplistic thesis—familiar from Herodotos (Histories II, 109) to Wittfogel's anti-Marxist vulgar Marxism—that mathematics was first created as a tool for early bureaucratic states. During the preceding decade I had looked at the "anthropology" of Babylonian mathematics under various perspectives, and the confrontation with the simplistic interpretation developed into a synthesis of these different approaches, each of which had been guided by a specific, more limited question.*

*In contrast to chapter 1, section 1, and its background in Høyrup (1980), both of which focus on the impact of teaching and only look at other facets through this mirror, Chapter 3 is thus an attempt to integrate all main aspects of the anthropology of early Mesopotamian mathematics as far as I understand them. I have strived, moreover, to substantiate and to add precision and detail to the sketch presented in Chapter 1. On the other hand, I include only as much technical mathematics as is necessary to elucidate the anthropological argument.*

*Obviously, the attempt at synthesis has led me to formulate certain views expressed in Chapter 1 (and elsewhere) differently, and to shift the emphasis at certain points. Furthermore, new epigraphic and archaeological material, as well as new interpretations of familiar sources, have come up during the 1980s. I will certainly not be aware of everything, especially not outside the domain of mathematical texts; nonetheless, what has come to my knowledge since 1980 weights heavily at several points.*

*Of special importance has been the series of Berlin Workshops on Concept Development in Babylonian Mathematics (four to date). As will be clear from the refer-*

ences, the synthesis draws extensively on work done by the members of this workshop, in particular on the works of Peter Damerow, Robert Englund, Jöran Friberg, Hans Nissen, and Marvin Powell.

 I have dedicated the study to the memory of Kenneth O. May, who in 1974 commented upon my first amateurish attempt at broad historical syntheses that although he agreed with my general thesis and found the generalizations plausible, what was needed were specific examples in which the interaction between mathematics and other phases of culture was "traced out *and* verified in detail" *(his emphasis). I hope that the essays contained in this volume, not least the present one, would have met his standards.*

# TIMETABLE

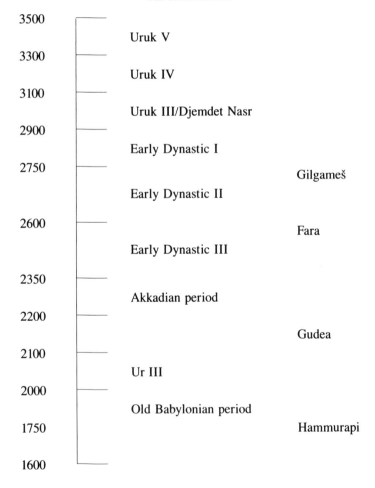

| 3500 | |
| | Uruk V |
| 3300 | |
| | Uruk IV |
| 3100 | |
| | Uruk III/Djemdet Nasr |
| 2900 | |
| | Early Dynastic I |
| 2750 | Gilgameš |
| | Early Dynastic II |
| 2600 | Fara |
| | Early Dynastic III |
| 2350 | |
| | Akkadian period |
| 2200 | Gudea |
| 2100 | |
| | Ur III |
| 2000 | |
| | Old Babylonian period |
| 1750 | Hammurapi |
| 1600 | |

## 1. Mathematics and the Early State

In his famous and somewhat notorious book *Oriental Despotism*, Karl Wittfogel (1957:29f.) presented a simple thesis connecting the first development of mathematics and astronomy with the rise of the early "Oriental" state—*viz.*, that the state was "hydraulic," i.e., developed in order to plan large-scale irrigation, and that mathematics and mathematical astronomy were created for that purpose:

(A)   The need for reallocating the periodically flooded fields and determining the dimension and bulk of hydraulic and other structures provide continual stimulation for developments in geometry and arithmetic. . . . Obviously the pioneers and masters of hydraulic society were singularly well equipped to lay the foundations for two major and interrelated sciences: astronomy and mathematics.

As a rule, the operations of time keeping and scientific measuring and counting were performed by official dignitaries or by priestly (or secular) specialists attached to the hydraulic regime. Wrapped in a cloak of magic and astrology and hedged with profound secrecy, these mathematical and astronomical operations became the means both for improving hydraulic production and bulwarking the superior power of hydraulic leaders.

This thesis is in fact widely held, though often in less outspoken and rigid form. As also observed by Wittfogel, it was already proposed by Herodotos to explain the presumed Egyptian origin of geometry. My reason to take Wittfogel's very explicit statement as my starting point is that it exposes the problematic nature of the conventional thesis so clearly. If we concentrate on Mesopotamia, Wittfogel is wrong on all factual accounts (Egypt would come out no better):

• Irrigation systems only became a bureaucratic concern (and then only in certain periods) many centuries after the rise of statal bureaucracy (which took place in the later fourth millennium[1]). No doubt the irrigation economy provided the surplus needed to feed the bureaucracy; but it was taken care of locally, and often by kin-based communities (as it often is even in today's Iraq).[2]
• Old Babylonian mathematical texts (c. 1700 B.C.) deal with construction of irrigation works, but only with the need for manpower, the wages to be paid, and the volume of earth involved. The dimensions of the constructions were not determined mathematically.

- Neither the sacred nor the secular calendar was ever involved in irrigation planning in Mesopotamia.
- Mathematical astronomy was only created about 3000 years after the rise of the state, and was concerned with the moon and the planets, i.e., it was irrelevant for irrigation planning.
- Even astrology is a late invention. Only in the first millennium are bureaucratic computation and occult endeavors of any sort connected through a common group of practitioners.

The easy version of the connection between the rise of the state and the development of mathematics (in Mesopotamia and elsewhere) is thus an illusion. In order to approach the issue in a profitable way we will have to ask some apparently trite questions: what *is* a state, and what *is* mathematics—if we are to discuss the two entities in the perspective of the Bronze rather than the Atomic Age.

## 2. The Early State and Its Origin

In his book, Wittfogel (1957: 383–86) points to two classical approaches to the problem of early state formation—both due to Friedrich Engels. Engels summarizes the thesis of *Die Ursprung der Familie, des Privateigentums und des Staats* as follows (*MEW* XXI, 166f.):

(B)    Da der Staat entstanden ist aus dem Bedürfnis, Klassengegensätze im Zaum zu halten, da er aber gleichzeitig mitten im Konflikt dieser Klassen entstanden ist, so ist er in der Regel Staat der mächtigsten, ökonomisch herschenden Klasse, die vermittelst seiner auch politisch herrschenden Klasse wird und so neue Mittel erwirbt zur Niederhaltung und Ausbeutung der unterdrückten Klasse.

In *Anti-Dühring*, on the other hand, he considers the state as "Verselbständigung der gesellschaftlichen Funktion gegenüber der Gesellschaft," which then, as the opportunity presented itself, changed from servant to master, whether "als orientalischer Despot oder Satrap, als griechischer Stammesfürst, als keltischer Clanchef u.s.w.," but where it shall still be remembered that "der politischen Herrschaft überall eine gesellschaftsliche Amtstätigkeit zugrunde lag" (*MEW* XX, 166f.).

Both points of view are present in the standard references of modern political anthropology. According to Morton Fried's *Evolution of Political Society,* the state arises as "a collection of specialized institutions and agencies, some formal and other informal, that maintain an order of stratification" (Fried 1967: 235), where a "stratified society" itself is understood as one "in which members of the same sex and equivalent age status do not have equal access to the basic resources that sustain life" (Fried 1967: 186)—i.e., in a generalized sense, a class society. Elman Service, on the other hand, sees statal organization as the result of a quantitative and often gradual development from "relatively simple hierarchical-bureaucratic chiefdoms, under some unusual conditions, into much larger, more complex bureaucratic empires" (1975: 306). The chiefdom itself is a hierarchical organization legitimized by social *functions* wielded by the chief for common benefit (according to Service, mostly functions of a redistributive nature) in a theocratic frame of reference, where "economic and political functions were all overlaid or subsumed by the priestly aspects of the organization" (Service 1975: 305).

Another oft-quoted contributor to the general debate should be singled out for relevance to the following. Robert Carneiro, arguing (1981: 58) that "what a chief gets from redistribution proper is esteem, not power," observes (Carneiro 1981: 61) that

(C)     As long as a chief merely returns everything he has been handed, he gains nothing in wealth or power. Only when he begins to keep a large part of it, sharing with his retainers and supporters but not beyond that, does his power begin to augment.

But the power of a chief to appropriate and retain food does not flow automatically from his right to collect and redistribute it. Villagers freely allow a chief to equalize each family's share of meat or fish or crops through redistribution because they benefit from it. But they will not willingly suffer the same chief to keep the lion's share of food for himself. Before doing this, he must acquire additional power, and that power must come from some other source.

Power, then, depends on the ability of the chief to transform redistribution proper (where the chief retains only a small percentage of what passes through his hands) into *tribute* or *taxation,* where he keeps a large part for himself and for the "core of officials, warriors, henchmen, retainers, and the like who will be personally loyal to him and through whom he can issue orders and have them obeyed" (61). The origin of this transformation Carneiro sees in *warfare resulting from population pressure.* Warfare is the reason that early class societies consist of three and not just two classes (Carneiro 1981: 65):

(D)     The two classes that are added to a society as it develops are a lower class and an upper class, and the rise of these two classes is closely interrelated. The lower class [ . . . ] consists initially of prisoners who are turned into slaves and servants. At the same time, however, an upper class also emerges, because those who capture and keep slaves, or have slaves bestowed upon them, gain wealth, prestige, leisure and power through being able to command the labor of these slaves.

Even though considering the transition "from autonomous villages, through chiefdoms and states, to empires" as a continuous process (67), Carneiro finally finds it useful to distinguish the state (69, quoting Carneiro 1970: 733) as

(E)     an autonomous political unit, encompassing many communities within its territory and having a centralized government with the power to draft men for war or work, levy and collect taxes, and decree and enforce laws.

Though illustrated by references to ethnographic and historical material, the theories cited here are general theories. During the last 15 to 20 years they have been tried out by specialists on a large number of single cases, which has provided many insights into the applicability of the concepts involved and into the historical variability of the diverse processes to which the theories make appeal. It would lead too far to discuss them in general,[3] and I shall only quote two of special relevance for the Mesopotamian case. First, in a discussion of Archaic Greece Runciman (1982: 351) distinguishes "the emergence of a state from nonstate or stateless forms of social organization" by "these necessary and jointly sufficient criteria":

(F)     Specialization of governmental roles; centralization of enforceable authority; permanence, or at least more than ephemeral stability, of structure; and emancipation from real or fictive kinship as the basis of relations between the occupants of governmental roles and those whom they govern.

Second, working on Mesopotamian and Iranian material Henry T. Wright and Gregory A. Johnson (1975: 267) formulate a description focusing "on the total organization of decision-making activities rather than on any list of criteria," defining a state

(G)     as a society with specialized administrative activities. By "administrative" we mean "control," thus including what is commonly termed "politics" under administration. In states as defined for purposes of this study, decision-making activities are differentiated or specialized in two ways. First, there is a hierarchy of control in which the highest level involves making decisions about other, lower-order decisions rather than about any particular condition or movement of material goods or people. Any society with three or more levels of decision-making hierarchy must necessarily involve such specialization because the lowest or first-order decision making will be directly involved in productive and transfer activities and second-order decision-making will be coordinating these and correcting their material errors. However, third-order decision-making will be concerned with coordinating and correcting these corrections. Second, the effectiveness of such a hierarchy of control is facilitated by the complementary specialization of information processing activities into observing, summarizing, message-carrying, data-storing, and actual decision-making. This both enables the efficient handling of masses of information and decisions moving through a control hierarchy with three or more levels, and undercuts the independence of subordinates.

Unless "information," "data-storing," etc. are taken in a rather loose sense, societies traditionally regarded as indubitable states (such as Charlemagne's Empire) may well fall outside this definition. But in the Irano-Mesopotamian case the authors succeed in making it operational by means of sophisticated archaeology and through the application of geographical "central place theory." Furthermore, the specific definition of "control" involved may serve to distinguish the specific character of Irano-Mesopotamian state formation.

Control, indeed, may differ in kind—even control developed to the degree of vertical and horizontal specialization and division of labor described by Wright and Johnson. But if control and decision-making involve intense message-carrying and *data*-storing as the fundament for further decision-making, as was the case in Mesopotamia (cf. Section 3), then some means for accounting and the handling of data must develop together with the state—whether it be writing and numerical notations, something like the Andean *quipu*, or some third possibility. For this same reason, indeed, "archaeologists like[d] to use 'writing' as a criterion of civilization" (roughly synonymous with statal culture), as Gordon Childe pointed out in 1950,[4] while at the same time himself pointing to the equally important role of accounting (Childe 1950: 14). This brings us back to the problem of Mesopotamia.

## 3. The Rise of States in Southern Mesopotamia

The center of early Mesopotamian state formation was the southern-most part of Mesopotamia (Sumer); furthermore, for the whole period that I am going to consider in depth, the essential developments as far as mathematics is concerned took place in the Sumerian and Babylonian south to center—whence the heading. A description of the prehistoric development, however, cannot be circumscribed meaningfully to this area—already because most of the Sumerian territory was covered by water during most of the prehistoric period, but also because much larger areas were involved in parallel developments.

By 8000 B.C., permanent settlements had been established and agriculture and herding had become the principal modes of subsistence, although hunting, food-gathering, and fishing remained important subsidiaries well into historical times. Within the single settlements, social stratification may have developed around redistribution—needed precisely because of the combination of several complementary subsistence modes, cf. quotation (C). The single villages, however, were involved in no higher structures of settlement or redistribution—their very ecological localization shows that they were meant to live on their own, apart from participation in long-distance trade in obsidian and similar rare goods. This self-sufficiency holds good even for the rare large settlements such as Jericho, level B (seventh millennium), with at least 2000 inhabitants, and Çatal Hüyük (sixth millennium), with at least 5000, even though the internal social organization and stratification will probably have been much more complex here than in smaller settlements.[5]

In the sixth to fifth millennium, the paths followed by different parts of the Middle East diverged. In geographically suitable places such as the Susiana plain in Khuzestan (southwestern Iran), larger numbers of settlements can be seen to form interconnected systems, some of them possessing apparently central functions or positions (to judge, *inter alia,* from systematic size differentiations).[6] In the late fifth millennium, the city Susa had an area of some 10 hectares and was the center of a system of smaller settlements in Susiana. Central storerooms in what may be a sacred domain have been found in the city, and findings of seals and seal impressions in Susa and a neighboring small settlement bear witness of controlled delivery of goods from the small settlement to the center. But no traces of higher-level recording or summarizing occur in this archaeological layer (Wright and Johnson 1975: 273).

After a setback in population density,[7] the Early Uruk period (before the mid-fourth millennium) brought new growth. Susa had grown to 13 hectares and was the center of a three-level settlement system; ceramic ware was distributed from central workshops; and bitumen, chert, and alabaster were produced locally for exchange. In the following (Middle Uruk) period, the size of Susa doubled to 25 hectares and the city was internally differentiated; the settlement system became four-tiered. There is direct evidence for differentiated levels of administrative control (by means of seals, "tokens" and "bullae," cf. below), and perhaps already indirect evidence for the distribution of standardized grain rations to institution workers.[8] In the Late Uruk period (3500–3100 B.C.), the trend toward specialization and hierarchical control continued. Now, however, a similar level was reached in southern Mesopotamia, where Uruk became the dominant center. Susa, on the other hand, fell behind, and will be less interesting for the arguments of the following.[9]

The reason for this development is to be sought in climatic changes, which lowered the water level in the Gulf by some 3 meters after the mid-fourth millennium and diminished the rainfall in the area (Nissen 1983: 58–60). As a consequence, land that had been covered by salt marshes or had been inundated regularly by the rivers now became available for irrigation agriculture. Until then, settlements in southern Mesopotamia had been rather few, and not part of higher-level systems. Now, however, a larger settlement density (larger than anything known before in the Near East) and the creation of a noticeable surplus in agriculture became possible. The city Uruk (as large as 50 hectares in the Late Uruk period, and soon much larger still) became the center of a four-tiered settlement structure; the internal productive and administrative organization of the city was highly differentiated, vertically as well as horizontally; huge public works in the form of temple building were performed, workers as well as officials being paid in rations in kind; and outposts in northern Iraq as well as trading relations to Bahrain were established.[10]

The evidence for this development is twofold. Part of it is made up by the traditional archaeologists' array of settlement and building remains and of other artifacts. Part of it, however, consists of *carriers of meaning:* pictures carved in cylinder seals, on relief vases, and the like; and inscribed clay tablets, first with numbers only (in the preliterate Uruk V stratum, before 3300 B.C.) and then also with pictographic writing (in the "protoliterate" Uruk IV and Uruk III periods, 3300–2900 B.C.).

Even though there is an indubitable continuity from the Late Uruk script to the later Sumerian cuneiform, it is far from completely deciphered; cylinder seals, like all other pictures, are of course always ambiguous and polyvalent. Nonetheless, the combination of these carriers of social and linguistic meaning conveys a lot of information not available from earlier periods. Prominent facets of the picture that emerges are as follows:

- The city (and, as a consequence, the settlement system whose center it was) was under theocratic control. Its core was made up of a sacred terrain dominated by a number of large temples, which can only have been built because of the existence (and availability to the theocratic rulers) of a large agricultural surplus.

- Part of this surplus was apparently given as tribute—a famous temple vase shows a procession bringing offerings to the city goddess Inanna (reproduction and discussion in Nissen 1983: 113–15). But part of it must also have been extracted from laborers working directly on Temple domains, many of them most likely enslaved prisoners—apparently the most popular theme of the cylinder seals of high Temple officials shows vanquished and pinioned prisoners watched by a high (supreme?) official and being beaten up more or less explicitly (reproduction, select specimens, and discussion in Nissen 1986: 146–48).
- The ruling group of the city was constituted by the top officials in a hierarchy also encompassing lower officials and craftsmen's and workers' foremen (cf. below, on the "profession list"). All appear to have received rations in kind in a sort of quasi-redistributive system, and at least the highest officials received important allotments of land (Vaiman 1974: 20f.; whether this land was worked by personal servants or slaves or by "public" laborers is unclear).
- Quasi- (or pseudo-)redistributive features were also furthered by the lack of virtually all natural resources apart from pastures, agricultural land, fish, fowl, reed, and clay. All needs apart from these (in particular, i.e., those arising from temple building and equipment and the luxury needs of the governing group) depended on organized import and distribution.
- To keep track of tribute and other deliveries and of the products of public agriculture and herding, and also in order to calculate the rations of officials, workmen, and domestic animals, techniques for accounting and computation were developed (details follow). In the earliest (Uruk V) phase, tablets carrying only numerical/metrological inscriptions and seal impressions of responsible officials were employed. Whether used for accounting, as receipts, or as delivery notes, they could only be understood by somebody possessing full knowledge of the context of the transaction in question. In the next, terminal phase of the Late Uruk period (stratum IV, 3300–3100 B.C.), pictograms are written together with the numbers. Even though there is no full rendition of any spoken language, nor any attempt to render syntax, the tablets could now be used as supports for memory, and to summarize a whole series of transactions while tracking its course—especially because the tablets are written according to a fixed format for single transactions and totals. In the ensuing Uruk III or Jemdet Nasr phase (c. 3100 to 2900 B.C.), these formats grow more complex and more regular.[11]
- There is no doubt that the script was developed as an accounting and control device. Fully 85 percent of all written documents belong to the cat-

egory of economic tablets. The remaining 15 percent are made up by "lexical lists", apparently used for teaching purposes. A "profession list" describing the hierarchy of officials and professions turns up most frequently in the record. Other lists enumerate herbs; trees and wooden objects; dogs; fish; cattle; birds; place names; vessel types; and metal objects (see Nissen 1981). Literary and religious texts are as absent as monumental inscriptions.

• Nothing in the record suggests that general Temple functions, management of the Temple estate, and practical bookkeeping were separated. To the contrary, literacy (confined to the sole function of economic control) will probably have been too restricted for any full separation to have taken place (nor has a specific scribal function been identified in the profession list). As to the merging of priestly functions and Temple estate management, precisely the sanctification of originally redistributive functions will have made possible that transformation of redistribution into taxation that might otherwise have been impossible (cf. quotation [C]).

Whereas this much is fairly well established, other questions remain open—not only because the script is largely undeciphered, but also because of the nature of the written evidence. Three open questions are of some relevance for the present study.

First of all, the reach of statal domination is unclear. The profession list, as well as the location and immense size of temple buildings, tells us that the statal institution *par excellence,* irrespective of our choice of precise defining criteria, was the Temple. We know that the Temple bureaucracy had command of a large work force, that these workers, as well as a number of officials of varying rank, were supplied rations in kind. But we do not know how many of the workers were enslaved, nor whether there existed a stratum of peasants only loosely submitted to the Temple (paying, for instance, a limited tribute in the form of temple offerings or perhaps none at all, maybe and maybe not contributing corvée labor).[12] Temple accountants, after all, recorded transactions that regarded the Temple economy; they were not engaged in social statistics. Evidence from the third millennium suggests that free, kin-based peasant communities will have been an important part of the total social fabric.[13]

Second, we do not know the real make-up of the bureaucracy. Because we only know it from accounting and glyptics we may be inclined to see it as a suppressive and theocratic yet fundamentally Weberian bureaucracy. Ethnographically, however, this picture is highly improbable, and prosopographic studies of third-millennium material has given Marvin Powell (1986: 10) the impression that "the entire bureaucracy is knit together by an elaborate system of kinship, i.e., what we would call nepotism and influence."

Third, the specific organization of urban society, of the total settlement structure (not least concerning outposts like the town Habuba Kabira, built in northern Mesopotamia during Uruk V and then abandoned, and the relations to other administrative centers developing no later than Uruk III) and of most trades and handicrafts is unclear. Were traders Temple officials (in the mid-third millennium, some private venture must be presumed; see Adams 1974; 248)? Were the "chief," "junior chiefs," and "foremen" of the professions testified to in the profession list (see Nissen 1974: 12–14) really members of an all-encompassing hierarchy, or is the organization of the profession list a result of the particular and biased perspective of literate Temple bureaucrats? Is the appearance of the "chairman of the assembly" in the profession list an indication that a formerly primitive-democratic assembly of citizens had been subsumed under the Temple hierarchy, or is this just an expression of priestly wishful thinking?[14] Once again, third-millennium parallels suggest that the real situation was more intricate than the information that we are able to extract from the written documents.

These conclusions from third-millennium parallels may be combined with an observation made by Joan Oates (1960: 44–46): since both the essentials of temple groundplans in Eridu (one of the originally isolated settlements of the extreme south) and many other religious customs exhibit continuity since the fifth millennium, at least the culturally pivotal segment of the Late Uruk state building population appears to be autochthonous. The explosive increase in population after the mid-fourth millennium, on the other hand, is probably not to be ascribed to autochthonous breeding alone. Influx of new population segments, regimented somehow by the Temple institution (whose organization may have taken over much from the corresponding Susa institution) may have contributed to the creation of the three-class situation described by Carneiro (see quotation [D]): thanks to the surplus extracted from the Temple clients and subjects, the Temple staff could evolve into a new upper class, whereas the clients and possible enslaved workers made up the new lower class. Non-subject populations (be they autochthonous or immigrants) may have continued a traditional non-state existence with only limited submission to the statal institution.[15] For the very same reason, however, they will have been out of the administrative focus of the Temple managers. That accounting rationality which, as we shall see, contributed to the formation of mathematics, was only concerned with the relation between the Temple estate and its officials and dependants—and whatever the real complexities of state formation, the written record only reflects the pseudo-redistributive features of the situation.

As long as we restrict ourselves to the protoliterate period alone, however, all talk of the "real complexities" is first, pure speculation, and second, inane speculation. It is only given sense by the perspective of the following, "Early Dynastic" period (cf. Diakonoff 1969a: 178–80, and Powell 1978: 139).

## 4. City-States and Centralization

Apart from an initial lacuna of some 200 years in the written record, the source situation improves steadily and significantly during the following millennium. This has several reasons.

- First, the script evolved to the point where it is fairly well understood—both because of changes in the sign repertoire and because of incipient use of syllabic writing. Because of the latter development we even know that the language in use was now Sumerian, while we have no means to decide in which language the pictographs of the protoliterate period were told.[16]
- Second, writing was used much more broadly and more systematically. Around the mid-third millennium royal inscriptions, literary texts, and political and juridical documents (some of them involving communal and private land) turn up. Even the traditional genre, the economic texts, improves in coherence and systematization.
- Third, certain aspects of early third-millennium society are reflected in oral epics written down in the second half of the millennium.
- Fourth and finally, on a number of archaeological sites strata from the third millennium cover those from the late protoliterate period, for which reason the latter are poorly known.

The first 500 years after the protoliterate phase are known as the Early Dynastic period (ED). Its first part is characterized by continued population growth—around 2900 B.C. Uruk had grown to six square kilometers, half of Imperial Rome at its height—and by further diminishing rainfall and lowering water-level in the Gulf and hence also in the great rivers. Around the mid-third millennium, moreover, a new main branch of the Euphrates was formed. This had decisive consequences, as discussed in some detail by Nissen (1983: 141–48). What is important in the present connection is the development of a system of city-states, competing and often at war for the same water resources; and of kingly functions in these city-states, formally originating as Temple offices but in reality regents on their own and eager to stand forward in their inscriptions as benefactors and protectors of the temples of their cities and city gods (see the collection of royal inscriptions in Sollberger and Kupper 1971).

One of the Sumerian epics offers an interesting insight into the social structure, somewhat at cross-purposes with naive identification of State and Temple estate. In *Gilgameš and Agga* (ed. trans. Römer 1980) we are told that Agga, son of king Enmebaragesi, proceeds with his army from Kish to Uruk and delivers an ultimatum. King Gilagmeš of Uruk first tries to convince the council of elders of his city to fight back; he fails, and instead puts the matter before the council of "men" (*guruš*—capable of bearing arms? or commoners, if the "elders" are elder by status and not by age?), who agree with Gilgameš and entreat the aristocrats and mighty of the city to fight for Eanna, the city's temple established by An the heavenly god and "cared for" by the hero-king.

Most likely, the epic was only committed to writing toward the end of the third millennium; but since Enmebaragesi is a historic person (he has left an inscription, and belongs around the twenty-seventh century, in early ED II) the written text must build on fairly stable oral transmissions. Moreover, the conciliar institutions were definitely not as powerful toward the end of the millennium as presupposed by the text, and the term *guruš* had come to designate semi-enslaved laborers with no political influence or rights. The social situation delineated in the poem must therefore correspond to *some* historical reality.

That, however, is striking. Admittedly, Eanna is mentioned as the pride of the city—but definitely not as supreme owner or overlord. The affairs of the city are taken care of by the king in agreement with the two councils. The whole make-up reminds one more of the *Iliad* than of the managerial society intimated by the protoliterate archives. If the higher Temple officials are mentioned (and they probably are!), it is only as rich and powerful "first-class citizens," i.e., as aristocrats or "elders."

On the other hand, there is no doubt that the managerial tradition *was* very much alive, as testified by the continued and expanding use of the same script and the same accounting techniques as in Uruk IV–III, and by the persistent use of the familiar lexical lists. We are thus confronted with a truly *dual society,* as suggested above: one aspect can be described with some approximation as that "military democracy" that Engels portrays in *Der Ursprung der Familie.*[17] The other is the formally redistributive, functional state presupposed in his *Anti-Dühring*—and since these two complementary theories anticipate the main approaches of modern political anthropology we may conclude that the disagreements within this field correspond to the dialectic of real state formation.[18]

. . . At least to *the dialectic of real state formation as it happened in Mesopotamia.* The duality is, indeed, more obvious here than in many other cases (cf., however, Section 12 on parallels in medieval Europe). That is seen, for instance, if one compares the ways in which early Mesopotamian and other ancient monarchs made use of the techniques of literacy, once developed for accounting, to glorify themselves. Whereas most royal inscriptions of the ancient world boast of prowess and military success, until mid-ED III Sumerian royal inscriptions boast of temple building, of gifts given to the temple, of ceremonies performed, and of canal-building. Early Mesopotamian literacy was thus no transparent medium, but a strong ideological filter that would not allow certain utmost important aspects of the kingly function to be seen.[19]

Toward the end of the Early Dynastic period the temples and temple estates have come under the sway of the city kings, who treat them as their private property.[20] The existence of communal land is testified by sales contracts, but these always show that the land is sold to private individuals with high social status (high officials, members of the royal family), and often "at a nominal price" (Diakonoff 1969a: 177; cf. Powell 1978; 136f.). Since peasant clans will in any case only have sold their hereditary land when in distress or when submitted to severe pressure, we may conclude that this was probably the point where a state in Runciman's sense was establishing itself (cf. quotation [F], and note 15).

In the mid twenty-fourth century, as a next developmental step, the whole Sumerian region was then united into one territorial state by conquering kings, first by Lugalzagesi of Umma and soon afterward by the Semitic Sargon of Akkad. Powell (1978) sees this as a result of the conflicts arising, *inter alia,* from growing social and political tensions caused by the increase of private large-scale property—tensions that could not be released or held in check within the single city-state, in spite of attempts such as Uru'inimgina's "social reform".[21] From now on, the "despotic" territorial state or empire can be regarded as a rule in Mesopotamia, and the decentralized phases as interludes.

For reasons of obvious necessity, Sargon and his dynasty introduced more wide-ranging social controls than any predecessor, many of them further developments of the traditional accounting controls. Already the Early Dynastic radical changes in sociopolitical structure, however, had led to changes in the domain of written administration. Both phases of the transformation were reflected in the structure and practices of the environment responsible for this administration. The evolution that took place during the Sargonic reign continued trends established during the preceding two centuries while at the same time transforming them to the advantage of government.

The first step is testified in Fara (ancient Šuruppak) around the mid-third millennium. Here, for the first time, the scribes turn up in the administrative documents as a separate and hierarchically organized group, even provided with overseers and a "senior scribe" (Tyumenev 1969: 77); until then, the very term is absent from the sources—with the exception of one

Jemdet-Nasr tablet that shows that the profession is not hidden in one of the uninterpreted lines of the protoliterate profession list.

The reason for the emergence of the profession is probably straightforward: writing itself was used more widely for socially important purposes, apparently in connection with the beginning of the above-mentioned socio-economic transformations of ED III (see Powell 1978: 136f.). It is precisely in Fara that legal contracts, *viz.*, concerning the sale of land, turn up (see Krecher 1973, 1974). In Fara, too, a monetary function becomes visible for the first time (in Fara accomplished by copper, in later ED III by silver). Temple estate accounting, too, grew in extent and systematics. We seem to stand at the threshold dividing "ultra-limited literacy" from "limited literacy", to use a conceptual distinction proposed by John Baines (1988: 208).

As pointed out by Baines, limited literacy is really a *new* situation, with problems and possibilities of its own. First of all, this reflects itself in the education of the literate-to-be. Even though the old lexical lists were still in full use (but in decline after Fara), new types of school texts emerge, as it appears from Deimel's collection (1923; on p. 63 we find a student's drawing of the proud teacher); of special importance are the mathematical exercises, to which I shall return. Finally, the Fara period produced the beginning of *literary texts*, testified by fragments of a temple hymn and by the first proverb collection (Alster 1975; 15, 110). It seems that the scribes, once they had become a profession halfway on their own,[22] tried out the possibilities of the professional tools beyond their traditional scope (this will be even more obvious when we come to the mathematical exercises)—and a perusal of the tablets that the Fara scribe students produced suggests that they liked the enterprise: in many of the empty corners of tablets, irrelevant but nice drawings have been made, portraying teachers or deer or featuring complex geometrical patterns. One gets an immediate impression of enthusiasm for the freshness of scholarly activities similar to that reported from Charlemagne's Palace School in Aachen, or from Abelard's and Hugue of Saint Victor's "twelfth-century renaissance."

The trends beginning in Fara continue during ED III, during the Sargonic era, and during the post-Sargonic, decentralized twenty-second-century interlude. The number of legal contracts of many sorts keeps growing (cf. Bauer 1975); archives are used on many levels.[23] Systematic school teaching continues, though relatively few records (among which, however, mathematical exercises) survive. Writing becomes more phonetic and orderly already in ED III (Maurice Lambert [1952: 76] speaks of an outright *reform of writing* under Eannatum of Lagaš).[24] Even the creation of literary text continues, though with a change. No longer an expression of semi-autonomous scribal identity, hymns are written in the royal environment where they serve to demonstrate the king's affection for those temple institutions which had been subjected to royal authority, as discussed by William Hallo (1976: 184–86). Sargon's daughter Enḫeduanna may indeed be the first poet in world history known by name. Gudea, the most important ruler of Lagaš during the post-Sargonic decentralized interlude, appears to have had epics composed on command that transposed his own feats into the mythical past. Also in another respect is he seen parading as a culture hero: not only a temple builder in the abstract, like the kings of earlier inscriptions, Gudea has drafted the ground plan himself "in the likeness of Nisaba [the scribal goddess], who knows the essence of counting" (Cylinder A, **19**, 20–21, in Thureau-Dangin 1907: 110); he has also formed and baked the brick, brought precious materials from foreign countries, and performed all other crucial steps in person. Though the ruler of a city-state similar to those of former times and perhaps conscious of himself as a restorator of the order of old, Gudea no less than the Sargonides represents the tendency to make *inter alia* scribal culture subservient to a fundamentally secular power.

This is no less true in the following centralizing period, the so-called Third Dynasty of Ur or Ur III (not to be mixed up with Uruk III, a period named after an archaeological stratum in a different city), coinciding with the twenty-first century B.C.[25] The founding king, Urnammu, subdued the whole of southern Iraq, and undertook large building programs. Since relatively few written documents are known from his time, we have no detailed knowledge of his policies, nor from the first 20 years of his successor Šulgi. At that point, however, Šulgi instituted a military and administrative reform, and from then on huge amounts of administrative tablets exist. They uncover a centralized economy submitted to meticulous control. It is probably *not* true, as has been believed, that *all* land belonged to the state or to temple estates in practice controlled by the state; that *all* industry was governmental; that *all* merchants were *exclusively* government agents; nor that *all* manual work was done by semi-enslaved populations. But the very fact that these theses have been widely held shows that royal estates, governmental trade and governmental workshops, and even textile factories worked by slaves were all-important.[26] The precise booking of rations, work days, and of flight, illness, and death within the work force allotted to each overseer also reveals an extremely harsh regime. As pointed out by Robert Englund in his conclusive words (1990: 316), the understanding of working conditions conveyed by the administrative texts "kann vielleicht helfen, sich in den historischen Darstellungen des 3. Jahrtausends v. Chr. die Kosten der babylonischen Paläste und Statuen plastischer vorzustellen."

In this situation, whatever autonomy may have been left to communities and crafts will have been severely restricted. This is demonstrably true for scribal culture. The scribe, of course, was the pivot and, in principle, the hero of an administrative system the precision and scope of which Nikolaus Schneider (1940: 4) regarded as "überspitzt" even from his writing perspective within the National Socialist war economy. The scribal title was used as an honorific title of dignitaries in general (Falkenstein 1953: 128). Moreover, in one of the hymns glorifying King Šulgi he also presents himself as "a wise scribe of [the scribal goddess] Nisaba," a characteristic that stands as the culmination of a long series of images (trans. Klein 1981: 189, 191):

(H)  I, the king, from the womb I am a hero, [ . . . ], I am a fierce-faced lion, begotten by a dragon, [ . . . ], I am the noble one, the god of all the lands, [ . . . ], I am the man whose fate was decreed by Enlil, [ . . . ], I am Šulgi who was voluptuously chosen by Inanna [goddess of Uruk], I am a horse, waving its tail on the highway, [ . . . ], I am a wise scribe of Nisaba. Like my heroism, like my strength, my wisdom is perfected, its true words I attain, righteousness I cherish, falsehood I do not tolerate, words of fraud I hate!

Looking back at *Gilgameš* and *Agga* we observe that nothing is left of dual society. The world of kingly prowess and that of scribal administration (identified with wisdom and justice) are united in the same person, who boasts on both accounts in the same composition.

The so-called Ur-Nammu law code, which should in fact carry Šulgi's name (Kramer 1983; cf. Neumann 1989), shows a similar mixture in its prologue (ed., transl. Finkelstein 1969: 66–68). At the same time it elucidates the royal idea of justice, which on one hand involves metrological regularization and reform, on the other repeats the nice words (and the details!) of Uru'inimgina and Gudea too much in the manner of a literary topos to be really convincing (cf. Edzard 1974).

Two other Šulgi hymns (Sjöberg 1976: 172f.) tell about the king's purported time in the scribal school, and thus make clear which aspects of scribal cunning were central seen from the official perspective (which, we can be fairly sure in a society such as Ur III, was also the perspective communicated to the students): addition, subtraction, counting, and accounting according to one of the hymns; writing, field mensuration and drawing of plans, agriculture, counting, and accounting (and a couple of ill-understood subjects) according to the other.

Traditional topoi and nice handwriting apart, the idea of *justice* had been reduced to unified metrology and menaces against trespassers of royal regulations, and that of scribal art to functionality within the administrative apparatus. According to all evidence, scribes were taught in school to be proud of their function in the administrative machinery; no more place is left (in the official ideal) to professional autonomy than to communal primitive democracy. The higher level of literary (and, as we shall see, mathematical) creativity was in all probability the preserve of a "court chancellery" (*Hofkanzlei*, Kraus 1973: 23) where year names, royal hymns, politically suitable epic poems, and royal inscriptions were produced. On all accounts, the scribal art had been harnessed to a no-longer dual state—in trite practice, insofar as rank-and-file scribes are concerned; as a source for ideology, in the case of the elite.

## 5. Breakdown and Apogee

In spite of the immense role played by the scribes in Ur III, the problems associated with "limited literacy" appear to have been solved or suppressed. Scribal autonomous thought, as any autonomy except perhaps nepotism and appropriation of "public" property among the privileged, is absent from the sources. But the cost of bureaucratic control was too high, and the price of extensive building activities and an all-encompassing administrative network was a work force plagued by illness, death, and problems of flight—and even, if we are to believe indirect literary evidence, rebellious strikes.[27] Internal breakdown resulted,[28] followed by now-irresistible barbarian invasions and another interlude of decentralization, the beginning of the Old Babylonian period (2000 to 1600 B.C.).

One of the resulting smaller states (Isin) continued the Ur III system as best it could for a century, and has provided us with a school hymn describing the high points of the scribal art as embellished "writing on the tablets," together with use of "the measuring rod, the gleaming surveyor's line, the cubit ruler which gives wisdom",[29] not far from Šulgi's ideals, though without his emphasis on accounting. The other main successor state (Larsa) inaugurated a trend that was to culminate during the next phase of centralization, achieved by Hammurapi of Babylon (1793–1750).[30] On the whole, the system of state-controlled production was abandoned. Royal land was often (though not always) given to tenants instead of being organized as large estates run by servile labor, or it was assigned to officials or soldiers who leased it to farmers. Similarly, land belonging to wealthy city dwellers was often leased—and in general, private possession of large-scale landed property became common. (The survival of community-owned land is disputed; cf. Komoróczy [1978] versus Diakonoff [1971]).

Similarly, public foreign trade was replaced by private trade; at least one major city appears to have been run by the body of merchants with some autonomy (Oppenheim 1967). Royal workshops had probably been taken over by their managers at the breakdown of the Ur III system, and were now run privately; free laborers working for wages largely replaced the semi-enslaved workers receiving rations in kind. We even observe a kind of banking developing, conducted by members of an institution for unmarried noble-class women using their double kinship affiliation (to the real kin, and to the pseudo-kin of the institution) to bypass traditional obstacles to free trade in land (Stone 1982).

The activity of the latter institution testifies to the tendency to evade the constraints of communal traditions; it is also, on the other hand, one of many proofs that land—the all-decisive productive asset—was not exchanged on real market conditions (cf. Jakobson 1971). Individualism and monetary relations dominated the economy, but capitalism was far away. Anyhow, the new economic structure caused changes in many sociocultural spheres.

First, of course, business did not give up accounting and archives just because it was private. On the contrary, these spread to new social circles. Private letter-writing emerged, describing both private business and personal affairs—until then, only official letters had been known. Seals, hitherto insignia of officials, became tokens of private identity. And, of course, accountants and surveyors in private employment and street scribes writing down personal letters for pay appeared, as did freelance priests performing private religious rites.

Second, *individualism* itself took shape as a world view, manifesting itself not only in the private seal and the personal letter, but also in the religious sphere and in art. Whereas Ur III had consummated the transformation of the ordinary member of the primitive community into a subject of the state,[31] the Old Babylonian era made him reappear as a *private man*.

On the other hand, Old Babylonian society was still a royal state. The king was, as during many preceding centuries, the largest estate owner and directed many affairs, whereas local autonomies, when existing, were restricted. A new duality had thus evolved where, clearly, the "modern" aspect of society was the more vulnerable. Corresponding to the traditional royal aspect of society, the ancestral royal ideology also survived, and in fact got its most famous expression precisely in this time: the preface and postface to Hammurapi's law code (translated in Pritchard (ed.) 1950: 164–80), where the king appears as a sort of Bronze Age social democrat, assuring for his country affluence *and* justice. (The details of the text and the king's personality as it can be seen from his letters makes this look more honest than in Šulgi's comparable text).

The institution that connects this to the development of mathematics is the scribe school.[32] Before discussing the school itself, however, a brief remark should be made about language. Sumerian had been retreating as a spoken language already during Ur III, and maybe centuries before, as can be seen from the increasing dominance of Akkadian names. Official writing, it is true, persisted in Sumerian. In early Old Babylonian times, Sumerian was in all probability a dead language, and all non-scribal business was done in Akkadian (in the Babylonian dialect). Official writing, always by one scribe and meant to be read by another scribe, was still made with some recourse to Sumerian: at times full, and more or (often!) less grammatically correct Sumerian, at times staple Sumerian word signs used as abbreviations within otherwise Akkadian sentences. The Sumerian literary tradition, moreover, was transmitted in the scribal school, though increasingly in bilingual versions.

As to the school itself, its situation reflected that of the general economy. Some schools have been found within palace precincts, and may hence be regarded as official institutions. Others, however, have been located in living areas for scribes; they can hardly have been anything but private enterprise (Lucas 1979: 311f. presents a survey). In both cases, however, the students were trained for similar, "notarial," accounting, and "engineering" functions, i.e., for key positions in general social practice in private or official business.[33] Evidently, the *sine qua non* for any scribe was to master the practical skills needed to perform these tasks.

Besides these skills, however, future scribes were taught to be proud of their profession. A number of texts have survived that were used in the school to inculcate professional pride. They tell us about the curriculum, but they also tell us which part of the curriculum was central for professional pride. The picture gained from these texts stands in significant contrast to actual scribal functions.

First, indeed, the continuation of the Sumerian tradition beyond Hammurapi's time is, as formulated by Kraus (1973: 28), "das größte Rätsel, welches der altmesopotamische Schreiber uns aufgegeben hat." Scribes had to learn Sumerian because other scribes used Sumerian! Even more paradoxical, scribal school students were expected to speak the dead language with good pronunciation. Tradition alone will not do (though even the survival of traditions requires a motivation on the part of their carriers and hence an explanation), since the scribal school tradition appears to take a fresh start in the early Old Babylonian period. (All the texts formulating its ideology belong to the second millennium.)

Sumerian simply, however, is not the culmination of the scribal art. According to the "Examination Text A",[34] the accomplished scribe must know everything about bilingual texts; he must know occult writings, and occult meanings of signs in Akkadian as well as Sumerian; he must be familiar with the concepts of musical practice, and he must understand the distorted idiom of a variety of crafts and trades. Into the bargain then comes mathematics, to which we shall return. All that, as a totality, has a name (of course Sumerian): nam-lú-ulù, "humanity" (van Dijk 1953: 23–26; Sjöberg 1973: 125).

True enough, the phenomenon has some similarity both to the practice of legalese and to the worst aspects of Modern humanism as a self-aggrandizing device for bureaucrats and court servants. Instead of making analogies, however, we may try to formulate an explanation starting from a more precise analysis of the Babylonian concept itself. We may then notice that *everything has to do with scribal practice,* but scribal practice transposed from the region of practical necessity into that of virtuosity. What appears from other didactical texts is that the scribe is expected to be proud, not of accomplishing his actual tasks but *of his identity and ability as a scribe.*

This connects scribal ideals to both aspects of contemporary general ideology. First, the scribal function as a whole was by tradition a public function. If the king was to guarantee affluence and justice, who but the scribe was to do the job? On the other hand, the scribe was also an individual, a private man. In order to assure oneself of being something special, a *human being par excellence,* it was of course desirable to view oneself as the one who gives the king prudent advice, and this is in fact part of scribal boasting (Landsberger 1960: 98). But there was not much satisfaction in pointing to trite, everyday scribal activities, i.e., to the actual ways to "guarantee affluence and justice." After all, phonetic Akkadian could be written with some 80 cuneiform signs. Everybody would be able to learn that. But everybody would not attain a level of virtuosity. Scribal professional pride needed something really difficult as its foundation; but the difficulties had to *belong at least formally to the territory of scribal tasks* if they were to serve *professional* pride. This, according to all evidence, is the reason for the specific configuration of Old Babylonian scribal "humanism", and for its appearance as *art pour l'art.*

Another characteristic of the "examination texts" and related didactical texts should be mentioned before we leave the subject. In contrast to the picture presented by the Fara school texts they always appear to reflect a rather suppressive ambience—ever-recurrent in an early text (known as "Schooldays") where the schoolboy tells his experience of the day are the words "caned me" (Kramer 1949: 205). In "Examination Text A" the student stands back as an ignorant dumbfounded by the teacher. Admittedly, it *is* the teacher who speaks through the text. But the double-bind situation it suggests is still psychologically informative. The message seems to be that the scribe should be proud of being a scribe, but only privately; on service he should be a humble functionary who knows his place. Scribes were to be servants, not rulers, and in reality, rarely advisors of those in power. The scribe was to keep balance between actual loyalty and personal autonomy. His situation may have been similar to that of a medieval clerk. Yet Renaissance humanism was as far ahead as capitalism; the Old Babylonian scribe was, after all, closer to the Fara scribe testing for the first time the possibilities of his professional tools than he was to Benvenuto Cellini.

## 6. Mathematics

"The state" as a concept turned out to be subject to more dispute than presupposed by Wittfogel, my initial punching bag. What about mathematics?

Nowadays, of course, we know the meaning of the term inside our own world—at least until we are asked about borderline cases such as accounting, engineering computation, magic squares, or structuralist grammar. Well within the border we have a cluster of indubitably mathematical practices, disciplines, and techniques, cohering through shared use or investigation of abstract, more or less generalized number or space or of other abstract structures.

Many single elements of this cluster can be traced far back in time, and can be found in non-literate contexts, often at quite advanced levels. Currently, the term *ethnomathematics* is used about these elements when found in non-literate cultures (Ascher and Ascher 1986). It is important to notice, however, that "ethnomathematics," no less than "mathematics," is *our* concept. The inhabitants of Malekula in Vanuatu (formerly New Hebrides) would hardly have recognized the bunch of elements of their culture classified by us as "mathematical" as one entity. Their "kinship group theory" belongs more closely together with the kinship and marriage customs in general than with the drawing of closed patterns, which on the other hand belongs with the relation and passage between life and death.[35] Counting and the geometry of house-building will belong to still other domains.

Non-literate populations visited by modern ethnographers are not identical with the ancestors of ancient civilizations; but it is a fair assumption that the mathematical techniques and practices of the latter constituted something similar in structure (or rather, lack of *own* structure) to ethnomathematics. Similarities may well have gone much further—as we shall see, graphs similar to those of Malekula were familiar in the ancient Near East. If we are going to look for mathematics as *one* entity we may thus choose between two options: either we define one specific domain (traditionally number and counting) as being *their* mathematics, which will allow us to postulate the existence of mathematics far back into an indefinite past; or we may decide (as I intend to do) that the distinctive characteristic of mathematics as *one* entity is the *coordination* of several abstracting practices.

The choice of coordination as the defining feature does not free us from all arbitrariness. It is still a question, for example, whether counting and addition are one or two practices; if they are two, the introduction of addition is already mathematics, since it cannot be done in isolation from counting. So, I shall end up by defining the transition to mathematics as the point where *preexistent and previously independent* mathematical practices are coordinated through a minimum of at least intuitively grasped understanding of formal relations. Remaining ambiguities I shall accept as an unavoidable ingredient of human existence.

### 7. From Tokens to Mathematics

The earliest mathematical technique that can be attested in the Near East is represented by small objects of burnt clay found as far back as the late ninth millennium B.C. and still present in the protoliterate period.[36] From early times, a variety of shapes are found: spheres, rods, cones, circular disks, more rarely other shapes. Many types are found in two sizes, and in certain cases the objects are marked by various incisions. During the fourth millennium, the number of shapes and of extra varieties created through multiple incision proliferates violently.

Because of continuity with later metrological notations (on which below), the objects must be tokens, i.e., *tangible symbols* for other objects—normally goods of economic importance, it appears. Obviously, the tokens constitute a *system* of symbols, used all over Iran, Iraq, Palestine, and Turkey.

The emergence of the system appears to coincide with the change to agricultural subsistence (Schmandt-Besserat 1986: 254). Agriculture itself, of course, will have had no need for symbolization, nor will barter of grain for obsidian (or whatever exchange can be imagined). The most plausible suggestion for the function of the token system is supplied by the excavation of a fifth millennium site (Tell Abada) in east-central Iraq (Jasim and Oates 1986: 352). Tokens are found in several places; yet groups of varied tokens (e.g., eight spheres, four cones, one disc, one rod) contained together in vessels are found only in one place, but there repeatedly: in the most important building of the village, which, according to a number of infant burials may have had religious functions, but whose many rooms show it not to be a mere shrine (or temple). Most likely, it was also a communal storehouse, the heart of a religiously sanctified redistributive system that was moving toward taxation in favor of responsible personnel, and within which the tokens have served for accounting (Schmandt-Besserat 1986: 268f.).

This interpretation is supported by other evidence. Tell Abada is not the only place where the tokens turn up in non-residential buildings (Schmandt-Besserat 1986: 254.). Moreover, tokens (or, rather, prestige versions of tokens made in stone) are also found as high-status grave goods from the sixth millennium onward, for example, in the fourth millennium site Tepe Gawra (near Ninive)—in the grave apparently possessing the highest status six stone spheres constitute the total deposit (Schmandt-Besserat 1986: 255.). Admittedly, Jasim and Oates (1986: 351f.) mention this as an argument for non-accounting functions of the objects; more plausible, however, is Schmandt-

Besserat's explanation (1986: 269 and, in more detail, 1988: 7f.) that the occurrence of tokens in the deposits of high-status burials reflects a high-status position for those who administered by means of tokens while living; their presence in infant graves in Tepe Gawra and elsewhere, furthermore, suggests that the manipulation of tokens was (or belonged with) a hereditary function (as burial deposits in children's graves are normally taken by archaeologists as evidence for hereditary social ranking).[37]

Because of later continuity the meaning of certain tokens can be interpreted. So, a disk marked with a cross appears to stand for a sheep (and two disks for two sheep). Most, however, are uninterpreted or only tentatively interpreted, whereas the principles involved are subject to only limited doubt. They can be illustrated by Schmandt-Besserat's suggestion that a small cone stands for a specific measure of (i.e., a specific type of basket or jar containing) grain, a small sphere for another, larger measure/container, and large cones and spheres for still larger measures (Schmandt-Besserat 1986: 268). Other types might signify other staple products (dried fruit, oil, wool, . . . ). We observe that the marked disk stands for both quality (sheep) and quantity (one) at the same time; the same holds for the cone if representing the grain contained by a specific container type. There is no symbol for abstract number or for volume as such. Since the containers for grain and for oil were different, volume concepts had to be specific. Measure only exists as "natural measure," and number only as concrete number.[38]

The fourth millennium proliferation of the number of token types corresponds to the need of the more highly organized economy of social systems like that of the Susiana plain. New commodities had to be handled, and those of old to be followed in more detail (from later evidence we may guess, for instance, that "sheep" would be differentiated into ewes, rams, and male and female lambs). In addition, the tokens were now used as "delivery notes" for goods sent from the periphery to Susa, enclosed in sealed containers made of clay "bullae".[39]

A disadvantage of the sealed bulla as a bill of lading was that it had to be broken in order to be "read". A solution, however, was at hand: before the tokens were put into the bulla they were pressed into its surface, each leaving a clearly visible impression. The observation that thereby the enclosed tokens had become superfluous will have called forth another step: the replacement of the hollow bulla by a flattened lump of clay where the impressions could be made (by tokens or, rather, by styli able to make similar impressions) and over which the cylinder seal could be rolled. These are the first genuine clay tablets, normally known as "numerical tablets"; like the bullae, they are found in Susa and the Susian orbit as well as in Habuba Kabira, the Uruk V outpost (those of Uruk are found in rubbish heaps and cannot be dated).[40]

As carriers of information, the numerical tablets had an important advantage over the bullae: their surface could be structured, first by distinguishing the four edges of an approximately square tablet and next by dividing the surface into compartments through incised lines. Another advantage was discovered in Uruk IV: through pictographs, *quality* could be separated from, or added to, *quantity*. A drawn circle with a cross was used to indicate sheepness, and impressions looking like pictures of small and large cones and spheres were used to indicate the number of sheep.[41]

The whole development from the introduction of bullae with impressions of tokens and seals to the creation of the pictographic script was evidently coupled to the development of a complex society and to the needs of statal administration for more precise controls, as it was delineated above. It was no consequence of state formation *per se:* as pointed out already, the control involved in state formation need not be bureaucratic control. But the development was a consequence of *state formation as it actually happened in the Sumero-Susian area,* and we may assume that it was the age-old connection between sanctified unequal redistribution and token accounting that made bureaucratic control a natural corollary of the further change of the redistributive system toward taxation.

Improvement of bookkeeping is an improvement of a mathematical technique, which was thus an effect of state formation. But bookkeeping alone does not constitute mathematics.

On the other hand, mathematics *did* emerge in the process, and even in the form of multiple coordination. First, we may look at the metrological sequences and number systems used in the texts. These were first analyzed thoroughly by Jöran Friberg (1978), whose preliminary results have now (on the whole) been confirmed and expanded through computer analysis as part of the Berlin Uruk project (Damerow and Englund 1987).

The first thing to be observed about these systems is that counting is still concrete. In fact, although the basic signs (varied through combination in various ways and addition of strokes) are pictures of the small and large spheres and cones,[42] a number of different systems are in use, with different relations between the visually identical signs.

First, there are *two sequences for counting.*[43] One (the "sexagesimal system") starts by a small cone (1), continues by a small circle (10), a large cone (60), a large cone with an impressed small circle (600), a large circle (3600), and culminates with a large circle with an impressed small circle. This system, characterized by its systematic shift between the factors 10 and 6, is used to count slaves, cattle, tools made from wood or stone, vessels (standing for a specific measure of their customary content), and probably lengths.

The other main counting system (the "bisexagesimal system," with units in the ratios 1: 10: 60: 120: 1200: 7200, i.e., successive factors 10, 6,

2, 10, 6) is used to count products related to grain (rations? bread?) and certain other products.

Besides, *three metrological sequences* have been identified. One is used for capacity measures for grain. If the basic unit is $B$ (a small cone), the next are 6 $B$ (small circle), 60 $B$ (large circle), 180 $B$ (large cone), and 1800 $B$ (large cone with inscribed small circle)—the factor sequence is thus 6, 10, 3, 10. We observe that both order and ratios differ from those of the sexagesimal number system.

Another metrological sequence (testified only in Uruk III/Jemdet Nasr) is used for areas. It was still in use in far later times, which allows us to interpret the small cone as an iku (c. 60 m × 60 m). Then follows a small cone with inscribed small circle (6 iku), a small circle (18 iku), a large circle with inscribed small circle (180 iku) and a large circle (720 iku) (factor sequence 6, 3, 10, 6).

A third metrological sequence is of unidentified use.

Obviously, all sequences are based on the principle of bundling, which demonstrates that principles derived from counting were applied to the regularization of natural measures. Apart from that (admittedly important) step, however, the plurality of sequences and the absence of any system in the succession of the same symbols and in the sequence of ratios is hardly a proof that the career of mathematics had begun.

This beginning, however, is demonstrated by closer investigation of features not yet mentioned. First, what I have just described is just one part of the sequences, from the basic unit upwards. This is the part whose signs derive from the old token system, and which may therefore be of indefinitely older age—even though it is not implausible that the counting notations and the area notation were fresh creations, taking over the symbols of the grain system and adapting them to the actual bundling steps of the verbal counting systems and to the area metrology in use (on areas, see below). The other part consists of fractional sub-units, which are positively new. In the counting sequences, the first sub-unit (½, in the sexagesimal system, and in specific contexts perhaps ⅒) is symbolized by the small cone turned 90° clockwise, which would of course make no sense for freely rattling tokens. In the grain system, a first step is made in a similar fashion,[44] producing ⅕ $B$ ( = $C$). In a second step downward, ⅟ₙ $C$ (n = 2, 3, 4, probably 5, and possibly higher values) is symbolized by n small cones arranged in a rosette. (No area units below the *iku* are attested, but this may well be because such smaller units do not occur in allocations of land—our only epigraphic evidence for area metrology). This involves a knowledgeable application of "inverse" counting to metrological innovation, and must thus be characterized as *mathematics*.

Another metrological innovation based on mathematical premeditation pertains to the calendar—more precisely, one of the calendars.[45] Until much later, indeed, the timekeeping calendar is a lunisolar calendar, whose months are on the average 29½ days, shifting between 29 and 30. Of these months there are 12 to a year, and about every three years an intercalary month is inserted in order to adjust the year to the tropical and agricultural year. In Ur III accounting, months of changing length were of course unacceptable, and a system was employed where the overseer was responsible for pressing 30 days' worth of work out of each worker per month, irrespective of its real length, and got food and fodder rations for his workers and animals according to the same principle. Now, through meticulous analysis of certain protoliterate herding texts, Robert Englund has been able first to confirm an interpretation of the timekeeping notation proposed by Vaiman (1974) on intuitive grounds, and second to show that the Ur III administrative calendar was in reality a protoliterate invention and practice.

The notation combines the pictogram showing a sun half-raised above the horizon with strokes (counting the years), ordinary sexagesimal numbers (months), and sexagesimal numbers turned 90° clockwise (days). Already for the reason that these distinctions only make sense when the symbols are fixed in clay will this be a fresh invention of the protoliterate period. The free, creative manipulation of several sexagesimal counting systems demonstrates mental independence of context-bound counting and ability as well as resolution to combine different elements of mathematical thought in order to create an adequate tool.[46]

Similarly, even the creation of a counterfactual calendar in order to attain mathematical regularity can be seen legitimately as an exhibition of coordination, *viz.*, between bureaucratic organization and mathematical thought. It will also involve at least an intuitively based decision that the rounding error was not larger that acceptable. On both accounts, the administrative calendar is thus evidence of genuine mathematics as defined previously.

All this had to do with the complex of counting, metrology, and accounting. A final observation involves geometrical practice in the network.

We have as yet no direct evidence that the area of a rectangular field was calculated from its length and width—none of the texts that appear to indicate lengths and widths contain area information. But two pieces of indirect evidence can be found. First, the same area system (or at least an area system with the same sequence of factors) is known from later times to be strongly geared to the length unit.[47] Thus, the basic area unit is the sar, which is the square of the fundamental unit of length (the nindan or "rod", equal to c. 6 m), but whose name (presumably meaning "garden plot"; Powell 1972: 189–93) suggests an independent origin as a "natural unit." The iku itself is a square ešé (the ešé, meaning "rope," being equal to 10 nindan). Further on in the sequence, the bur ( = 18 iku), again appears to have originated as a natural unit.

This suggests that the system emerged from a mathematical process of normalization, where natural seed or irrigation measures were redefined in terms of length units, thus stabilizing the system, as Powell points out (p. 177)—and since the upper end of the sequence is already present in Uruk III (where no area units below the *iku* are testified but may still have existed), the redefinition must have taken place already by then.

The other piece of indirect evidence is a protoliterate tablet referred to by Damerow and Englund (1987: 155 n. 73). It deals with a surface of which the two (identical) lengths and the two (slightly different) widths are told. Calculating the area by the "agrimensor formula"[48] one finds a nice round value: 10 times the highest area unit, that is, c. 40 km$^2$. The implausibly large value tells us that we have to do with a school exercise, and the improbability to hit upon the round value by accident suggests that the exercise was constructed so as to achieve it, and thus that the area had to be calculated as done in later times.

Area measurement is not the only element of geometrical practice attested in the protoliterate period. Already the ground plan of the late preliterate "Limestone Temple" (E. Heinrich 1982: 74 and Abbildung 114), perhaps even two fifth-millennium temples (Heinrich, 32 and Abb. 71, 74), possess a regularity that suggests architectural construction. Remains of a ground plan left under an early Uruk IV (or possibly late Uruk V) temple, moreover, shows that it was carefully laid out by colored string (Heinrich 1938: 22, cf. 1982: 63, 66). One of the many different groups of experts present in protoliterate Uruk must hence have been architects skilled in practical geometrical construction[49]—and since only "official" prestige buildings suggest the existence of a geometrical plan, they must have worked exclusively for the Temple.

We can also be reasonably sure that the planning of buildings and of building enterprises will have involved computation of brickwork and manpower requirements. First, a culture that defines a specific administrative month for the sake of fodder calculations would hardly take the enormous costs of prestige building just as they came. Second, the evidence for precise geometrical layout coupled to the standard brick demonstrates that calculations could be made, as indeed they were in later times; it is plausible that this was even the idea behind the mutual adjustment of standards. If so, however, the computation of areas and volumes from linear dimensions will have arisen already in the architects' sphere, and the gearing of field-area measurement to measures of length will also have involved the architectural branch of practical geometry. Protoliterate mathematics will already have coordinated number and metrical space—both, of course, as practical concerns and not as abstract fields of interest.[50]

The formation of mathematics as a relatively coherent complex was thus concomitant with the unfolding of the specific Uruk state. Is that to say that it was a direct consequence of statal bureaucratic rationality—sort of modified and attuned Wittfogel thesis, mechanistic-functionalist though on revised premises? Hardly. Other early bureaucratic states have existed without bringing similar results,[51] and bureaucratic management of agriculture would probably have been better served by natural measures (as suggested by the changes in Babylonian metrology after the mid-second millennium). Bureaucracy itself does not demand the type of coherence inherent in the Uruk formation of mathematics. What is involved, we might say with Weber, is a *particular* spirit of bureaucracy, one tempted by intellectual and not by merely bureaucratic order. We also find it expressed in the lexical lists, which are more than a means of teaching the script: they also provide an ordered cosmos, and a cosmos of a specific sort. Putting wooden objects together in one category, vessels in another, and so on, amounts to what Luria (1976: 48ff.) labels "categorical classification," in contradistinction to his "situational thinking."[52] Still, the lists *are* a means for teaching, and thus a vehicle not only for literacy but also for the "modern," abstracting mode of thought—precisely the mode of thought preferring mathematical coherence to situationally adequate seed measures, and the like. The latter part of their message will have supported, and have been supported by, the development of the main administrative tool: the clay tablet with its ordered formats.[53]

Insofar as the emergence of mathematics is to be ascribed to a particular Uruk variant of the bureaucratic spirit, this spirit was thus interacting intimately with, and largely a consequence of, the school organization of teaching (whose typical features we already encountered in a mathematical exercise). If a complex process is to be reduced to a simplistic formula, the emergence of mathematics was called forth neither by technical needs or by the bureaucratic organization nor by writing *per se,* but *only through the interaction of these with each other and with that school institution that provided recruits and technical skills to the bureaucracy.*

## 8. Trends in Third-Millennium Mathematics

As long as the Sumerian city-states remained dual societies, mathematics was on the same side as writing and bureaucracy. Throughout the third millennium, therefore, the career of mathematics runs parallel to that of expanding bureaucratic systems, spreading literate activities and improved writing. In so far as all this was a simple continuation of the trends inherent in the protoliterate state, mathematics too was a continuation.

Let us first look at metrology. One may wonder that no metrological sequence for weights has been mentioned (unless, of course, the unidentified sequence contains weight units)—especially in view of the fact that metal smelting is actually attested in Uruk (Nissen 1974: 8–11). But technical activities of this sort were not the concern of accounting, and whatever the craftsmen have done was not committed to writing and thus subjected to mathematical regularization.[54]

Later, when copper and silver acquired monetary functions, on the other hand, weight became an accounting concern *par excellence*. In the beginning of ED III, thus, the weight system *is* well attested. A consequence of this late development of weight metrology is a high degree of mathematical systematization (see Powell 1971: 208–11) in the shape of "sexagesimalization," adoption of the fixed factor 60 from the principal (in ED III the *only*) counting system, in analogy with what had already happened in the protoliterate creation of the calendar notation. Starting from the top, a "load" (some 30 kg, the Greek "talent") is divided into 60 mana, each again subdivided into 60 gin (the later *šekel*). The gin is subdivided in *še,* "barleycorns," which in real life weigh much too little to fit another sexagesimal step; but $180 = 3 \cdot 60$ še to a gin agrees fairly well with real barley.

Sexagesimalization was not the preserve of the weight system. In general, when pre-existent systems were extended, it was done "the sexagesimal way." So, for instance, $60^3$ and $60^4$ were added to the counting sequence; the gin was transferred from weights to other systems in the generalized sense of $\frac{1}{60}$; and established systems were expanded upward through multiplication of the largest traditional unit by sexagesimal counting numbers. This development is most straightforwardly explained as the natural consequence of the situation that mathematics was *already present* as a coherent way of thought, both in actu and as impetus and challenge, carried by continuing school teaching.

Another perceptible trend is parallel to that of centralized reforms of writing and bureaucratic procedures (and, though only on the ideological level, to the recurrent idea of a "social reform"): intentional and methodical changes of metrology in order to facilitate bureaucratic procedures. This is of course analogous to the protoliterate introduction of the administrative calendar; the instance that is best certified in the pre-Ur III period is the Sargonic introduction of a new capacity measure in the order of the barrel, the "gur of Akkad" of 30 ban $= 300$ sila ( $\approx 300$ l) instead of the current gur of 24 ban $= 240$ sila and the Lagaš gur of 144 sila (see Powell 1976: 423, for a discussion of the advantages of the new unit in connection with computations of rations).

A third trend, finally, is akin to the appearance of literary texts, and like literary text it begins in the Fara period, concomitantly with the emergence of the scribal profession as a separate group. We might speak of a first instance of *pure mathematics,* namely, of mathematical activity performed in order to *probe the possibilities* of existing concepts and techniques and neither for immediate use in practice nor for plain training of skills to be used in practice.

The evidence is constituted by the oldest mathematical exercises after those of the protoliterate period, which could only be distinguished from real-world accounting and mensuration by the occurrence of round and implausibly but not impossibly large numbers and by the lack of the name of an official carrying responsibility for the transaction (Friberg 1990: 539). One of the Fara problems (Jestin 1937, #188; unpublished analysis by Jöran Friberg) is *almost* of the same type, with the difference that now the area involved is rather *impossibly large.* Two other Fara texts (Jestin 1937, #50 and #671) require that the content of a silo containing 2400 "great gur," each of 480 sila, be distributed in rations of 7 sila per man (the correct result is found in #50: 164, 571 men, and a remainder of 3 sila; the solution of the other tablet is wrong or at best uncompleted—analysis of the two texts and of the method used in Høyrup 1982). A fourth text (analyzed by Jöran Friberg 1986: 16–22), comes from the Syrian city Ebla (whose mathematics was avowedly taken over from the Sumerians) and is presumably of slightly later date. It deals with the successive division of 100, 1000, 10,000, 100,000 and 260,000 by 33 (concretely: if 33 persons get 1 gubar of barley, how much barley do you count out for 100, 1000, 10,000 and 260,000 persons?)

Apart from being division problems and from the "impossibly large" numbers of rations dealt with, the three last problems have one decisive thing in common: the divisors are *irregular,* they fit the metrologies and number systems used *as badly as possible* (Ebla spoke a Semitic language and had decimal number words, but combined these in writing with the Sumerian sexagesimal system; 33, of course, is irregular on both accounts). As Jöran Friberg (1986: 22) puts it,

(I)    the fact that three of the four oldest known mathematical problem texts[55] were concerned with exactly the same kind of "nontrivial" division problems must be significant: the obvious implication is that the "current fashion" among mathematicians about four and a half millennia ago was to study non-trivial division problems involving large (decimal or sexagesimal) numbers and "non-regular" divisors such as 7 and 33.

A number of school exercises dating between the Fara period and Ur III (mostly Sargonic) have been identified (see Powell 1976). Some of them are characterized by the occurrence of "impossibly large" numbers, for example, a field long enough to stretch from the Gulf to central Anatolia. There is no trace, however, of continued interest in "pure mathematics"—which, in view of the striking statistics cited by Friberg, must be significant. As literary creativity, once a scribal exploration of the possibilities of a professional tool, was expropriated by the royal court as a political device, so also mathematical exploration appears to have vanished from a school more directly submitted to its bureaucratic function in a society losing its traditional dual character. Two verifiable forces survived as determinants for the development of "school-and-bureaucracy mathematics": sexagesimalization and systematization governed by the dynamics of internal coherence; and regularization determined by the requirements of bureaucratic efficiency.

A small and isolated tablet found on the floor of a Sargonic temple suggests that a third force *may* possibly have operated outside the school-and-bureaucracy complex; more on this in connection with the Old Babylonian development, to which it is connected (note 69).

### 9. The Paramount Accomplishment of Bureaucracy

Waning duality dwindled further in Ur III, the school-and-bureaucracy complex reached a high point, and so did bureaucratic and accounting rationality. No wonder, then, that Ur III brought about the culmination of the tendencies of late ED and Sargonic mathematics.

We already encountered Šulgi's administrative reform, and we remember that metrological reform was presented as a cornerstone in his establishment of "justice." Another, mathematically more decisive, part of the administrative revolution was the development of the conceptual and technical tools for the many calculations inherent in the reform.

First of all, a new number notation was created as a final outcome of the process of sexagesimalization: the sexagesimal *place value system,* which permitted indefinite continuation of numbers into the regions of large and small. The idea had been in the air for several centuries, as demonstrated first by the generalized use of the gin in the sense of $\frac{1}{60}$, and next also by the particular idiom of a late Sargonic school exercise discussed by Powell (1976: 427), where a "small gin" is introduced for $\frac{1}{60}$ of $\frac{1}{60}$. But precisely the use of *names* for the fractional powers shows that the system was not positional, and was not extendable *ad libitum*. We can thus be fairly sure that the introduction of place value does not antedate Ur III.[56]

The Mesopotamian place value notation was a pure floating-point system, with no indication of absolute place (in likeness of a slide rule); it could thus only be used for intermediate calculations—in accounting, one sixth of a workday, for example, still had to be designated "10 gin" (Powell 1976,

421) to avoid misunderstandings (and so it was—*loc. cit.*). For this reason, only very few indubitably Ur III tablets carry indisputable place value numbers,[57] though some do (one instance is discussed in Powell 1976, 420).

The important point about the place value notation is not the possibilities it offers in additive and subtractive accounting, where the disadvantage of a double number system will have outweighed the ease of writing that it brought about. It lies in the multiplicative possibilities of the system to surmount the conflict between mathematical and technical rationality (as discussed in connection with the tendency of protoliterate scribes to prefer mathematical coherence to practical orientation), and to do this more radically than could be done by changes in the metrological system. If a platform had to be built to a certain height and covered by bricks and bitumen, for example, changes in length measures could not be made that at the same time would facilitate manpower calculations for the earth- and brickwork, the computation of bricks to be used, and the consumption of bitumen. But once the place value system was available, tables could do the trick. A metrological table could be used to transform the different units of length into sexagesimal multiples of the nindan. A table of "constant factors" would tell the amount of earth carried by a worker in a day, the number of bricks to an area unit, and the volume of bitumen needed per area unit. With these values at hand everything was a question of sexagesimal multiplications and divisions, which again were facilitated by recourse to tables, this time tables of multiplication and of reciprocal values. The conflict between "natural" and "mathematical" measure was solved similarly in other domains, and so well solved that supplementary technical measures could be introduced *ad libitum*, as indicated by an apparent proliferation of brick systems. This was the great advantage of the system in a society where the scribes were financially responsible overseers of all sorts of productive activities.

It is a fair guess that the place value system was probably invented with the purpose to solve these problems, but since we do not possess the memoirs of the inventor we cannot know.[58] What can be known is that other highly adequate place value systems are known historically to have spread at a snail's pace, in processes taking hundreds of years or even longer. If the invention was not made in Šulgi's think tank (something like the administrative department of Kraus' conjectured *Hofkanzlei*), a central decision must at least have been made to propagate the system through the scribe school, which must then have been under centralized control (as one would guess anyhow, given the character of Ur III society and Šulgi's interest in a school teaching what his scribes needed).

Much the same could be said about other aspects of the administrative system, especially of the introduction of a system of *balanced accounts,* at times with automatic cross-checking.[59] The school provided the administration with accountants and calculators whose collective competence has hardly been equalled by any comparable body west of China before the eighteenth or nineteenth century (A.D., for once!). Judged on the purely utilitarian premises inherent in the Šulgi hymns cited already, the Ur III school did everything that could be done.

It is remarkable, then, that no trace whatsoever is left of non-utilitarian mathematical interests from the period. Not only are texts lacking, which in itself proves nothing, since no school texts at all from the period have been identified. More decisive: an investigation of the mathematical terminology of the subsequent Old Babylonian period shows that terms used for current operations of utilitarian calculation are Sumerian; the key terms of the non-utilitarian branches, on the other hand, are Akkadian, and the oldest non-utilitarian texts formulate even the additive and subtractive operations (for which current Sumerian terms existed, of course) in Akkadian—with the exception of the finding of reciprocals and the extraction of square roots, which referred to tables in the Ur III tradition, and the traditional Sumerian terms for which were even adopted as loanwords and provided with Akkadian declination.[60] According to all evidence, Ur III thus managed to bring its scribes to a high level of mathematical competence without engendering any sort of pure-mathematical interest, i.e., any intellectually motivated investigation of the possibilities of professional tools beyond the needs of current business—in contrast to the situation in Fara, where much more modest competence did call forth "pure" investigation. Borrowing an expression from a classical discussion of other aspects of the Mesopotamian intellect (von Soden 1936), Ur III demonstrates "Leistung und Grenze" of the early bureaucratic state as a promoter of mathematical development.

## 10. The Culmination of Babylonian Mathematics

The vast majority of Mesopotamian genuine mathematical texts come from the Old Babylonian period. Before Marvin Powell and Jöran Friberg began their work, almost nothing was known from the third and fourth millennia, and no system whatsoever had been noticed in the meager material (even the connection between the Ur III administration and the creation of the sexagesimal system was only suggested as a conjecture by Powell in 1976). From the 1300 years separating the Old Babylonian from the Seleucid period, again practically nothing was known (since the, Jöran Friberg has located a few items). Finally, a small number of texts with Seleucid dating had been published. No wonder, then, that the Old Babylonian period was considered the culmination of Babylonian mathematics, which in histories of mathematics was simply identified with this climax.

Much of this is of course a consequence of the source situation. Since there is some though not full continuity from Old Babylonian to Seleucid mathematics, something *must* have existed in the intermediate years. Yet today, when at least a sketchy picture of the state of the mathematical art in the early and the intermediate period can be made, Old Babylonian mathematics is enforcing its particular character upon us in more real terms: never before, and never after, was comparable depth and sophistication achieved in ancient Mesopotamian mathematics. Even the source situation seems to reflect realities and not merely the random luck of excavators and illegal diggers: after the Old Babylonian period the institutional focus for the production of sophisticated mathematics disappeared.

Why is that? What *was* the make-up of Old Babylonian mathematics? And what was its purpose?

First of all, Old Babylonian, quite as much as third-millennium mathematics, spells *computation*. All texts compute something, they *never prove* in Euclidean manner, and they only *explain* through *didactical discussion of specific examples* of computation.

Many computations are purely utilitarian, of course. The texts are scribe school texts (the teacher's copies, not students' solutions, as most of the pre-Ur III texts that have come down to us); and graduate scribes, as we remember, would normally go into notarial jobs, where they needed little but accounting mathematics, or into engineering-like occupations, where a wider range of practical geometry and similar topics would be required.[61] Utilitarian mathematics was thus a continuation of Ur III mathematics, involving sexagesimal calculation, the use of the tables of metrological conversion and of "constant factors," knowledge of accounting and surveying procedures and of computational techniques at the level of the rule of three, and familiarity with the computation of areas and (occasionally fairly intricate) volumes.[62] All this, in fact, is found, often in complex combinations as in "real scribal life," where the manpower needed to dig a trench and carry off the dirt was more interesting than its volume.

Just as important in school, however, were non-utilitarian computations, to judge from the statistics of extant texts. Dominating in this field was a domain traditionally denoted "algebra" by historians of mathematics, and which is in fact homomorphic with second- and higher-degree equation algebra of the medieval and Modern epoch. The designation can be argued to be problematic, because a literal reading of the terms of the Old Babylonian discipline indicates that it does not deal with number but with areas (quite literally: with fields), and because a close investigation demonstrates that the methods used were indeed sort of "naive" (i.e., reasoned but not explicitly demonstrative) cut-and-paste geometry.[63]

Many problems belonging to this category look fairly abstract. For instance, we may be given the sum of the length and the width of a rectangular field and the sum of the area and the excess of the length over the width, and then be asked to compute the length and the width (AO 8862, in *MKT* I, 108f., cf. interpretation in Høyrup 1990: 309ff.). In this case, only the remark that "I went around it" tells that the person stating the problem speaks of a real field; other problems are even more deprived of the smell of real life. Still others, however, attach themselves directly, for example, to military engineering practice, as may be illustrated by this example[64]:

7.  Of dirt, 1,30,0 (sar), gán. A city inimical to Marduk I shall seize.
8.  6 (nindan) the (breadth of the) fundament of the dirt. 8 (nindan) should still be made firm before the city wall is attained
9.  36 (kùš) the peak (so far attained) of the dirt. How great a length
10. must I stamp in order to seize the city? And the length behind
11. the *ḫurḫurum* (the vertical back front reached so far?) is what? You, detach the igi of 6, the fundament of the dirt—0;10 you see. Raise 0;10 to
12. 1,30,0, the dirt—15,0 you see. Detach the igi of 8—0;7,30 you see.
13. Raise 0;7,30 to 15,0—1,52;30 you see. Double 1,52;30—
14. 3,45 you see. Raise 3,45 to 36—2,15,0 you see. 1,52;30
15. make surround—3,30,56;15 you see. 2,15,0 from 3,30,56;15
16. tear out—1,15,56;15. What is the equilateral? 1,7;30 you see.
17. 1,7;30 from 1,52;30 tear out—45 you see, the elevation of the city wall.
18. ½ of 45 break—22;30 you see. Detach the *igi* of 22;30—0;2,40.
19. Raise 15,0 to 0;2,40—40, the length. Turn back, see 1,30,0, the dirt. Raise 22;30,
20. ½ of the elevation, to 40, the length—15,0 you see. Raise 15,0 to 6—
21. 1,30,0 you see, 1,30,0 is the dirt. The method.

To a first glance, this looks like a slightly idealized piece of engineering mathematics: a siege ramp formed like a right triangular prism is to be constructed, and we know certain parameters concerning the structure and have to find the others. (The minor blunder than an already given value is asked for again, instead of another that is actually found, is due to an editor-copyist's mixup with other problems on the same configuration—one follows on the same tablet).

A second look, however, changes everything. The construction has already started; we already know how much dirt is going to be used for the ramp, as well as the height already reached and the remaining distance. But we do not know the intended total length or final height of the ramp, nor the length of the part built so far! The outcome, after intricate geometrical considerations, is a problem of the second degree.

Evidently, such a problem would never present itself to a military engineer in real life. In fact, no single second-degree (or higher) problem in the Babylonian mathematical corpus—and there are lots of them—corresponds to a situation that could be encountered in practice, nor can any be imagined within the Babylonian horizon. And yet, such artificial problems were highly popular (the same unfinished ramp, for example, turns up in another tablet making use of a somewhat different terminology and thus probably produced in a different school).[65] Definitely, mathematics needed not be applicable in order to acquire high status within the curriculum—*if only it looked applied,* as the above puzzle from the military engineers' wonderland.

This may look a paradox at a first glance. Why should evidently "pure" mathematics be disguised as applied? Neugebauer (1954: 780), obviously disgusted, speaks of "educational artificiality which fancies it is making simple geometrical problems more appealing by using practical examples containing unreal assumptions." Why should pure mathematics be restricted to computation? And why on earth should a school for future clerks, managers, and engineers make so much of the training of useless skills?

The answers have to do with the position of the scribal profession and the role of the scribal school. Like the writing of phonetic Akkadian, accounting mathematics and trite computation of prismatic volumes were too uncomplicated to serve as the foundation for professional pride. In order to demonstrate *virtuosity,* Akkadian had to be supplemented by Sumerian and secret writing, and the volume computation had to be turned around into a second-degree puzzle. Higher "algebra" was thus the expression of scribal "humanism" corresponding to the numerate aspect of the scribal vocation (and a choice expression), as Sumerian was the expression corresponding to the literate aspect. The important thing about second-degree "algebra" was not that it could not be used; the distinctive characteristic was that it was *complex,* i.e., *non-trivial.* The situation repeats that of the Fara scribes on a higher level, whose investigation of the possibilities of writing produced the first literary texts, and whose comparable experiments with their computational tools produced "pure" division problems.

But virtuosity had to be *scribal* virtuosity in order to serve *professional* pride (which would of course be the only sort of pride at which a scribal school could aim). Therefore, even complex mathematical problems should belong at least in form to the category of scribal problems. Though "pure" *in substance,* scribal mathematics was by necessity *applied in form.*

Strictly speaking, furthermore, the numerate aspect of the scribal venture was not *mathematical* in a general sense but *computational.* The virtuoso scribe had to be a virtuoso in *finding the correct number.* Pure *mathematics* in the sense that we have derived from the Greeks was not open as an option. Only *pure computation* would make the day.[66]

Finally, the scribe was a *practitioner,* no philosopher or teacher. In Babylonia as everywhere else, the main thing for a practitioner is to be able to handle his methods aptly and correctly. In mathematics at the Old Babylonian level, this requires more than a modicum of understanding.[67] But in all vocational training then as now, apt and correct handling of methods is learned primarily through *systematic training supported by explanation,* not vice versa—as it was once formulated, you do not extinguish a fire by lecturing on the nature of water. Though transmission of *methods* was the central *aim* of the school, the *solution of* (adequately selected) *problems* was thus necessarily the *central teaching mode*—as, again, real, practical problem solution was the ultimate purpose of the training of *utilitarian* methods.

Nothing thus remains of the supposed paradox when it is seen in the light of Old Babylonian scribal humanism. But another important characteristic persists that should be discussed. The "unfinished ramp" illustrated the "humanist" character beautifully, through exaggerating features that are present yet less conspicuous in other problems. But exaggeration is, already by definition, untypical, and so also in this case. The text is so much of a riddle that we can almost hear the hidden wording of lines 9–10 as "[ . . . ] *Tell me, if you are a clever scribe,* how great a length must I stamp in order to seize the city?" Most texts, however, are much more terse and, more important, the majority of those which contain several problems are fairly or even highly systematic (with the exception of some late Old Babylonian anthology texts—the "unfinished ramp" is known from precisely these). The riddle shows the family likeness between Old Babylonian "pure computation" and that "recreational mathematics" which, before it became a column in newspapers and mathematics teachers' journals, was a "pure," virtuoso outgrowth of practitioners' "oral mathematics".[68] But systematization is of course foreign to any genre of campfire riddles, mathematical as well as non-mathematical. The systematics of the Babylonian mathematical texts reflect that school system in which they were composed, and that tendency to establish "bureaucratic order" even in the intellectual realm that had characterized it since the protoliterate period.[69] If many of the distinctive "humanistic" characteristics of Old Babylonian mathematics must be explained with reference to the particularities of the carrying school institution as a relatively autonomous institution in an individualistic culture, its overall character of *mathematics* was still guaranteed by the traditional character of the school as developed in interplay with the bureaucratic state.

The dependence of Old Babylonian mathematics on the school-and-bureaucracy complex and its characteristic double-bind conditioning, on the other hand, was also the factor that effectively inhibited the emergence of theoretical mathematics of the Greek kind. As I have formulated it elsewhere (1990: 337), the scribal school was "only moderately inquisitive and definitely not critical." This befitted the education of future "humble officials knowing their place" yet proud of their social status. In later times similar institutions would provide fairly suitable vehicles for the transmission of the ultimate outcome of the Greek mathematical endeavor; yet the dawn of the

Greek endeavor itself was too dependent on non-suppressive critical discussion to be within the reach of a scribal school culture.

## 11. Devolution

I shall finish my discussion of Mesopotamia by some cursory and undocumented remarks on the state, the scribal profession, and the development of mathematics after the end of the Old Babylonian period, before passing to some even more brief comparative observations.

The end of the Old Babylonian epoch inaugurated the dissolution of much of the complex that, according to the previous sections, had shaped and even engendered Mesopotamia mathematics.

First of all, it was the end of that sort of state which, since its emergence as a pseudo-redistributional organization, had guarded the pretense to be the upholder of justice and affluence (in spite of often contrary realities). The end of the Old Babylonian epoch was brought about by the Hittites, who sacked Babylon, after which a warrior people (the Kassites) took over power in the Babylonian area. They exploited it, not by taxation (however vaguely) disguised as redistribution but by direct extorsion, as conquerors would mostly do until the advent of the more sophisticated methods of the Modern era, taking over part of the land, allying themselves with the autochthonous upper class and pressing tribute from a re-communalized peasant class.

City life, on that occasion, did not disappear *completely*—but the proportion of town to country dwellers reverted to the level of the preliterate Middle Uruk period.

The scribe school disappeared. Administration and scribes were still needed, but scribes were from now on trained as apprentices inside their "scribal family."

The self-asserting individualism of the Old Babylonian period disappeared. The particular scribal expression of Old Babylonian individualism, "humanism," disappeared, too. Instead, scribal pride was founded on the membership in an age-old *tradition*. That cloak of magic and secrecy which Wittfogel ascribes to the bureaucrats of the managerial and functionalist state is in fact a product of the intellectual crisis caused by its breakdown.

Finally, even mathematics disappeared—at least from the archaeological horizon. But realities are involved, too. Techniques, of course, survived. But the few texts of later times suggest that the integrity of mathematics as a subject on its own disappeared—whereas Old Babylonian mathematical texts would contain *nothing but* mathematics, things were now mixed up. "Pure" scribal interest in mathematics disappeared, it seems. The evidence suggests, indeed, that second-degree algebra, even though it turns up again in a few Seleucid texts, survived in a practitioners' (surveyors' and/or architects') rather than in the scribal environment.[70] Moreover, the evolution of metrology suggests that technical mathematical skills declined. As explained before, the routines and procedures associated with the place value system overcame the conflict between mathematical and technical rationality, thus making the use of "natural measures" unnecessary. From the Kassite era onward, however, the metrological system changes; field measures keyed to the squared length unit, for example, are replaced by seed measures. Apparently, technical efficiency was no longer compatible with "mathematical efficiency," i.e., coherence and simplicity.

In the Late Babylonian epoch (from c. 600 B.C. onward), finally, mathematics reappears above the horizon. Its practitioners are no longer primarily scribes, i.e., accountants and engineering managers; instead, they designate themselves as "exorcists" (*āšipu*) or "priests" (*šangû*). The latter title, oddly enough, coincides with the Sumerian sanga, who was not only a priest but also a manager of temple estates and a teacher in the Fara school; the Late Babylonian *šangû*-mathematician, however, was no practical manager, like his protoliterate predecessor, but an astrologer.

The astrologer-priests who created Late Babylonian mathematical astronomy performed technical wonders, no doubt. Their skill in developing interpolation schemes and six-place (sexagesimal places!) reciprocal tables is impressive. But if we understand mathematics as "a coherent way of thought, both *in actu* and as impetus and challenge," then the high point of Mesopotamian mathematics was reached in Hammurapi's Bronze Age and never again.

## 12. Supplementary Comparative Observations

A possible test of the plausibility of the theses advanced here on the connections between the specific process of state formation and development and the emergence and shaping of mathematics would be cross-cultural comparison. Implicitly, of course, much of the analysis *is* already cross-culturally based, through the use of theoretical tools sharpened on non-Mesopotamian whetstones. In the present appendix I shall only point to two possibilities of explicit comparison.

First, of course, Egypt suggests itself, as every time a mirror is to be held up to Mesopotamia.[71] State formation in Egypt was roughly contemporary with that of Uruk, presumably slightly later. Its background, however, was more explicitly in agreement with Carneiro's warfare model. Pharaoh united Egypt through conquest, and courtly art demonstrates that he was proud of that. The early Egyptian state was not built on any redistributive pretext or ideology.[72]

Writing was also roughly contemporary, and presumably slightly later. But until late in the Old Kingdom, literacy was extremely restricted, and not before the Middle Kingdom, i.e., in the outgoing third millennium, did a scribal school arise.

If no other forces were present that could nourish the process, we should thus expect the development of mathematics as a coherent whole and, especially, of "pure" orientations, to be much slower than in Mesopotamia. As far as it can be judged from the meager evidence from the Old Kingdom, this seems indeed to be the case. First, some generosity is already required to see the development of the Egyptian unit fraction system as evidence of a "pure" orientation; but even if that is granted one will have to observe that the unit fraction system seems only to be created *as a system* in the Middle Kingdom, in the wake of the new scribe school institution. No other branch of Egyptian mathematics can at all be considered nonutilitarian. Second, it is questionable how far the unification of single techniques into a coherent whole had developed before it was definitely brought about by the unit fraction system.

Whereas ancient Egypt is the ever-recurrent mirror in which Mesopotamian development is seen, medieval Western Europe is rarely mentioned as an analogue. In one important aspect, however, the medieval West *is* relevant, *viz.*, as a dual society. If Gilgameš shares essential features with the Homeric kings, he can also be compared to a Frankish warrior king. The Church, on the other hand, shares with the Sumerian Temple the status of a purported institutionalization of the common good; to a large extent, its incomes derived from benevolent gifts (often compulsory, it is true, and in the case where the gift was a nobleman's donation of land with appurtenant peasants, it could only be made productive through continued compulsion; but these details are irrelevant for my present purpose, and nobody knows whether realities were much different in Sumer). The interesting thing is that *literacy was until the High Middle Ages the exclusive ally of the ecclesiastical "Temple institution"*; except for a few dreamers of learning such as Charlemagne and Otto III, one can describe the history of central medieval learning without ever presenting the feudal power in person. "Who was the feudal lord who donated land for the Cluny monastery? It doesn't matter: feudal lords did that sort of thing!" "Who pacified the French core areas in the late tenth century to the extent that cathedral schools could revive? No single lord or king; it was part of a general trend visible in many places"; and so on. And vice versa of course: in the *Poema de Mio Cid*, the tale of this most Christian hero of the Spanish *reconquista*, the role of the Church is as secondary as in *Gilgameš and Agga*. Societal duality is thus a recurrent historical type in state formations not yet fully satisfying Runciman's criteria (quotation F), and literacy and learning belong with the institutionalization of alleged general interest, not with the warrior-robber lordship in its interaction with a prestate, communal or kinship-based sector.

The medieval parallel can be pursued further, into the High Middle Ages. Then, of course, duality was reabsorbed, and royal centralization was well served by literate clerks. But at the same time the environment of learning, rapidly growing and therefore less directly subject to the "Temple" institution, went through a process of intellectual emancipation, first in the "twelfth-century renaissance" and then in the universities. But scholars remained *clerks*, first because of the general sociocultural and the particular institutional context, second because of the future social position of most university students. As in the Old Babylonian scribal school, though less strictly, the traditional binding to the "Temple" and the actual nexus to scribal (notarial and cameralistic) functions in existing society set limits to the tendencies toward intellectual enfranchisement.

# 4

## The Formation of "Islamic Mathematics": Sources and Conditions

In memory of
George Sarton

### Introduction

*As the reader will immediately notice, the following essay shares the initial quotation from Aristotle's* Metaphysics *with Chapter 2. A couple of other quotations are also repeated. This is symptomatic of a kind of symmetry relation between the two investigations: Chapter 2 explored how pre-Modern practitioners' (subscientific) mathematics was socially as well as cognitively* separate from, *and not to be understood as an* application of, *scientific mathematics. The present study, for its part, examines the first historical process in which the two were brought into a mutually fecundating relationship without either of them losing its autonomy and legitimacy; it locates the first appearance of this "miracle," no less decisive for the emergence of the Modern and contemporary scientific-technological complex than the Greek creation of autonomous theory, in the Islamic Golden Age.*

*Many factors will have been involved in this metamorphosis. Economic affluence and amalgamation of cultures formed an indispensable background for that general cultural outburst which makes us speak of a Golden Age. But affluence, amalgamation, and ensuing gilding of the age had presented themselves at earlier occasions without producing comparable effects on the organization of knowledge. Instead of tracing the material roots of the cultural upsurge I try in what follows to uncover the distinctive features of eighth- to twelfth-century Islamic society that made its scholars pursue goals which either had not appealed to or had been inaccessible to their ancient precursors. As far as the earlier part of the period is concerned, an open-minded fundamentalist religious world-view, in combination with the absence of stable institutions, and in particular the absence of effective religious authority, appears to have been decisive; in later times, the development of stable institutions carrying on the integrative attitudes that had materialized in the early fluent situation made these attitudes effective.*

*A first, rather crude, version of the investigation was read at the George Sarton Centennial Conference in Ghent in 1984, where Dr. Hosam Elkhadem had invited me to speak on "any topic of [my] choice" related to the history of Muslim medieval science. I took advantage to recast it in 1987, when I received an invitation to contribute to the newly started journal* Science in Context. *Apart from the usual adjustment of the reference system and the introduction of cross-references, the essay as it presents itself here is on the whole identical with the version published in 1987—with the exception of those passages where intended editorial corrections of style and language had distorted the meaning of the text.*

## Biographical Cues

Based mainly upon DSB; GAS; GAL; *Suter (1900); Sarton (1927); Sarton (1931); and Dodge (1970).*

*ʿAbbasid dynasty   Dynasty of Caliphs, in effective power in Baghdad from 749–50 to the later tenth/earlier eleventh century, formally continued until 1258.*

Abraham Bar Ḥiyya (Savasorda)   Fl. *before 1136. Hispano-Jewish philosopher, mathematician, and astronomer.*

Abū Bakr   *Early ninth century? Otherwise unknown author of the* Liber mensurationum.

Abū Kāmil   Fl. *late ninth and/or early tenth century. Egyptian mathematician.*

Abū Maʿšar   *787–886. Eastern Caliphate. Astrologer.*

Abūʾl-Wafāʾ   *940–997/98. Eastern Caliphate. Mathematician and astronomer.*

Al-Bīrūnī   B. *973, d. after 1050. Astronomer, mathematician, historian, and geographer from Khwārezm.*

Al-Fārābī   C. *870–950. Eastern Caliphate. Philosopher.*

Al-Fazārī   Fl. *second half of the eighth century. Eastern Caliphate. Astronomer.*

Al-Ghazzālī   *1058–1111. Eastern Caliphate. Theologian.*

Al-Ḥajjāj   Fl. *later eighth to early ninth century. Eastern Caliphate. Translator of the* Elements *and of the* Almagest.

Al-Ḥaṣṣār   Fl. *twelfth or thirteenth century, probably in Morocco. Mathematician.*

Al-Jāḥiẓ   C. *776–868/69. Iraq. Muʿtazilite theologian; zoologist.*

Al-Jawharī   Fl. c. *830. Eastern Caliphate. Mathematician and astronomer.*

Al-Karajī   Fl. c. *1000. Eastern Caliphate. Mathematician.*

Al-Khayyāmī   *1048(?)–1131(?). Iran. Mathematician, astronomer, philosopher.*

Al-Khāzin   D. *961/971. Iran. Mathematician and astronomer.*

Al-Khazinī   Fl. c. *1115–1130. Iran. Astronomer, theoretician of mechanics and instruments.*

Al-Khuwārizmī   Fl. c. *980. Khwārezm. Lexicographer.*

Al-Khwārizmī   *Late eighth to mid-ninth centuries. Eastern Caliphate. Mathematician, astronomer, geographer.*

Al-Kindī   C. *801 to c. 866. Eastern Caliphate. Philosopher, mathematician, astronomer, physician, etc.*

Al-Māhānī   Fl. c. *860 to c. 880. Eastern Caliphate. Mathematician, astronomer.*

Al-Maʾmūn   *786–833. ʿAbbasid Caliph 803–833, ardent* muʿtazilite, *patron of* awāʾil *learning.*

Al-Nadīm   C. *935 to 990. Baghdad. Librarian, lexicographer.*

*Al-Nasawī    Fl. 1029–1044. Khurasan, Eastern Caliphate. Mathematician.*

*Al-Nayrīzī    Fl. early tenth century. Eastern Caliphate. Mathematician, astronomer.*

*Al-Qalaṣādī    1412 to 1486. Spain, Tunisia. Mathematician, jurisprudent.*

*Al-Samawʾal    Fl. mid-twelfth century, Iraq, Iran. Mathematician, physician.*

*Al-Umawī    Fl. fourteenth century. Spain, Damascus. Mathematician.*

*Al-Uqlīdisī    Fl. 952/953, Damascus. Mathematician.*

*Al-Yaʿqūbī    Fl. 873–891. Eastern Caliphate. Shiite historian and geographer.*

*Anania of Širak    Fl. first half of the seventh century. Armenia. Mathematician, astronomer, geographer, etc.*

*Banū Mūsā ("Sons of Mūsā")    Three brothers, c. 800 to c. 875, Baghdad. Mathematicians, translators, organizers of translation.*

*Ḥunayn ibn Isḥāq    808–873. Eastern Caliphate. Physician, translator.*

*Ibn Abī Uṣaybiʿa    1203/4–1270. Syrian lexicographer and physician.*

*Ibn al-Bannāʾ    1256–1321. Morocco. Mathematician, astronomer.*

*Ibn al-Haytham    965 to c. 1040. Iraq, Egypt. Mathematician, astronomer.*

*Ibn Khaldūn    1332–1406. Maghreb, Spain, Egypt. Historian, sociologist.*

*Ibn Qunfudh    D. 1407/8. Algeria. Jurisprudent, historian. Commentator to ibn al-Bannāʾ.*

*Ibn Ṭāhir    D. 1037. Iraq, Iran. Theologian, mathematician, etc.*

*Ibn Turk    Earlier ninth century. Turkestan. Mathematician.*

Ikhwān al-Ṣafāʾ    *("Epistles of the Brethren of purity") Tenth century* Ismāʿīlī *encyclopedic exposition of philosophy and sciences.*

*Kamāl al-Dīn    D. 1320. Iran. Mathematician, mainly interested in optics.*

*Kūšyār ibn Labbān    Fl. c. 1000. Eastern Caliphate. Astronomer, mathematician.*

*Māšāʾallāh    Fl. 762 to c. 815. Iraq. Astrologer.*

*Muḥyiʾl-Dīn al-Maghribī    Fl. c. 1260 to 1265, Syria and Iran. Mathematician, astronomer, astrologer.*

*Naṣīr al-Dīn al-Ṭūsī    D. c. 1214. Iran. Astronomer, mathematician.*

*Qayṣar ibn Abīʾl-Qāsim    1178–1251. Egypt, Syria. Jurisprudent, mathematician, technologist.*

*Qusṭā ibn Lūqā    Fl. 860 to 900. Eastern Caliphate. Physician, philosopher, translator.*

*Rabīʿ ibn Zaid    Fl. c. 961. Bishop at Cordoba and Elvira, astrologer.*

*Severus Sebokht    F. mid-seventh century. Syrian bishop. Astronomer, commentator on philosophy.*

*Thābit ibn Qurra    836–901. Eastern Caliphate. Mathematician, astronomer, physician, translator, etc.*

*ʿUmar ibn al-Farrukhān    762–812. Eastern Caliphate. Astronomer, astrologer.*

*ʿUmayyad dynasty    Dynasty of Caliphs, 661–750.*

*Yūḥannā al-Qass    Fl. first half of tenth century? Mathematician, translator of mathematics.*

## 1. Introducing the Problem

When the history of science in prehistoric or Bronze Age societies is described, what one finds is normally a description of technologies and of that kind of practical knowledge which these technologies presuppose. This state of our art reflects perfectly well the state of the arts in these societies: They present us with *no specific, socially organized, and systematic search for and maintenance of cognitively coherent knowledge* concerning the natural or practical world—i.e., with nothing like our own scientific endeavor.

The ancestry of that specific endeavor is customarily traced back to the "Greek miracle," well described by Aristotle (*Metaphysica* 981$^b$14–982$^a$1, translated by Ross 1928):

> At first he who invented any art whatever that went beyond the common perceptions of man was naturally admired by men, not only because there was something useful in the inventions, but because he was thought wise and superior to the rest. But as more arts were invented, and some were directed to the necessities of life, others to recreation, the inventors of the latter were naturally always regarded as wiser than the inventors of the former, because their branches of knowledge did not aim at utility. Hence when all such inventions were already established, the sciences which do not aim at giving pleasure or at the necessities of life were discovered, [ . . . ]
>
> So [ . . . ], the theoretical kinds of knowledge [are thought] to be more the nature of Wisdom than the productive.

The passage establishes the fundamental distinction between "theoretical" and "productive" knowledge, between "art" and "science," and thus the break with those earlier traditions where knowledge beyond the useful was carried by precisely those groups which possessed the highest degree of useful knowledge.[1] Inherent though not fully explicit is also a fairly absolute social and cognitive separation of the two. Though obvious deviations from this ideal can be found in several ancient Greek scientific authors (some of whom we shall mention in the following), Aristotle's discussion can be regarded as a fair description of the prevailing tendency throughout Greek antiquity.

On the other hand, it is definitely not adequate as a description of Modern or contemporary attitudes toward the relation between science and technology (which we are often disposed to regard as "applied science"[2]). We are thus separated from the Bronze Age organization of knowledge not only by a "Greek miracle" but also by at least one later break, leading to the acknowledgment of the practical implications of theory. Customarily we locate this break in the Late Renaissance, and regard Francis Bacon as a pivotal figure.

A first aim of the present paper is to show that the break took place earlier, in the Islamic Middle Ages, which first came to regard it as a fundamental epistemological premise that problems of social and technological practice can (and should) lead to scientific investigation, and that scientific theory can (and should) be applied in practice. Alongside the Greek we shall hence have to reckon an *Islamic miracle*. A second aim is to trace the circumstances that made medieval Islam produce this miracle.

I shall not pursue these two aims in broad generality, which would be beyond my competence. Instead, I shall concentrate on the case of the mathematical sciences. I shall do so not as a specialist in Islamic mathematics[3] but as a historian of mathematics with a reasonable knowledge of the mathematical cultures connected to that of medieval Islam, basing myself on a fairly broad reading of Arabic sources in translation. What follows is hence a tentative outline of a synthetic picture as it suggests itself to a neighbor looking into the garden of Islam; it should perhaps best be read as a set of questions to the specialists in the field formulated by an interested outsider.

From this point of view, the mathematics of the Islamic culture[4] appears to differ from its precursors by a wider scope and a higher degree of integration. It took up the full range of interests in all the mathematical traditions and cultures with which it came in contact, "scientific" as well as "subscientific" (a concept I shall discuss subsequently); furthermore, a significant number of Islamic mathematicians mastered and worked on the whole gamut from elementary to high-level mathematics (for which reason they tend to see the former *vom höheren Standpunkt aus* [from the higher vantage point], to quote Felix Klein). Even if we allow for large distortions in our picture of Greek mathematics introduced by the schoolmasters of late antiquity (cf. Toomer 1984: 32), similar broad views of the essence of mathematics appear to have been rare in the mature period of Greek mathematics; those who approached it tended to miss either the upper end of the scale (like Hero) or its lower part (like Archimedes and Diophantos, the latter with a reserve for his lost work on fractions, the *Moriastica*).

The "Greek miracle" would not have been possible had it not been for the existence of intellectual source traditions. If we restrict ourselves to the domain of the exact sciences, nobody will deny that Egyptian and Babylonian calculators and astronomers supplied much of the material (from Egyptian unit fractions to Babylonian astronomical observations) that was so radically transformed by the Greek mathematicians.[5] Equally certain is, however, that the Egyptian and Babylonian cultures had never been able to perform this transformation, which was only brought about by specific social structures and cultural patterns present in the Greek *polis*.[6] Similarly, if we want to understand the "miracle" of Islamic mathematics, and to trace its unprecedented integration of disciplines and levels, we must also ask both for the sources that supplied the material to be synthesized and for the forces and structures in the culture of Islam which caused and shaped the transformation—the formative conditions.

## 2. Scientific Source Traditions: The Greeks

In Section 1, a dichotomy between "scientific" and "subscientific" source traditions was introduced. I shall return to the latter and discuss why they must be taken more seriously into account than normally done. At first, however, I shall concentrate on the *scientific* sources—more distinctly visible—and first of all on the most visible of all, to medieval Islamic lexicographers as well as to modern historians of science[7]: Greek mathematics.

That this source was always regarded as having paramount importance will be seen, for example, from the *Fihrist (Catalogue)* written by the tenth century Baghdad court librarian al-Nadīm.[8] The section on mathematics and related subjects contains the names and known works of 35 pre-Islamic scholars. Twenty-one of these are Greek mathematicians (including writers on harmonics, mathematical astronomy, and mathematical technology). All the others deal with astrology (in the narrowest sense, it appears) and Hermetic matters (four of these belong to the Greco-Roman world, six to the Assyro-Babylonian orbit, and four are Indians). Thus, not a single work on mathematics written by a non-Greek, pre-Islamic scholar was known to our tenth-century court librarian,[9] who would certainly be in a good position to know anything there was to know.

Central to the Greek tradition, as it was taken over by the Islamic world, were the *Elements* and the *Almagest*. Together with these belonged, however, the "Middle Books," the *Mutawassiṭāt:* The "Little Astronomy" of Autolycos, Euclid, Aristarchos, Hypsicles, Menelaos, and Theodosios; the Euclidean *Data* and *Optics;* and some Archimedean treaties.[10] Even Apollonios and a number of commentators to Euclid, Ptolemy, and Archimedes (Pappos, Hero, Simplicios, Theon, Proclos, Eutocios) belong to the same cluster.[11]

Somewhat less central are the Greek arithmetical traditions, whether Diophantos or the Neopythagorean current as presented by Nicomachos (or by the arithmetical books of the *Elements,* for that matter). Still, all the works in question were of course translated; further work on Diophantine ideas by al-Karajī and others is well testified,[12] and even though Nicomachean arithmetic was, according to Ibn Khaldūn, "avoided by later scholars" as "not commonly used [in practice]" (*Muqaddimah* VI,19), it inspired not only Thābit (Nicomachos' translator) but other scholars too.[13] Finally, the treatment of the subject in encyclopedic works demonstrates familiarity with the concepts of Pythagorean and Neopythagorean arithmetic.[14]

Also somewhat peripheral—*yet definitely less peripheral than they were for those Byzantine scholars whose selection of works to be studied and hence to survive has created our picture of Greek mathematics*—are the subjects that we might characterize tentatively as "technological mathematics" (al-Fārābī speaks of *ʿilm al-ḥiyal*, "science of artifices"—Palencia 1953, Arabic p. 73) and its cognates: optics and catoptrics, "science of weights," and non-orthodox geometrical constructions (geometry of movement, geometry of fixed compass-opening). They are well represented, for example, in works by Thābit, the Banū Mūsā, Qusṭā ibn Lūqa, ibn al-Haytham, and Abū'l-Wafāʾ—detailed discussions would lead too far astray.

### 3. Scientific Source Traditions: India

The way al-Nadīm mentions the Indians reflects the way the Indian inspiration must have looked when seen from the Islamic end of the transmission line, even though he misses (and is bound to miss) essential points. Indian mathematics, when it reached the Caliphate, had, according to all available evidence, become anonymous: The ideas of Indian trigonometry were adopted via Siddhantic astronomical works and *zījes* based fully or in part on Indian sources[15]; Islamic algebra was untouched by Indian influence—which it would hardly have been, had the Islamic mathematicians had direct access to great Indian authors in the style of Āryabhaṭa and Brahmagupta.[16]

Below the level of direct scientific import, some influence of Indian algebra is plausible. This is indicated by the metaphorical use of *jidhr* ("root, stem, lower end, stub" etc.) for the first power of the unknown. Indeed, this same metaphor (which can hardly be considered self-evident, especially not in a rhetorical, non-geometric algebra—cf. Section 6) is found already around 100 B.C. in India (Datta and Singh 1962: II, 169). In all probability, however, this borrowing was made via practitioners' subscientific transmission lines, to which we shall return; furthermore, the ultimate source for the term need not have been Indian.

Apart from trigonometry, the main influence from Indian mathematics is the use of "Hindu numerals." If the Latin translation of al-Khwārizmī's introduction of the system is to be believed (and it probably is[17]), he only refers it to "the Indians." So already does Severus Sebokht in the mid-seventh century (fragment published and translated in Nau 1910: 225f.). The earliest extant "algorism" in Arabic, that of al-Uqlīdisī from the mid-tenth century, is no more explicit. Most of its references are to "scribes" or "people of this craft"—evidently, local users of the technique are thought of; there is no explicit reference to the origin of the craft, nothing more precise than "Indian reckoners" (trans. Saidan 1978: 45, 104, 113). In addition, the dust-board so essential for early "Hindu reckoning" was known under a Persian, not an Indian name (*takht*—Saidan 1978: 351). Finally, the methods of indeterminate equations and combinatorial analysis (which both are staple goods in Indian arithmetical textbooks) are not found with the early Islamic expositions of Hindu reckoning (even though examples of indeterminate equations can be found in textbooks based on finger reckoning).[18] So the Islamic introduction of Hindu reckoning can hardly have been based on direct knowledge of "scientific" Indian expositions of arithmetic. Like trigonometry, it appears to derive from contact with practitioners using the system. Some inspiration for work on the summation of series may have come from India. Apart from the chessboard problem (to which we shall return), the evidence is not compelling, and proofs given by al-Karajī and others may be of Greek as well as Indian inspiration.

The two scientific source traditions were mainly tapped directly through translations from Greek and Sanskrit. To some extent, however, the mathematics of Indian astronomy found its way through Pahlavi, whereas elementary Greek astronomy may have been diffused through both Pahlavi and Syriac.[19] Neither of these secondary channels of transmission appears to have been scientifically creative, and they should probably only be counted among the scientific source traditions insofar as we distinguish "scientific" (e.g., astronomical and astrological) practice from "subscientific" (e.g., surveyors', master builders', and calculators') practice.

### 4.  Subscientific Source Traditions: Commercial Calculation

This brings us to the problems of subscientific sources, which we may initially approach through an example. The last chapter in al-Uqlīdisī's arithmetic is entitled "On Doubling One, Sixty-Four Times" (Kitāb al-Fuṣūl . . . IV, 32, trans. Saidan 1978: 337). Evidently, we are confronted with the chessboard problem, to the mathematical solution of which already al-Khwārizmī had dedicated a treatise,[20] and whose appurtenant tale is found in various Islamic writers from the ninth century onwards.[21] Al-Uqlīdisī, however, states that "This is a question many people ask. Some ask about doubling one 30 times, and others ask about doubling it 64 times," thereby pointing to a wider network of connections. In the mid-twelfth century, Bhāskara II asks about 30 doublings in the Līlāvatī (trans. Colebrooke 1817: 55); so does Problem 13 in the Carolingian collection Propositiones ad acuendos juvenes, ascribed to Alcuin (ed. Folkerts 1978). A newly published cuneiform tablet from the eighteenth century B.C.[22] contains the earliest extant version of the problem. Like the chessboard problem, it deals with grain, and its 30 doublings correspond to the 30 cases of a current game board; but like the Carolingian problem, it avoids speaking of "doubling" or "multiplication by 2," telling instead that "to each grain/soldier comes another one."

The problem belongs in the category of "recreational problems," defined by Hermelink (1978: 44) as "problems and riddles which use the language of everyday but do not much care for the circumstances of reality"—to which we may add the further observation that an important aspect of the "recreational" value of the problems in question is a funny, striking, or even absurd deviation from these circumstances. With good reason, Stith Thompson includes the chessboard doublings in his *Motif-Index of Folk Literature* (1975: V, 542 [Z 21.1]). Seen from a somewhat different perspective, we may look at recreational mathematics as a "pure" outgrowth of the teaching and practice of practitioners' mathematics (which, in the pre-Modern era, spells computation). It does not seek mathematical truth or theory; instead, it serves the display of virtuosity.[23]

Other recreational problems share the widespread distribution of the repeated doublings. Shared problem types (and sometimes shared numbers) and similar or common dress connect the arithmetical epigrams in Book XIV of the *Anthologia graeca*,[24] Anania of Širak's arithmetical collection from seventh century Armenia,[25] the Carolingian *Propositiones,* part of the ancient Egyptian Rhind Papyrus, and ancient and medieval problem collections from India and China.[26] They turn up without fancy dress in Diophantos' *Arithmetica,* and recur in medieval Islamic, Byzantine, and Western European problem collections. The pattern looks very much like the distribution of folktales (even to the extent that Diophantos' adoption of the material can be seen as a parallel to the literate adoption of folktale material). The geographical distribution is also roughly congruent with that of the Eurasian folktale (*viz.,* "from Ireland to India"—Stith Thompson 1946: 13ff.). This, however, can only be regarded as a parallel, not as an explanation. First of all, the recreational problems cover an area stretching into China, beyond the normal range of Eurasian folktales[27]; second, mathematics can be entertaining only in an environment that knows something about the subject. The predominant themes and techniques of the problems in question point to the community of traders and merchants interacting along the Silk Road, the combined caravan and sea route reaching from China to Cadiz.[28]

"Oral mathematics" cannot be encountered *in vivo* in the sources. Like folktales before the age of folklorists, it has normally been worked up by those who took the care to write it down, adoption entailing adaption.[29] Generally they were mathematicians, who at least arranged the material systematically, and perhaps gave alternative or better methods for solution or supplied a proof. In a few cases, however, they added a description of the situation in which they found the material. So Abū Kāmil in the preface to his full mathematical treatment of the indeterminate problem of "the hundred fowls,"[30] which he described (German trans. Suter 1910: 100) as

a particular type of calculation, circulating among high-ranking and lowly people, among scholars and among the uneducated, at which they rejoice, and which they find new and beautiful; one asks the other, and he is then given an approximate and only assumed answer, they know neither principle nor rule in the matter.

A similar aggressive description of reckoners who

strain themselves in memorising [a procedure] and reproduce it without knowledge or scheme [and others who] strain themselves by a scheme in which they hesitate, make mistakes, or fall in doubt

is given by al-Uqlīdisī in connection with the continued doubling asked by "many people" (trans. Saidan 1978: 337).

It is precisely this situation that has distorted the approach to the sub-scientific traditions. The substratum was anonymous and everywhere present, and its procedures deserve the designation of recipes rather than methods. Every mathematician, when inspired from it, would therefore have to employ his own techniques to solve the common problems (or at least have to translate the recipes into his own theoretical idiom)—Diophantos would use rhetorical algebra, the Chinese *Nine Chapters on Arithmetic* would manipulate matrices, and the *Liber abaci* would find the answer by means of proportions. We should hence not ask, as commonly done, whether Diophantos (or the Greek arithmetical environment) was the source of the Chinese or vice versa. There was *no specific source: The ground was everywhere wet.*

Besides supplying problems and procedures, the merchants' and book-keepers' community appears to have provided Islamic mathematics with two of its fundamental arithmetical techniques: The peculiar system of fractions, and the "finger reckoning."

The system of fractions is built up by means of the series of "principal fractions" 1/2, 1/3, . . . 1/10 (the fractions that possess a name of their own in the Arabic language) and their additive and multiplicative combinations (described, for example, in Saidan 1974: 368, and in Juschkewitsch 1964: 197ff.; cf. Høyrup 1990b). The system has been ascribed to Egyptian influence and to independent creation within the territory of the Eastern Caliphate.[31] It turns out, however, that already some Old Babylonian texts use similar expressions—for example, "the third of $X$, and the fourth of the third of $X$" for $5/12 X$.[32] So we are really confronted with an age-old system and at least a common Semitic usage—but, for example, the formulation of Problem 37 of the Rhind Papyrus suggests in fact a common *Hamito*-Semitic usage, a usage that had already provided the base on which the Egyptian scribes developed their learned unit fraction system around the end of the third millennium B.C.[33] Since the same "fractions of fractions" are also used in those of the arithmetical epigrams of the *Anthologia graeca* which have a dress connected to the trade routes and to technologies of Near Eastern origin (see Høyrup 1990b: 297–99), they appear to have spread over the whole Near East and the Eastern part of the Roman Empire in late antiquity, and thus to have been well rooted in the commercial communities all over the region covered by the Islamic expansion; further evidence of this is their use in arithmetical textbooks written for merchants, accounting officials, and the like in the earlier Islamic period (cf. Section 14).

The use of "principal fractions" and "fractions of fractions" appears to coincide with that of "finger reckoning," another characteristic method of Islamic elementary mathematics. It was referred to as *ḥisāb al-Rūm wa'l-ʿArab* (Saidan 1974; 367 [misprint corrected]), "calculation of the Byzantines and the Arabs." A system related to the one used in medieval Islam had been employed in ancient Egypt.[34] Various ancient sources refer to the symbolization of numbers by means of the fingers (see Menninger 1957: II, 11–15), without describing, it is true, the convention employed. But since the very system used in Islam was described around A.D. 700 in Northumbria by Bede (*De temporum ratione*, cap. I—edited by Jones 1943: 179–81), who would be familiar with descendants of ancient methods rather than with the customs of Islamic traders, we may safely assume that the ancient system was identical with both.

### 5. Subscientific Source Traditions: Practical Geometry

So far we have dealt with what appears to have been a more or less shared tradition for practitioners of bookkeeping and commercial arithmetic (ḥisāb, to use the Arabic term). Another group possessing a shared tradition (for practical geometry) was made up of surveyors, architects, and "higher artisans."[35]

In the case of this subscientific geometry, we can follow how the process of mathematical synthesis has begun long before the Islamic era. Indeed, various ancient civilizations had had their specific practical geometries. The (partly) different characters of Egyptian and Babylonian practical geometry have often been noticed.[36] The melting pots of the Assyrian, Achaemenid, Hellenistic, Roman, Bactrian, and Sassanian empires mixed them up completely,[37] and through the Heronian corpus some Archimedean and other improvements were infused into the practitioners' methods and formulas.[38] This mixed and often disparate type of calculatory geometry was encountered locally by the mathematicians of Islam, who used it as a basic material while criticizing it—just as they encountered, used, and criticized the practices of commercial and recreational arithmetic.

## 6. Algebra and its Alternative

As pointed out in Section 3, Islamic algebra was in all probability *not* inspired from Indian scientific algebra. Detailed analysis of a number of sources suggests instead a background in the subscientific tradition—or, indeed, in two different subscientific traditions. I have published the arguments for this elsewhere,[39] and here shall therefore present the results of the investigation only briefly.

*Al-jabr* was performed by a group of practitioners engaged in *ḥisāb* (calculation) and spoken of as *ahl al-jabr* (algebra people) or *aṣḥāb al-jabr* (followers of algebra). The technique was purely rhetorical; a central subject was the reduction and resolution of quadratic equations—the latter by means of standardized algorithms (analogous to the formula $x = \frac{1}{2}b + \sqrt{(\frac{1}{2}b)^2 + c}$ solving the equation $x^2 = bx + c$, etc.) unsupported by arguments, the rhetorical argument being reserved for the reduction. Part of the same practice (but possibly not understood as covered by the term *al-jabr*) was the rhetorical reduction and solution of first-degree problems.

As argued in Section 3, part of the characteristic vocabulary suggests a subscientific (but probably indirect) connection to India. An ultimate connection to Babylonian algebra is also inherently plausible, but not suggested by any positive evidence; in any case the path that may possibly have led from Babylonia to the early medieval Middle East must have been tortuous, since the methods employed in the two cases are utterly different.

The latter statement is likely to surprise, since Babylonian algebra is normally considered to be either built on standardized algorithms or on oral rhetorical techniques. A detailed structural analysis of the terminology and of the distribution of terms and operations inside the texts shows, however, that this interpretation does not hold water; furthermore, it turns out that the only interpretation of the texts that makes coherent sense is geometric—the texts have to be read as (naive, non-apodictic) constructional prescriptions, dealing really, as they seem to do when read literally, with (geometric) squares, rectangles, lengths and widths (all considered as *measured* entities); they split, splice and combine figures so as to obtain a figure with known dimensions in a truly analytic though completely heuristic way. Only certain problems of the first degree (if any) are handled rhetorically, and no problems are solved by blind standard algorithms.[40]

This is quite certain for Old Babylonian algebra (c. seventeenth century B.C.)—here, the basic problems are thought of as dealing with rectangles, and they are solved by naive-geometric "analysis." A few tablets dating from the Seleucid period and written in the Uruk environment of astronomer-priests contain second-degree problems, too. They offer a more ambiguous picture. Their garb is geometric, and so apparently is the method (though synthetic rather than analytic); but the geometric procedure is obviously thought of as *an analogy to a set of purely arithmetical relations* between the unknown magnitudes.

Having worked intensively with Babylonian texts for some years I was utterly amazed to discover accidentally their distinctive rhetoric (character-ized by fixed shifts between present and past tense, and between the first, second, and third person singular) in the medieval Latin translation of a *Liber mensurationum* written by an unidentified Abū Bakr. The first part of this *misāḥa* (surveying) text contains a large number of problems similar to those known from the Old Babylonian tablets: A square plus its side is 110; in a rectangle, the excess of length over width is 2, and the sum of the area and the four sides is 76; and so on. The problems, furthermore, are first solved in a way strikingly reminiscent of the Old Babylonian methods (although the matter is obscured by the absence of a number of figures alluded to in the text); a second solution employs the usual rhetorical reductions and solutions by means of standard algorithms. The second method is spoken of as *aliabra*, evidently a transliteration of *al-jabr*. The first usually goes unlabeled, being evidently the standard method belonging to the tradition; in one place, how-ever, it is spoken of as "augmentation and diminution"—apparently the old splicing and splitting of figures.

A precise reading of the text in question leaves no reasonable doubt that its first part descends directly from the Old Babylonian "algebra" of mea-sured line segments (the second part contains real mensuration in agreement with the Alexandrinian tradition). Once this is accepted as a working hypoth-esis, a number of other sources turn out to give meaningful evidence. The geometrical "proofs" of the algebraic standard algorithms given by al-Khwārizmī and ibn Turk will have been taken over from the parallel naive-geometric tradition; Thābit's Euclidean proof of the same matter is therefore really *something different*, which probably explains his silence with regard to these hitherto presumed predecessors; and so forth.

Especially interesting is Abū'l-Wafā''s report of a discussion between (Euclidean) geometers on one hand and surveyors and artisans on the other in his *Book on What Is Necessary from Geometric Construction for the Artisan* (10.xiii; Russian trans. Krasnova 1966: 115). He refers to the "proofs" used by the latter in questions concerned with the addition of figures; these proofs turn out to be precisely the splitting and splicing used by Abū Bakr and in the Old Babylonian texts. This confirms a suspicion already suggested by the appearance of the "algebra" of measured line segments in a treatise on men-suration: this "algebra" belonged to the practitioners engaged in subscien-tific, practical geometry, and was hence a tradition of surveyors, architects, and higher artisans. *Al-jabr,* on the other hand, was carried by a community of calculators, and was considered part of *ḥisāb,* Abū Kāmil seems to tell us (see note 39).

## 7. Reception and Synthesis

These different source traditions had already merged to some extent in pre-Islamic times. The development of a syncretic practical geometry was discussed previously, and the blend of (several sorts of) very archaic surveying formulae with less archaic recreational arithmetic in the *Propositiones ad acuendos juvenes* was also touched on.[41] Still, merging, and especially critical and creative merging, was no dominant feature.

From the ninth century onward, it came to be the dominant feature of Islamic mathematics. The examples are too numerous to be listed, but a few illustrations may be given.

A modest example is the geometric chapter of al-Khwārizmī's *Algebra.* As shown by Solomon Gandz,[42] it is very closely related to the Hebrew *Mišnat ha-Middot,* which is a fair example of pre-Islamic syncretic practical geometry, or at least a very faithful continuation of that tradition.[43] Al-Khwārizmī's version of the same material is not very different; but before treating of circular segments precisely in the way it is done in the *Mišnat ha-Middot* he tells that the ratio $3\frac{1}{7}$ between perimeter and diameter of a circle "is a convention among people without mathematical proof" (cf. note 38); he goes on to inform us that the Indians "have two other rules," one equivalent to $\pi = \sqrt{10}$ and the other to $\pi = 3.1416$[44]; finally he gives the exact value of the circular area as the product of semiperimeter with semidiameter, together with a heuristic proof.[45] So not only are the different traditions brought together, but we are also offered a sketchy critical evaluation of their merits.

If the whole of al-Khwārizmī's *Algebra* is taken into account, the same features become even more obvious. The initial presentation of the *al-jabr* algorithms is followed by their geometrical justification, by reference to figures inspired by the "augmentation-and-diminution" tradition, but which are more synthetic in character, and for the sake of clear presentation make use of Greek-style letter formalism.[46] A little further on, the author attempts his own extrapolation of the geometrical technique in order to prove the rules of rhetorical reduction.

The result is still somewhat eclectic—mostly so in the chapter on geometry. Comparison and critical evaluation do not amount to real synthesis. But in the work in question, and still more in the total undertaking of the author, an effort to make more than random collection and comparison of traditions is clearly visible. Soon after al-Khwārizmī, furthermore, other authors wrote more genuinely synthetic works. One example, in the same field as al-Khwārizmī's naive-geometric proofs of the *al-jabr* algorithms, is Thābit's treatise on the "verification of the rules of *al-jabr*" by means of *Elements* II.5–6 (ed., trans. Luckey 1941); another, in the field of practical geometry, is Abū'l-Wafā"'s *Book on What Is Necessary from Geometric Construction for the Artisan,* where methods and problems of Greek geometry (including, it now appears, Pappos' passage on constructions with restricted and constant compass opening—see Jackson 1980) and Abū'l-Wafā"'s own mathematical ingenuity are used to criticize and improve upon practitioners' methods, but where the practitioners' perspective is also kept in mind as a corrective to otherworldly theorizing.[47] These examples could be multiplied *ad libitum.* Those already given, however, will suffice to show that the Islamic synthesis was more than bringing together methods and results from the different source traditions; it included an explicit awareness of the difference between theoreticians' and practitioners' perspectives and of the legitimacy of both, as well as acknowledgement of the possible relevance and critical potentiality *of each* when applied to results, problems, or methods belonging to the other. Whereas the former aspect of the synthesizing process, though much further developed than in other ancient or medieval civilizations, was not totally unprecedented, the latter *was* exceptional (cf. also Section 13).

Apart from a violent cultural break and an ensuing cultural flowering (which of course explains much, but unspecifically), what accounts for the creative assimilation, reformulation, and (relative) unification of disparate legacies as the "mathematics of the Islamic world?" And what accounts for the specific character of Islamic mathematics as compared, for example, with Greek or medieval Latin mathematics?

## 8. "Melting Pot" and Tolerance

I shall not pretend to give anything approaching an exhaustive explanation. Instead, I shall point to some factors that appear to be important, and possibly fundamental.

On a general level, the "melting-pot effect" was an important precondition for what came about. Within a century from the *Hijra,* the whole core area of Medieval Islam had been conquered, and in another century or so the most significant strata of the Middle Eastern population were integrated into the emerging Islamic culture[48]; this—and also the movements of single scholars as well as those of larger population groups, especially toward the Islamic center in Baghdad—broke down earlier barriers between cultures and isolated traditions and offered the opportunity for "cultural learning." The religious *and cultural* tolerance of Islam was also important here. Muslims were of course aware of the break in history created by the rise of Islam, and in the field of learning a distinction was maintained between *awā'il*, "pristine" (i.e., pre-Islamic) and Muslim/Arabic "science" (i.e., *'ilm*, a term largely congruent with Latin *scientia* and perhaps better rendered as "field of knowledge"). Since, however, the latter realm encompassed only religious (including legal), literary, and linguistic studies,[49] the complex societal setting of learning in the mature Islamic culture prevented the development of attitudes such as the ancient Greek overall contempt for "barbarians." Furthermore, the rise of people whose roots were in *different* older elites to elite positions in the Caliphate may have precluded that sort of cultural exclusiveness which came to characterize Latin Christianity during *its* phase of cultural learning (where, even when most open to Islamic and Hebrew learning, it took over only translations and practically no scholars, and showed little interest in translating works with no relation to the "culturally legitimate" Greco-Roman legacy[50]).

Whatever the explanations, Islam remained freer of ethnocentrism and culturocentrism than many other civilizations.[51] Because of this tolerance, the intellectual and cognitive barriers that were melted down in the pot were not immediately replaced by new barriers, which would have blocked up the positive effects of cultural recasting. It also permitted Islamic learning (both in its initial phase and later) to draw on the service of Christians, Jews, and Sabians, and on Muslims rooted in different older cultures—for example, Māšā'allāh the Jew, Thābit the heterodox Sabian, Ḥunayn ibn Isḥāq the Nestorian, Qusṭā ibn Lūqā the Syrian Christian of Greek descent, Abū Maʿšar the heterodox, pro-Pahlavi Muslim from the Hellenist-Indian-Chinese-Nestorian-Zoroastrian contact point at Balkh, and ʿUmar ibn al-Farrukhān al-Ṭabarī the Muslim from Iran. In later times, the conversion from Judaism of al-Samaw'al late in life appears to have been unrelated to his scientific career. (Cf. the biographies in *DSB*). Still later, the presence of Chinese astronomers collaborating with Muslims at Hulagu's observatory at Maragha underscores the point (see Sayılı 1960: 205–7).

Still, the "melting pot effect" and tolerance were only preconditions—"material causes," in a quasi-Aristotelian sense. This leaves open the other aspect of the question: Which "effective" and "formal" causes made medieval Islam scientifically and mathematically creative?

### 9. Competition?

It has often been claimed that the early ninth-century awakening of interest in pre-Islamic (*awā'il*) knowledge[52] "must be sought in the new challenge which Islamic society faced" through the "theologians and philosophers of the religious minorities within the Islamic world, especially the

Christians and Jews" in "debates carried on in cities like Damascus and Baghdad between Christians, Jews and Muslims," the last being "unable to defend the principles of faith through logical arguments, as could the other groups, nor could they appeal to logical proofs to demonstrate the truth of the tenets of Islam" (thus the formulations of Seyyed Hossein Nasr [1968: 70]).

One problematic feature of the thesis is that "we have very little evidence of philosophical or theological speculation in Syria [including Damascus] under the ʿUmayyad dynasty," as observed by De Lacy O'Leary (1949: 142), otherwise close to the idea of stimulation through intellectual competition. Another serious challenge to the thesis is the fact that Islamic learning advanced well beyond the level of current Syriac learning in a single bound (not least in mathematics). Here the cases of the translators Ḥunayn ibn Isḥāq and Thābit should be remembered. Both began their translating activity (and Thābit his whole scientific career) in the wake of the ʿAbbasid initiative. Most of their translations and other writings were in Arabic; what they made in Syriac was clearly a secondary spin-off, and included none of their mathematical writings or translations (unless some minor work be hidden among Thābit's "various writings on astronomical observations, Arabic and Syriac"—Suter 1900: 36). Ḥunayn's translations, in particular, show us that by the early ninth century A.D. the Syriac environment was almost as much in need of broad Platonic and Aristotelian learning as the Muslims[53] (cf. also Section 11). True, if we return from philosophy to mathematics, a Provençal-Hebrew translation from A.D. 1317 of an earlier Mozarabic treatise claims that an Arabic translation of Nicomachos' *Introduction to Arithmetic* was made from the Syriac before A.D. 822.[54] This testimony, even if reliable, is however of little consequence: Veritable understanding of scientific mathematics is certainly not a necessary condition for interest in Nicomachos, nor is it the inevitable outcome of even profound familiarity with his *Introduction*. Equally little can be concluded from the existence of a (second-rate) Syriac translation of Archimedes' *On the Sphere and Cylinder*, since it may well have been prepared as late as the early ninth century and thus have been a spin-off from the ʿAbbasid translation wave.[55]

A possible quest for intellectual competitiveness will thus hardly do, and definitely not as sole explanation of the scientific and philosophical zeal of the early ninth-century ʿAbbasid court and its environment. Similarly, the ʿAbbasid adoption of many institutions and habits from the Sassanian state and court and the concomitant peaceful reconquest of power by the old social elites[56] may explain the use of astrologers in the service of the court, for example, at the famous foundation of Baghdad. But then it does not explain why Islamic astrologers were not satisfied with the *Zīj al-Šah* (the astronomical tables prepared for the Sassanian Shah Yazdijird III), with its connection to the past of the reborn Sassanian elite. An attempt to be *still more splendid* than the Sassanian Shah could explain that the Caliph made his scholars tap Siddhantic and Ptolemaic astronomy directly from the sources, instead of using only the second-hand digests known from the Pahlavi astrologers (see Pingree 1973: 35). Continuity or revival of elites and general cultural patterns are, however, completely incapable of explaining a sudden *new* vigor of scientific culture in the Irano-Iraqian area—a truly qualitative jump. In particular, they are unable to tell why a traditional interest in the astronomico-astrological applications of mathematics should suddenly lead to interest in mathematics *per se*—not to speak of the effort toward synthesizing separate traditions.

### 10. Institutions or Sociocultural Conditions?

Sound sociological habit suggests that one look for explanations at the institutional level. Yet a serious problem presents itself to this otherwise reasonable "middle range" approach to the problem (to use Robert K. Merton's expression[57]): The institutions of Islamic learning were only in their swaddling clothes in the early ninth century A.D., if born at all. In this age of fluidity and fundamental renewal, Islamic learning formed its institutions quite as much as the institutions formed the learning.[58] In order to get out of this closed circle of pseudo-causality we will hence have to ask why institutions became shaped the way they did. The explanations should hold for the whole core area of medieval Islam and should at the same time be specific for this area.

Two possibilities suggest themselves: Islam itself, which was shared as a cultural context even by non-Muslim minorities and scholars; and the Arabic language. Language can be ruled out without hesitation. Admittedly, the general flexibility of Semitic languages (especially the rich verbal system with its complex of generalized aspect and voice, and its vast array of corresponding nominal derivatives) makes them well suited both to render accurately foreign patterns of thought and to serve as the basis for the development of autochthonous philosophical and scientific thought. But Syriac and other Aramaic dialects are no less Semitic than Arabic; they had been shaped and ground for philosophical use over centuries, and much of the Arabic terminology was in fact modeled upon the Syriac (see Pines 1970: 782). The Arabic tongue was an adequate medium for what was about to happen, but it replaced another medium that was just as adequate. Language cannot be the explanation. Islam remains.

Of course, the explanation need not derive from Islam regarded as a system of religious teachings—what matters is Islam as a specific, integrated social, cultural, and intellectual complex. From this complex some important factors can be singled out.

## 11. Practical Fundamentalism

One factor is the very character of the complex as an integrated structure—i.e., those implicit fundamentalist claims of Islam which have most often been discussed with relation to Islamic law as

> the totality of God's commands that regulate the life of every Muslim
> in all its aspects; it comprises on an equal footing ordinances regarding
> worship and ritual, as well as political and (in the narrow sense) legal
> rules, details of toilet, formulas of greeting, table-manners, and sick-
> room conversation. (Schacht 1974: 392)

True, religious fundamentalism in itself has normally had no positive effects on scientific and philosophical activity, and it has rarely been an urge toward intellectual revolution. In ninth-century Islam, however, fundamentalism was confronted with a complex society in transformation; religious authority was not segregated socially as a "Church"—ḥadīth (traditions, cf. note 58) and Islamic jurisprudence in general were the province mainly of persons engaged in practical life, be it handicraft, trade, secular teaching, or government administration (see Cohen 1970). Furthermore, a jealous secular power did its best to restrain the inherent tendencies of even this stratum to get the upper hand.[59]

Fundamentalism combined with the practical involvement of the carriers of religious authority may have expressed itself in the recurrent tendency in Islamic thought to regard "secular" knowledge (scientia humana, in the medieval Latin sense) not as an alternative to Holy Knowledge (kalām al-dīn, "the discourse of Faith,"[60] scientia divina) but as a way to it, and even to contemplative truth, ḥāl (sapientia, in yet another Latin approximate parallel). Psychologically, it would be next to impossible to regard a significant part of the activity of "religious personnel" as irrelevant to their main task,[61] or as directly irreverent (as illustrated by the instability generated by scandalous popes (and other Church hierarchs) and leading to the Reformation revolts; in Islam, the ways of the Umayyad caliphs provided as effective a weapon for Khārijite radicals and for the ʿAbbasid take-over as that given to Anabaptists and to Lutheran princes by the Renaissance popes). That integration of science and religious attitude was not just a Mutakallimūn's notion but was shared to a certain extent by active mathematicians is apparent from the ever-recurring invocation of God in the beginnings and endings of their works (and the references to Divine assistance interspersed inside the text of some works[62]).

The legitimization of scientific interests through the connection to a religion that was fundamentalist in its theory and bound up with social life in its practice may also by analogy have impeded the complete segregation of pure science from the needs of daily life without preventing it from rising above these needs. This generated not only the phenomenon that even the best scientists would occasionally be concerned with the most practical and everyday applications of their science,[63] but also the general appreciation of theory and practice as belonging naturally together—cf. Section 7 and especially Section 13.

The plausibility of this explanation can be tested against some parallel cases. One of these is that of Syriac learning, which belonged in a religious context with similar fundamentalist tendencies. Syrian Christianity, however, was carried by a Church, i.e., by persons who were socially segregated from social practice in general, and Syriac learning was carried by these very persons. The custodian and non-creative character of Syriac learning looks like a sociologically trite consequence of this situation—as Schlomo Pines (1970: 783) explains,

> pre-Islamic monastic Syriac translations appear to have been undertaken mainly to integrate for apologetic purposes certain parts of philosophy, and perhaps also of the sciences, into a syllabus dominated by theology. In fact great prudence was exercised in this integration; for instance, certain portions of Aristotle were judged dangerous to faith, and banned.

Other interesting cases are found in twelfth- and thirteenth-century Latin Christianity. Particularly close to certain ninth-century Islamic attitudes is Hugh of Saint Victor, the teacher and rationalistic mystic from the Paris school of Saint Victor (cf. Chenu 1974). He was active during the first explosive phase of the new Latin learning (like the Islamic late eighth and early ninth century A.D., the phase where the *Elements* was translated), in a school that was profoundly religious and at the same time bound up with the life of its city. The sociological parallels with ninth-century Baghdad are striking. Also striking are the parallel attitudes toward learning. In the propaedeutic *Didascalicon*,[64] Hugh pleads for the integration of the theoretical "liberal arts" and the practical "mechanical arts"; his appeal to "learn everything, and afterwards you shall see that nothing is superfluous"[65] permits the same wide interpretation as the Prophet's saying, "Seek knowledge from the cradle to the grave"[66]; and he considers Wisdom (the study of which is seen in I.iii as "friendship with Divinity") a combination of moral and theoretical truth and practitioners' knowledge[67]—one can hardly come closer to al-Jāḥiz' formula as quoted in note 60.

Hugh was, however, an exception already in this own century. The established Church, as represented by the eminently established Bernard of Clairvaux, fought back; even the later Victorines demonstrate through their teaching that a socially segregated ecclesiastical body is not compatible with synthesis between religious mysticism, rationalism, and an open search for all-encompassing knowledge. As a consequence, the story of twelfth- and thirteenth-century Latin learning can on the whole be read as a tale of philosophy as potentially subversive knowledge, of ecclesiastical reaction, and of an ensuing final synthesis where the "repressive tolerance" of Dominican domesticated Aristotelianism blocked the future development of learning.[68] It is also a tale of segregation between theoretical science and practitioners' knowledge: lay theoretical knowledge gained a subordinated autonomy, but only by being cut off from the global world-view[69] and concomitantly also from common social practice. Autonomy thus entailed loss of that unsubordinated, mutually fecundating integration with practical concerns that was a matter of course in Islam.[70]

## 12. Variations of the Islamic Pattern

In the preceding pages, Islam was regarded as an undifferentiated whole or a broad average. Such generalizations have their value. Still, looking at specific religious currents or theological schools might give us some supplementary insight into the mechanisms coupling faith, learning, and social life, or at least give the picture the shades of life.

One current in which the coupling between the "discourse of philosophy" and the "discourse of Faith" was strong was the *Mu'tazila*. According to earlier interpretations, admittedly, the Mu'tazilite attitude to philosophy should have been as lopsided as that of Syriac monasticism[71]; but from Heinen's recent analysis it appears that the *Mu'tazila* in general did not derive the Syro-Christian sort of intellectual censorship on philosophy from the theological aims of *kalām* (cf. note 60). Under al-Ma'mūn, who used Mu'tazilism in his political strategy, the inherent attitudes of this theological current were strengthened by the ruler's interest in clipping the wings of those traditionalists whose fundamentalism would lead them to claim possession of supreme authority, concerning knowledge first of all, but implicitly also in the moral and political domain (cf. note 59 and surrounding text). Among the *Ismā'īlī* there was an equally strong (or stronger) acceptance of the relevance of the *awā'il* knowledge for the acquisition of Wisdom, even though the choice of disciplines was different from that of the *Mutakallimūn* (cf. the polemic quoted in note 71): To judge from the *Ikhwān al-ṣafā'*, neoplatonic philosophy, Harranian astrology, and the Hermeticism of late antiquity were central subjects (see Marquet, "Ikhwān al-ṣafā"'); but the curriculum of the *Ismāʿīlī* al-Azhar *madrasah* in Cairo included philosophy, logic, astronomy, and mathematics (see Fakhry 1969: 93), i.e., central subjects of ancient science. The same broad spectrum of religiously accepted interests (to which comes also Jabirian alchemy) can be ascribed to the Shiite current in general. In the Ašʿarite reaction to Muʿtazilism, and in later Sunna, the tendency was to emphasize fundamentalism and to reject non-Islamic philosophy, or at least to deny its relevance for Faith. Accordingly, the curriculum of the Sunni Nizamiyah *madrasah* in Baghdad included, alongside the traditionalist disciplines (religious studies, Arabic linguistics, and literature) only "arithmetic and the science of distributing bequests" (Fakhry 1969: 93)—the latter being in fact a "subdivision of arithmetic," as ibn Khaldūn explains.[72]

In the long run, Muʿtazilism lost out to Sunnism, and in the very long run the dominance of traditionalist Sunnism (and of an equally traditionalist Shiism) was probably one of the immediate causes that Islamic science lost its verve. During the Golden Age, however, when institutionalization was still weaker than a vigorous and multidimensional social life, the attitudes of the formulated theological currents reflected rather than determined ubiquitous dispositions (cf. also note 66). Muʿtazilism was only the most clear-cut manifestation of more general tendencies, and the reversal of the ʿAbbasid Muʿtazilite policy in 849 did not mean the end of secular intellectual life in the Caliphate nor of the routine expression of religious feelings in the opening and closing sentences of scholarly works. Furthermore, through Sufi learning, and in the person of al-Ghazzālī, secular knowledge gained a paradoxical new foothold—more open, it is true, to quasi-Pythagorean numerology than to cumulative and high-level mathematics,[73] but still able to encourage more serious scientific and mathematical study. It appears that the Sufi mathematician ibn al-Bannāʾ did in fact combine mathematical and esoteric interests (see Renaud 1938); and even though al-Khayyāmī's Sufi confession may be suspected of not reflecting his inner opinions as much as his need for security (see Youschkevitch and Rosenfeld, "Al-Khayyāmī," and Kasir 1931: 3f.), his claim that mathematics can serve as part of "wisdom"[74] must either have been an honest conviction or (if it was meant to serve his security) have had a plausible ring in contemporary ears. At the same time, the two examples show that the role of Gnostic sympathies was only one of external inspiration: Ibn al-Bannāʾ's works are direct (and rather derivative) continuations of the earlier mathematical and astronomical tradition (see Vernet, "Ibn al-Bannāʾ") and al-Khayyāmī's treatise was written as part of the running tradition of metamathematical commentaries to the *Elements,* and in direct response to ibn al-Haytham. Gnostic sympathies might lead scholars to approach and go into the mathematical traditions, but they did not transform the traditions, nor did they influence how work was carried out within traditions.

### 13. The Importance of General Attitudes: The Mutual Relevance of Theory and Practice

Analysis according to specific religious currents is hence not to be taken too literally when the shaping of mathematics is concerned. As long as religious authority was not *both* socially concentrated and segregated *and* in possession of scholarly competence (as it tended to be in thirteenth-century Latin Christianity), the attitudes of even dominant religious currents and groups could influence the internal development and character of learning only indirectly, by influencing overall scholarly dispositions and motivations. (What it could do directly without fulfilling the two conditions was to strangle rational scholarship altogether; such things happened, but they affected the pace and ultimately the creativity of Islamic science, which is a different issue). If only scholars could find a place in an institution under princely protection (of type "library with academy," i.e., *Dār al-ʿilm* and the like, or an observatory) or covered by a religious endowment (of type *madrasah*, hospital and so forth), the absence of a centralized and scholarly competent Church permitted them to work in relative intellectual autonomy, as long as they kept within the limits defined by institutional goals[75]: princes, at least, were rarely competent to interfere with learning by more subtle and precise means than imprisonment or execution.[76] The general attitude—that mathematics *qua* knowledge was religiously legitimate and perhaps even a way to Holy Knowledge, and that conversely, the Holy was present in the daily practice of this world—could mold the disposition of mathematicians to the goals of their discipline; but even a semi-Gnostic conception of rational knowledge as a step toward Wisdom appears not to have manifested itself as a direct claim on the subjects or methods of actual scientific work—especially not as a claim to leave traditional subjects or methods.

Accordingly, in Islamic mathematical works explicit religious references are normally restricted to the introductory dedication to God, the corresponding clause at the end of the work, and perhaps passing remarks invoking his assistance for understanding the matter or mentioning his monopoly on supreme knowledge. Apart from that, the texts are as secular as Greek or medieval Latin mathematics. It is impossible to see whether the Divine dedications in Qusṭā ibn Lūqā's translation of Diophantos[77] are interpolations made by a Muslim copyist, or they have been written by the Christian translator with reference to a different God—they are quite external to the rest of the text. The ultimate goals of the activity were formulated differently from what we find in Greek texts (*when* a formulation is found at all in Greek texts—but cf. the initial quotation from Aristotle). Irrespective of his personal creed, the Islamic mathematician would not be satisfied by staying at the level of immediate practical necessities—he would go beyond these and produce something higher, *viz.*, principles, proofs, and theory; nor would he, however, feel that any theory, however abstract, was *in principle* above application,[78] or that the pureness of genuine mathematics would be polluted by possible contact with more daily needs. Several exemplifications were discussed in Section 7 (al-Khwārizmī, Abū'l-Wafā') and Section 11, note 63 (al-Uqlīdisī, ibn al-Haytham). An example involving a nonmathematician (or rather a philosopher not primarily mathematician) is al-Fārābī's chapter on ʿilm al-ḥiyal, the "science of artifices" or "of high-level application" (see end of Section 2). We should of course not be surprised to find the science of *al-jabr waʾl-muqābalah* included under this heading—algebra *was* already a high-level subject when al-Fārābī wrote. But even if we build our understanding of the subject on Abū Kāmil's treatise, we might well feel entitled to wonder at seeing it intimately connected to the complete ancient theory of surd ratios, including both that which Euclid gives in *Elements* X and "that which is not given there" (Spanish trans. Palencia 1953: 52). More expressive than all this is, however, the preface to al-Bīrūnī's trigonometrical *Book on finding the Chords in the Circle . . .* , which I quote in extensive excerpts from Suter's German translation (1910a—emphasis added):

You know well, God give you strength, for which reason I began searching for a number of demonstrations proving a statement due to the ancient Greeks concerning the division of the broken line in an arbitrary circular arc by means of the perpendicular from its centre, and which passion I felt for the subject [ . . . ], so that you reproached [?] me my preoccupation with these chapters of geometry, not knowing *the true essence of these subjects, which consists precisely in going in each matter beyond what is necessary.* If you would only, God give you strength, observe the aims of geometry, which consists in determining the mutual relation between its magnitudes with regard to quantity, and [if you would only observe] that it is in this way that one reaches knowledge of the magnitudes of *all things measurable and ponderable found between the centre of the world and the ultimate limits of perception through the senses.* And if you would only know that by them [the geometrical magnitudes] are meant the [mere] forms, detached from matter [ . . . ]. Whatever way [the geometer] may go, through exercise will he be *lifted from the physical to the divine teachings,* which are little accessible because of the difficulty to understand their meaning, because of the subtlety of their methods and the majesty of their subject, and because of the circumstance that not everybody is able to have a conception of them, *especially not the one who turns away from the art of demonstration.* You would be right, God give you strength, to reproach me, had I neglected to search for these ways [methods], and used my time where an easier approach would suffice; of if the work had *not arrived at the point which constituted the fundament of astronomy,* that is to the calculation of the chords in the circle and the ratio of their magnitude to that supposed for the diameter [ . . . ].

Only in God the Almighty and All-wise is relief!

By going beyond the limits of immediate necessity and by cultivating the abstract and apodictic methods of his subject, the geometer hence worships God—but only on the condition that (like God, we may add) he cares for the needs of (astronomical) everyday existence.

Al-Bīrūnī's formulation is unusually explicit, perhaps reflecting an unusually explicit awareness of current attitudes and their implications. Normally these attitudes stand out most obviously in comparison with texts of similar purpose or genre from neighboring cultures. Particularly gratifying in this respect is the field of technical literature. As mentioned in note 70 in the case of astronomy, the prevailing tendency in Latin learning was to opt for the easy way by means of simplifying, nonapodictic compendia. Highly illustrative are also the various Anglo-Norman treatises on estate management. One such treatise was compiled on the initiative of (or even by) the learned Robert Grosseteste[79]; yet it contains nothing more than common sense and rules of thumb. Not only was the semi-autonomous playing field granted to philosophical rational discussion in the thirteenth-century compromise not to encroach on *sacred* land; neither should it divert the attention of *practical* people and waste their time. In contrast, a handbook on "commercial science," written by one Šaykh Abū'l-Faḍl Jaʿfar ibn ʿAlī al-Dimišqī somewhere between 870 and 1174 A.D., combines general economic theory (the distinction between monetary, movable, and fixed property) and Greek political theory with systematic description of various types of goods and with good advice on prudent trade.[80] Knowledge of the fine points of trade was, just like mathematics or any other coherently organized, systematic knowledge, considered a natural part of an integrated world-view covered by *al-Islām.* In agreement with the basic (fundamentalist but still non-institutionalized) pattern of this world-view, the theoretical implications of applied knowledge were no more forgotten than the possible practical implications of theory.

The use of theory to improve on practice looks like a fulfillment of the ancient Heronian and Alexandrinian project.[81] It has not been totally inconsequential—the acceptance of $\pi = 22/7$ was discussed already; "Archimedes' screw" and Alexandrinian military and related techniques were also reasonably effective (see Gille 1980). On the whole, however, the project had proved beyond the forces of ancient science and of ancient society—for good reasons, we may assume, since fruitful application of theory presupposes a greater openness to practitioners' specific problems and perspective than current in Greek science.[82] Thus, as was claimed in the introductory chapter, the *systematic* theoretical elaboration of applied knowledge was a specific creation of the Islamic world. It was already seen in the early phase of Islamic mathematics, when the traditions of "scientific" and "subscientific" mathematics were integrated. The great synthetic works in the vein of al-Khwārizmī's *Algebra* or Abū'l-Wafā''s *Book on What Is Necessary From Geometric Construction for the Artisan* were already discussed (Section 7), as was their occasionally eclectic character. One step further was taken in cases where a problem taken from the subscientific domain was submitted to theoretical investigation on its own terms (i.e., not used only as inspiration for an otherwise independent investigation, as when Diophantos takes over various recreational problems and undresses them in order to obtain pure number-theoretical problems). Thābit's Euclidean "verification of the rules of *al-jabr*" was mentioned in the same section, and Abū Kāmil's preface to his investigation of the recreational problem of "hundred fowls" was quoted in Section 4.[83] A final and decisive step occurred when the results of theoretical investigation were adopted in and transmitted through books written for practitioners. A seeming first adoption of Thābit's *Verification* is found in Abū Kāmil's *Algebra*. At closer inspection, however, the reference to Euclid is rather ornamental, and the Euclidean proposition referred to is "demonstrated" by naive geometry. But in Abraham bar Ḥiyya's (Savasorda's) *Collection on Mensuration and Partition* the actual argument can only be understood by somebody who knows his Euclid by heart.[84]

## 14. The Institutionalized Cases: (1) Madrasah and Arithmetical Textbook

The two most interesting cases of infusion of theory into an inherently practical mathematical tradition appear to be coupled not only to the general dispositions of Islamic culture but also to important institutions that developed in the course of time. The first institution that I shall discuss is the writing of large-scale, reasoned arithmetical textbooks. Al-Uqlīdisī's work was already discussed (Section 4 and note 63); in his introductory commentary to the translation Saidan describes a number of other works that have come down to us.[85] In the early period, two largely independent types can be described: The "finger-reckoning type" and the "Hindi type," using verbal and Hindu numerals, respectively. The two most important finger-reckoning books are those of Abū'l-Wafā' (description in Saidan 1974) and of al-Karajī (trans. Hochheim 1878). The two earliest extant Hindi books are those of al-Khwārizmī (extant only in Latin translation) and al-Uqlīdisī. Among lost works on the subject from the period between the two, al-Kindī's treatise in four sections can be mentioned.[86] Later well-known examples are Kūšyār ibn

Labbān's explanation of the system for astronomers (trans. Levey and Petruck 1965) and al-Nasawī's for accounting officials.[87]

After the mid-eleventh century, it becomes difficult to distinguish two separate traditions. Whereas al-Nasawī, when examining around A.D. 1030 earlier treatises on his subject, would still (according to his preface) restrict the investigation to Hindi books, his contemporary ibn Ṭāhir inaugurated an era where the traditions were combined, writing himself a work presenting "the elements of hand arithmetic and the chapters of *takht* ["dust-board," i.e., Hindi—JH] arithmetic" together with "the methods of the people of arithmetic" (apparently his section 6 on Greek theoretical arithmetic) and the "arithmetic of the *zīj*" (sexagesimal fractions).[88]

The same combination is found again in the Maghrebi arithmetical tradition as we know it from works of al-Ḥaṣṣār (reported extensively in Suter 1901), ibn al-Bannāʾ (ed. trans. Souissi 1969), al-Umawī[89] and al-Qalaṣādī (ed., trans. Souissi 1988), and through ibn Khaldūn's report.[90] It is interesting in several respects, not least for its systematic development of arithmetical and algebraic symbolism,[91] the former of which was taken over by Leonardo Fibonacci in the *Liber abaci*.[92]

The early writers of large arithmetic textbooks appear to have been relatively independent of each other. During the initial phase of synthesis, they collected, systematized, and reflected upon current methods and problems of one or the other group of practitioners, and occasionally they used earlier treatments that were accessible as books. The first source appears to have been used in al-Uqlīdisī's work, whereas al-Nasawī quotes the written works he has consulted. The integration of finger- and Hindi reckoning, on the other hand, appears to depend upon the more continuous teaching tradition of the *madrasah*. Already ibn Ṭāhir is reported to have taught at the mosque (Saidan, "Al-Baghdādī"), and, as mentioned in Section 12, the only non-"traditional" subject permitted at the Baghdad Nizamiyah *madrasah* was arithmetic. In the Maghreb tradition, ibn al-Bannāʾ was taught and himself became a teacher of mathematics and astronomy at the *madrasah* in Fez (Vernet, "Ibn al-Bannāʾ"). This makes it inherently plausible that even al-Ḥaṣṣar, upon whose works he commented, had relations with the *madrasah* (at the very least his works must have been used there). Ibn al-Bannāʾ's network of disciples also appears to cohere through the social network of *madrasah* learning. As a writer of commentaries to al-Bannāʾ, Al-Qalaṣādī must be presumed to belong to the same context (cf. Saidan, "Al-Qalaṣādī"). Finally, even al-Umawī was active as a teacher in Damascus. The theoretical elevation of the subject of arithmetic was hence not only a product of the general dispositions of Islamic culture; according to all evidence it was also mediated by the *madrasah*, which in this respect came to function as an embodiment and as an institutional means of fixation of these same attitudes.

That theoretical elevation of this practical subject requires a specific explanation becomes evident if we compare the Islamic tradition with the fate of its "Christian" offspring: The *Liber abaci*, which carried the elevation of practical arithmetic to its apex. This is not to say that Leonardo's book was a cry in the desert. Its algebra influenced scholarly mathematics in the fourteenth century—for example, Jean de Murs (see G. l'Huillier 1980, *passim*); it is also plausible that it inspired Jordanus de Nemore.[93] Part of the material was also taken over by the Italian "abacus schools" for merchant youth. The scholars, however, took over only specific problems and ideas, and the abacus teachers only the more elementary, practically oriented facets of the work. Western Europe of the early thirteenth century possessed no institution that could appreciate, digest, and continue Leonardo's work. Only in the fifteenth century do similar orientations turn up once again—apparently not without renewed relations to the Islamic world.[94]

### 15. The Institutionalized Cases: (2) Astronomy and Pure Geometry

The other case of an entire tradition integrating theoretical reflection and investigation into a branch of practical mathematics is offered by astronomy. Because of its ultimate connection to astrology, astronomy was itself a practical discipline[95]; the *mathematics* of astronomy was, of course, practical even when astronomy itself happened to be theoretical.

In the Latin thirteenth through fifteenth centuries, this practical aim of the mathematics of astronomy had led to reliance on compendia, as observed in note 70. In Islam, however, astrology was the occasion for the continuing creation of new *zījes* and for the stubborn investigation of new planetary models. Islam was not satisfied with using good old established models like the *Zīj al-Šah* and the *Zīj al-Sindhind* in the way the Latin late Middle Ages went on using the *Theorica planetarum* and the *Toledan Tables* for centuries.

Astronomy can even be seen to have been the main basis for mathematical activity in medieval Islam. This appears from even the most superficial collective biography of Islamic mathematicians. Their immense majority is known to have been active in astronomy.[96] Since astronomy (together with teaching at levels up to and including that of the *madrasah*, which hardly required anything more advanced than large arithmetic and mensuration textbooks) was the most obvious way for a mathematician to earn a living, one is forced to conclude that the astronomer's career involved quite serious work on mathematics, at times serious work *in* mathematics.

This is also clear from al-Nayrīzī's introductory explanation to his redaction of the al-Hajjaj version of the *Elements:* here it is stated that "the discipline of this book is an introduction to the discipline of Ptolemy's *Almagest*" (Latin trans. Besthorn and Heiberg 1897: I,7). Later, the same connection was so conspicuous that the Anglo-Normal writer John of Salisbury could observe in 1159 (*Metalogicon* IV.vi, trans. McGarry 1971: 212), that "demonstration," i.e., the use of the principles expounded in Aristotle's *Posterior Analytics,* had

> practically fallen into disuse. At present demonstration is employed by practically no one except mathematicians, and even among the latter has come to be almost exclusively reserved to geometricians. The study of geometry is, however, not well-known among us, although this science is perhaps in greater use in the region of Iberia and the confines of Africa. For the peoples of Iberia and Africa employ geometry more than do any others; they use it as a tool in astronomy. The like is true of the Egyptians, as well as some of the peoples of Arabia.

So, it was not only the factual matter of the *Elements* that was reckoned part of the astronomical curriculum. According to the rumors that (via the translators?) had reached John of Salisbury, the geometry of astronomy was also concerned with the metamathematical aspects and problems of the *Elements.*

In the initial eager and all-devouring phase of Islamic science (say, until al-Nayrīzī's time, i.e., the early tenth century A.D.), the general positive appreciation of theoretical knowledge may well have laid the foundation both for the extension of astrology into the realm of high-level theoretical astronomy and for the extension of astronomy into that of theoretical mathematics. Down-to-earth sociology of the astronomers' profession may be a supplementary explanation of the continuation of the first tradition: The importance of the court astronomer (and, in case it existed, of the court observatory) could only increase if astronomy was a difficult and inaccessible subject. But even if this common-sense sociology is correct, it is not clear why intricacy should be obtained via the integration of metamathematics, the difficulties of which would only be known to the astronomer himself, and which would therefore hardly impress his princely employer. Why then should the integration survive for so long?

It appears, once more, that the original positive appreciation of (mathematical) theoretical knowledge was materialized institutionally, in a relatively fixed curriculum for the learning of astronomy. This curriculum began (as stated by al-Nayrīzī) with the *Elements,* and it ended with the *Almagest.* In between came the *mutawassiṭāt,* the "Middle Books" (cf. Section 2).

It is not clear to what degree this curriculum was fixed at different times. A full codification of the corpus of Middle Books is only known from the Naṣirean canon,[97] and the precise delimitation of the concept may have varied with time and place. Most remarkable are perhaps the indications that

Books I–II of the *Conics* may also have been considered normal companions of the *Elements* in the times of ibn al-Haytham (see Sesiano 1976: 189) and al-Khayyāmī (cf. note 11). It appears, however, that Ḥunayn ibn Isḥāq made a translation of the "Little Astronomy," which already served the purpose, and that Thābit had a similar concept.[98] Al-Nayrīzī too, we remember, appeared to have a fixed curriculum in mind.

So, from the ninth century A.D. onward it appears that astronomical practice and interest kept the focus upon pure and metatheoretical geometry not only because of a vague and general appreciation of the importance of theoretical knowledge[99] but also because of the institutional fixation of this appreciation. This does not imply that the long series of investigations of the foundational problems of the *Elements* were all made directly (or just presented) as astronomical *prolegomena*—the opposite is evident regarding al-Khayyāmī's *Discussion of Difficulties in Euclid* (cf. note 74) as well as Thābit's two proofs of the parallel postulate (both trans. Sabra 1968). Other metatheoretical investigations, however, were expressly written for recensions (*taḥrīr*) of the *Elements* for the introductory curriculum of astronomy—this is the case of Muḥyi'l-Dīn al-Maghribī's and Naṣīr al-Dīn al-Ṭūsī's proofs of the same postulate (see Sabra 1969: 14f. and 10, note 58, respectively). With utterly few exceptions, the authors of such metamathematical commentaries appear to have been competent in mathematical astronomy.[100]

## 16. A Warning

The foregoing might look like a claim that the global character and all developmental trends of Islamic mathematics can be explained in terms of one or two simple formulas. Of course this is not true. Without going into details, I shall point to one development that is puzzlingly different from those discussed previously: magic squares.[101] Their first occurrences in Islam are in the Jabirian corpus, in the *Ikhwān al-Ṣafāʾ*, and (according to ibn Abī Usaybiʿa) in a lost treatise from Thābit's hand. Various Islamic authors ascribe the squares to the semi-legendary Apollonios of Tyana,[102] or even to Plato or Archimedes. An origin in classical antiquity is, however, highly improbable: A passage in Theon of Smyrna's *On the Mathematical Knowledge Which is Needed to Read Plato* is so close to the idea that he would certainly have mentioned it had he heard about it[103]; but neither he nor any other ancient author gives the slightest hint in that direction. On the other hand, an origin in late Hellenistic or Sabian Hermeticism is possible, though still less probable than diffusion along the trade routes from China, where magic squares had long been known and used. This doubt notwithstanding, it is obvious that the subject was soon correlated with Hermeticism and *Ismāʿīlī* and related ideas. At least one mathematician of renown took up the subject—ibn al-Haytham, whose omnivorous habits we have already met on several occasions. Some progress also took place, from smaller toward larger squares and toward systematic rules for the creation of new magic squares. On the whole, however, the subject remained isolated from general mathematical investigations and writings. The exceptional character of ibn al-Haytham's work is revealed by an observation by a later anonymous writer on the squares, that "I have seen numerous treatises on this subject by crowds of people. But I have seen none which speaks more completely about it than Abū ʿAlī ibn al-Haytham" (quoted from Sesiano 1980: 188). The anonymous author combines the subject with arithmetical progressions in his own treatise; but integration into larger arithmetical textbooks or treatises seems not to have occurred. Islamic mathematics thus did not integrate *every* subject into its synthesis. Instead, magic squares appear to have conserved an intimate connection to popular superstition and illicit sorcery.[104]

It is not plausible that the exclusion of magic squares from the mathematical mainstream shall be explained by any inaccessibility to theoretical investigation—other subjects went into the arithmetical textbook tradition even though they were only known empirically and not by demonstration.[105] So the exclusion of magic squares from honest mathematical company must rather be explained by cultural factors: perhaps the subject did not belong within the bundle of recognized subdisciplines that had been constituted during the phase of synthesis; or its involvement with magic and sorcery made it a non-mathematical discipline[106]; or perhaps the involvement of mainstream mathematicians with practically oriented social strata made them keep away from a subject (be it mathematics or not) involved with sufi and other esoteric (or even outspokenly heretical) currents. I shall not venture into any definite evaluation of these or other hypotheses (even though ibn Khaldūn makes me prefer Nº 2), but conclude only that the place of magic squares in the culture of medieval Islam is *not* explainable in the same terms as the synthesis, the integration of practical mathematics and theoretical investigation, the development of the arithmetical textbook tradition, or the interest in the foundations of geometry. No culture is simple.

## 17. The Moral of the Story

The foregoing is hence not a complete delineation of medieval Islamic mathematics; nor was it meant to be. The purpose was to demonstrate that Islamic mathematics possessed certain features not present in any earlier culture (but shared with early Modern science) and to trace their causes. I hope that I have succeeded in demonstrating the existence of these features, and hence of an "Islamic miracle" just as necessary for the rise of our modern scientific endeavor as its Greek namesake, and to have offered at least a partial explanation of what happened.

This leaves us with a question of a different order: Was the integration of theory and advanced practice in Renaissance and early Modern Europe a legacy from Islam, or was it an independent but parallel development?

Answering this question involves us in the recurrent difficulty of diffusionist explanations. "Miracles" and other cultural patterns cannot be simply borrowed: they can only inspire developments inside the receiving culture. Even a piece of technology can only be borrowed if the receiver possesses a certain preparedness. The experience of cargo cults shows to which degree the recipient determines the outcome of even a seemingly technological inspiration, and investigations of any process of cultural learning will show us radical reinterpretations of the original message (and we may ask whether Charlemagne's identification of the Palace school of Aachen with the resurrection of Athenian philosophy was less paradoxical than any cargo cult).

We know the eagerness with which the European Renaissance tried to learn to the letter from ancient Rome and Greece—and we know the enormous extent to which the social and cultural conditions of Europe made it misunderstand the message. In contrast, no serious effort was made to understand the cultural messages of the Islamic world; on the contrary, great efforts were invested to prove that *such messages were morally wrong*. We can therefore be confident that no general cultural patterns or attitudes (even the attitudes toward rational knowledge and technology) were borrowed wholesale by Christian Europe. Nor was there any significant borrowing of institutions,[107] including those institutions which embodied the attitudes to knowledge. The only way Renaissance and early Modern Europe could learn from the "Islamic miracle" was through acquaintance with its products, i.e., through scholarly works and technologies it had produced or stamped. Because they were received in a society that was already intellectually and technologically mature enough to make an analogous leap, part of the "Islamic message" could be apprehended even through this channel. Primarily, however, Renaissance Europe developed its new integrative attitudes to rational and technological knowledge autochthonously; transfers were only of secondary importance.

This conclusion does not make the Islamic miracle irrelevant to the understanding of modern science. *First,* two relatively independent developments of analogous but otherwise historically unprecedented cultural patterns should make us ask whether similar effects were not called forth by similar causes. Here the sources of Islamic and Renaissance mathematics were of course largely identical (not least because Christian Europe supplemented the meager direct Greco-Roman legacy with translations from such Arabic works accessible in Spain, i.e., mainly works dating from the ninth century). These sources had not, however, been able to produce the miracle by themselves before the rise of Islam. Were there, then, any shared formative conditions that help us explain the analogous transformation of the source material?

Probably the answer is "yes." It is true that Western Christianity had been dominated in the High Middle Ages by a powerful ecclesiastical institution; moreover, after the twelfth century it could hardly be claimed to be fundamentalist. Yet precisely during the critical period (say, the period of Alberti, Ficino, Bruno, and Kepler) the fences of the Thomistic synthesis broke down, and rational knowledge came to be thought of both as a way to ultimate truths concerning God's designs and to radical improvements of practice. At the same time the ecclesiastical institution lost much of its force, both politically and in relation to the conscience of the individual; if anything, however, religious feelings were stronger than in the thirteenth century. It would therefore not be astonishing if patterns like the non-institutionalized practical fundamentalism of ninth-century Islam could be found among Renaissance scholars and higher artisans. It would also be worthwhile to reflect once more in this light upon the "Merton thesis" on the connections between Puritanism, social structure, and science.[108]

*Second,* the whole investigation should make us aware that there are no privileged heirs to the cultural "miracles" of the past. It is absurd to claim that "science, as we know it and as we understand it, is a specific creation of the Greco-Occidental world" (Castoriadis 1986: 264). On one hand, Greek "science" was radically different from "science as we know it and as we understand it." On the other, with relation to science (and in many other respects, too), it is no better (and no worse) to speak of a "Greco-Occidental" than of a "Greco-Islamic" world, and not much better to claim a "Greco-Occidental" than an "Islamo-Occidental" line of descent.

In times more serene than ours, these points might appear immaterial. If Europe wants to descend from ancient Greece and to be her heir *par excellence,* then why not let her believe it? Our times are, however, not serene. The "Greco-Occidental" particularity always served (and serves once again in many quarters) as a moral justification of the behavior of the "Occident" toward the rest of the world, going together with anti-Semitism, imperialism, and gunboat diplomacy. In theory it might be different, and the occidentalist philosopher just quoted finds it "unnecessary to specify that no 'practical' or 'political' conclusions should be drawn" from "our" privileged place in world history (Castoriadis 1986: 263 n. 3). It is, alas, *not* unnecessary to remind of Sartre's observation (1960: 28) that the "intellectual terrorist practice" of liquidation "in the theory" may all too easily end up expressing itself in *physical liquidation of those who do not fit the theory.*

As Hardy once told, "a science is said to be useful if its development tends to accentuate the existent inequalities in the distribution of wealth, or more directly promotes the destruction of human life." The ultimate drive of the present study has been to undermine a "useful" myth on science and its specifically "Greco-Occidental" origin—whence the dedication to a great humanist.

# 5

# Philosophy: Accident, Epiphenomenon, or Contributory Cause of the Changing Trends of Mathematics—A Sketch of the Development from the Twelfth Through the Sixteenth Century

To Andreas Jørgensen og Johannes Pedersen
and in memory of Mogens Pihl
friends and mentors

## Introduction

*The following essay resulted from an invitation to participate as a historian of mathematics in a symposium on Renaissance philosophy, concerned in particular with the role of the Aristotelian tradition in the genesis of the early Modern era. I was, as anticipated by the organizers, somewhat amazed by the invitation—and all the more so because my immediate feeling was that Aristotelianism (as well as formal philosophy in general) and mathematics have* no *close connection during the Renaissance and the earliest Modern period.*

*This* was *only an immediate feeling, since my familiarity with the sources stopped by the fourteenth century, and since the question was in any event not one I had considered deeply before, not even in connection with the High Middle Ages. But I was intrigued by the problem, and I decided to investigate in more detail whether my suspicion had a sound foundation.*

*What follows in therefore an investigation of the question* whether prevailing philosophies, inasfar as they are at all visible in the mathematical sources, have stamped (perhaps even determined) the ways in which mathematics developed from the twelfth through the sixteenth century, or whether they are purely epiphenomenal, maybe even irrelevant.

*This question, though inspired in its concrete form by my own amazement, relates to broader issues in the historiography of science. In terms of a simplistic scheme (familiar from the oral culture surrounding the history of science, misrepresenting reality as much as schemes in general, but still convenient like so many schemes—cf. the preface), three main models have been traditionally used to explain scientific development and change. According to one, scientists respond to the results of earlier science and to questions raised by these results ("internalism"); according to another, general (mostly technological) social needs are the moving force, and their absence a brake (one brand of "externalism"). The third approach (regarded as internalist by some and as externalist by others, and conventionally associated with Alexandre Koyré, whose carrying capacity is investigated in the following, looks into the general history of ideas, more specifically into the history of philosophy, for the causes that make scientists organize their search and shape their theories as they do.*

*The extent to which philosophies determine the casting of scientific work and thinking, as well as the answer to my specific question, is of course to some degree determined by the level on which* philosophy *is understood. For the sake of conceptual transparency I shall restrict the use of the word to the level of systematically organized thought, and exclude the loose sense of "attitudes" even when the attitudes in question* could be *expressed in terms of some philosophical system—an artisan is not to be labeled an* Aristotelian *just because he prefers empirical methods to a-priorism; I shall, however, also discuss the influences of proto- and quasi-philosophical attitudes and moods as well as the* relations *between philosophies and attitudes—thus exposing what appears to me to be misleading short-circuits in certain received interpretations of the history of mathematics through the history of philosophy.*

*As to the impact of further sociocultural factors, it should be obvious that major channels of influence for these are the philosophical and (since they lack scholarly obligation to well-belabored tradition and hence also internal rigidity) especially quasi-philosophical attitudes regarding the nature and purpose of mathematics and mathematical activity which they contribute to shape. Any historian who will not swear allegiance to internalism strict sensu, even (if such a person ever existed) the one who claims that technological needs or social structures are the ultimate and absolute determinants of everything, will therefore be obliged to take the level of philosophy and attitudes into account, at least as a compulsory mediating factor.*

*It should be no less obvious that philosophies and attitudes can only act through human minds, and thus only if they are accepted and shared. Since reception always implies a bargaining between what we are presented with and the conceptual and other cognitive tools already at our disposal, ground as they are through professional and non-professional experience, it is finally obvious that sociocultural factors* also *make themselves felt as co-determinants of* the *way philosophical and quasi-philosophical ideas are understood (and, we might say from our—dissimilar—horizon,* misunderstood) *by the actors. This latter question is included in the following discussion at a number of points, but not investigated systematically on its own.*

*In principle, reflections like these are of general and almost trivial validity. It depends on the sources, however, whether they can be ascertained in concrete detail or have to remain at the level of heuristic principles. As far as Old Babylonian mathematics was concerned, nothing more could be done from the mathematical texts that we know than to analyze the professional situation of their authors and target group, and to point out the structural similarities between general attitudes expressed more explicitly in other kinds of texts and the organization of mathematical thought. Similar, but even more restrictive conditions prevail when it comes to interpreting the anonymous mathematics of the subscientific traditions; even classical antiquity has left so little explicit material directly from the hands of mathematicians—Archimedes' letter to Eratosthenes, a few dedicatory prefaces where we mostly ignore the identity and position of the addressees, and where the very genre asks us to read with care—that direct verification of assumed influences of (for instance) Platonic philosophy or of more general moods in concrete cases is mostly excluded. Only the mature Islamic and Latin Middle Ages offer us an incipient direct insight, if not into what mathematicians thought then at least into what they wanted to say to their readers; only the European Renaissance, finally, presents us with a broad range of relevant sources.*

*That Chapter 5 differs in character and approach from Chapters 1 through 4 is thus no accident, and no mere consequence of an unexpected invitation.*

The following study is phenomenological in character, and in some way analogous to an archaeological surface survey. The method consists in a broad scanning of the sources for whatever evidence they might offer regarding explicit or implicit ideas about the nature, role, tasks, and legitimization of mathematical activity, reading them in relation to their context. Before we dive into the sometimes perplexing phenomena of some five centuries, however, a few prolegomena may be useful.

First: It will be convenient to discuss the problem in the grid constituted by conventional periodizations. A first period ("the twelfth century", in reality with vaguer limits) is (when seen from the point of view of the history of mathematics) dominated by the enthusiasm for Euclid and the *Almagest,* and ends when Aristotle becomes all-dominant in the *artes* (and so becomes the all-dominant problem for ecclesiastical authorities nervous about university curricula). The second period ("the thirteenth century") is that of assimilation of Aristotle, whereas the third ("the fourteenth century") presents us with a wealth of creative developments of Aristotelian philosophy, including mathematical developments. During the fourth period ("Early Renaissance", late fourteenth to mid-sixteenth century) mathematics and formalized philosophy live largely separate lives—and in the fifth period the foundation is laid *not least by developments related to mathematics* for the creation of the new philosophies of the seventeenth century.

Next: Three problems of method should be mentioned in advance, one practical and two of principle. One is the breadth of the "broad scanning." I have applied myself to cover at least essential sources representative of the most important currents from the whole period (but not of every major mathematician). Not everything I wanted has, however, been accessible to me; nor have I had the time to go into depth with everything that deserved it. Finally,

I may, of course, have overlooked important characters and tendencies unintentionally or have assessed them wrongly, in which case I will ask for correction rather than indulgence.

Two other problems concern the reality of entities regarded. Specific philosophies may have some historical coherence over a certain span of time, though even that can be problematic. *Attitudes, tendencies,* and *currents,* however, are elusive concepts, though indispensable if overall structures are to be perceived; their demarcations will by necessity be blurred, at times they will overlap, and it will often be impossible to claim that a specific author belongs to one current and only there. It should be kept in mind that currents and tendencies may at times represent *poles with relation to which authors can be seen to orient themselves,* rather than *classifications.*

One entity is of special importance for the study: mathematics. Is it justified to think of mathematics as something well-defined and possessing continuous existence from (at least) 1100 to 1700, or is this an anachronism, a piece of *whig history?* Isn't the idea of mathematics being a conserved entity in conflict with the obvious observation that the term covers something very different in the beginning and at the end of the period?

My answer to this Heraclitean point of view will be negative, which can be argued on several levels. I shall mention two: Socially, the actors themselves, those who generation for generation recreated the field, were convinced of both coherence and historical continuity; even a claim that previous generations had made barbaric or adulterated mathematics implied an acceptance that they made *mathematics.* Metaphysically, mathematics is characterized over the whole period by being an abstraction from sensible reality, dealing (as stated continuously from Augustine to Pascal) with a world created according to measure, number, and weight, and susceptible of some sort of argument or proof. Over the centuries the substance covered by this global characteristic would vary; but the existence of the category itself was constant.

## 1. The Twelfth Century

Medieval learning had inherited from antiquity the scheme of the Seven Liberal Arts and, quite as decisive, the idea that these arts constituted the apex of scientia humana. This idea was promoted not least by Isidore of Seville (c. 560–636), who remained an important authority throughout the Middle Ages. If we concentrate interest on the quadrivium part of the scheme (arithmetic, geometry, music, and astronomy), Isidore's attitude to the subject in his *Etymologies* is almost paradoxical: He is full of reverence for these important disciplines, but he knows next to nothing about them (if we define their contents according to the yardstick of the Alexandria school). All the same, the empty reverence proved important over the centuries: At every occasion where scholarly activity began burgeoning—be it Bede's (c. 673–735) Northumbria, Alcuin's (c. 735–804) Aachen, Hrabanus Maurus' (c. 776–856) Fulda, or Gerbert's (c. 930–1003) school in Rheims, mathematical subjects were among those cultivated to the extent and in the sense allowed by current conditions.[1]

Up to the end of the first millennium the interest in mathematical subjects is, it seems, mainly to be explained along these lines, as interest in *something in which the good scholar ought to be interested*, even though actual needs of ecclesiastical scholarly life made *computus* (Easter-reckoning) the only really living field from the seventh through ninth centuries.[2] The final result achieved was the re-establishment toward the end of the eleventh century of a complete *Latin quadrivium*, a cycle of mathematical disciplines considered to belong to and to round off the *scientia humana* level of Latin scholarship. Its high point was Boethius' *Arithmetic*. Geometry was represented by compilations of (pseudo-)practical geometry combining the surviving fragments of Boethius' translation of Euclid with material drawn from Roman agrimensors; it included the use of the so-called Gerbert abacus. Music was once again a mathematical theory of harmony (built on Boethius' translation of Nicomachos), after a dark interlude where it had dealt with actual song; and astronomy embraced, beyond the *computus*, also basic description of the celestial sphere and the astrolabe, and a little (very little!) astrology taken over from the Islamic world. Besides, various general expositions of the aims and authorities of the quadrivial arts were at hand (Martianus Capella's *Marriage of Philology and Mercury*, Cassiodorus's *Introduction to Divine and Human Readings*, and a variety of medieval compilations).

This was the foundation on which scholars had to build their understanding of mathematics, and with which the more ambitious among them became dissatisfied in the early twelfth century. To contemporary observers, this century was a bloom *par excellence* of the *artes*. Historians of philosophy would first of all think of its beginning as the period of Abelard and the inception of dialectics' supremacy. Abelard (1079–1142) himself, however, tells us indirectly that the quadrivium too was able to foster enthusiasm in the environment of young scholars, through the name he and Héloïse gave to their son: *Astralabius* (Abelard, *Historia calamitatum*, ed. Muckle 1950: 184f.).

More direct evidence is offered by the translators. A biography of Gherardo di Cremona (c. 1114–1187), the most prolific of all, tells that he was "educated from the cradle in the bosom of philosophy" and, dissatisfied with the limits of Latin studies, "set out for Toledo" to get hold of the *Al-*

*magest.* Having arrived he stayed there translating the Arabic treasures "until the end of life."[3] Another, anonymous, scholar pursued medical studies in Salerno until hearing that a Greek copy of the *Almagest* had arrived in Palermo; accordingly, he left for Sicily, started preparing himself by translating some minor works from the Greek, and finally translated Ptolemy's *Great Composition* (see Haskins 1924: 159–62). Adelard of Bath (fl. 1116–1142), finally, started writing in the tradition of the Latin quadrivium on the *Regulae abaci.* The treatise presents us with an overwhelmingly full discussion of this subject, referring to Boethius' *Arithmetic* and *Music,* to the traditional system of sub-units (the mutual multiplication of which gives occasion for many pages), to Gerbert, and of course to everything connected to the device itself.[4] Then he left home "to investigate the studies of the Arabs" (*Quaestiones naturales,* ed. Müller 1934: 4[32]), which resulted first in a metaphysical treatise *De eodem et diverso,*[5] and then in a work on *Quaestiones naturales* built in part on what he had learned from the "studies of the Arabs,"[6] and in a beautiful array of translations—including various astronomical treatises and at least one (probably two, possibly even three) translation of the *Elements.*[7]

The first effect of the mathematical translations was the completion of what I have called the *Christian quadrivium,* that quadrivial syllabus which Christian Latin Europe ("Christianity" understood as an ethnic rather than a religious identity) considered its legitimate heritage, because it completed the range of authors, works, and disciplines known (like Euclid, Ptolemy and others) by name from Isidore, Martianus Capella, and Cassiodorus; known (like optics) from Aristotle's works, of which most of those not translated before became available during the same century; or attributed by their titles to ancient authors (as the "science of weights" was attributed to Euclid). Islamic authors continuing these same disciplines were accepted as legitimate and necessary (though morally somewhat secondary) commentators and explanations of the same "Christian" quadrivium.

Another effect was the completion of *the total range* of mathematical subjects, which came to include two obviously non-"Christian" disciplines, namely *algebra* (and more elementary commercial calculation) and *algorism*—the latter meaning *calculation with Hindu numerals,* which was soon accepted as a useful and neutral tool by the environment adopting the new mathematics and astronomy.[8]

A total survey of the range of translations shows that mathematics played an important role (especially if astronomical and astrological works are counted as mathematics)—almost, perhaps fully, on a par with medical subjects and the concomitant *philosophia naturalis.* Another measure of the importance of the mathematical imports is supplied by traditionalist polemics against the new learning. In a *Sermon to the Purification of the Blessed Mary* from the late twelfth century, Étienne de Tournai complained that many Christians (and even monks and canons) endangered their salvation by studying.

> poetical figments, philosophical opinions, the [grammatical] rules of Priscian, the laws of Justinian, the doctrine of Galen, the speeches of the rhetors, the ambiguities of Aristotle, the theorems of Euclid, and the conjectures of Ptolemy. Indeed, the so-called Liberal Arts are valuable for sharpening the genius and for understanding the Scriptures; but together with the Philosopher they are to be saluted only from the doorstep.[9]

The poets, Priscian's grammar, the rhetors, and even Aristotle's discussion of sophisms belong to the traditional realm of the trivium; Justinian's Roman law must also be understood as an extension of the study of dialectical canon law and theology, and hence as a traditional subject that aroused a sudden, vigorous interest, to the dismay of Bernard of Clairvaux and his companions-in-arms. The really *new* learning is represented, we see, by Euclid, Ptolemy, Galen, and possibly (but probably not) by the "philosophical opinions." Broader interest in theoretical mathematics and in high-level astronomy (not necessarily followed by conspicuous competence) was evidently an important effect of the intellectual revolution in the twelfth-century scholarly environment, and not just a queer preference of the translators. As we can imagine, below this high level a still broader interest in less exacting mathematical subjects thrived.

At the very turn of the century, a monumental expression of the interest in mathematics was created outside the environment of the schools: Leonardo Fibonacci's (b. c. 1170, d. after 1240) *Liber abaci*[10] (written 1202, and containing somewhere between 250,000 and 300,000 words). The work is in principle an enormously extended algorism, a guide to the use of Hindu numerals not only for computation but for commercial calculation and algebra in general. *If* the work represented more than its author it would be evidence that the interest in mathematics had penetrated not only the schools but also the commercial environment of Northern Italy. To some extent this is certainly true; during the thirteenth century a system of lay commercial education developed in Northern Italy. The system was centered upon commercial calculation, as evident already from the name: the *abacus* school.[11] In its immensity, however, Leonardo's work is a personal achievement of its author, and (apart from a number of copies of the manuscript) nothing similar was made for centuries. Leonardo's genius, impressive as it is for the historian of mathematics, tells little about the conditions and intellectual climate of his environment, either during the twelfth or the thirteenth century.

How are we, however, to explain those more modest but still revolutionary developments that *were* characteristic of the twelfth century?

Let us first return to Adelard. His *Regulae abaci* reflect his background in the traditional Latin Arts, and the *De eodem et diverso* and the *Quaestiones naturales* demonstrate that he belongs in full right to the current of "twelfth century Platonism," with its inspiration from the *Timaios* and its interest in natural explanation.[12] This commitment is directly connected to the empirical and naturalist aspect of his translations, including his interest in astrology and natural magic. There is, however, no direct link from this very atypical brand of Platonism to pure geometry or mathematics in general. Instead, we must see it as expressing an uncritical climate of intellectual hunger, where *anything important* in relation to the lost intellectual heritage (as understood in the schools of the early twelfth century, and hence understood not as antiquarianism but as comprehension of the universe) was to be seized upon—especially such fundamental works as the *Almagest* itself and the *Elements*. Already Isidore and Augustine had quoted the Bible to the effect that "YOU made everything in measure and number and weight,"[13] from which Isidore concluded that

> By number, we are not confounded but instructed. Take away number from everything, and everything perishes. Deprive the world of computation, and it will be seized by total blind ignorance, nor can be distinguished from the other animals the one who does not know how to calculate.[14]

In the early twelfth century, Adelard and his fellows would see not only "computation" but the whole range of available mathematics as necessary if a world created in "measure and number and weight" were to be understood.

A related but even more open attitude is expressed by Hugue de Saint-Victor (c. 1096–1141). In *Didascalicon* VI.iii he exhorts, "learn everything, and afterwards you shall see that nothing is superfluous."[15] The immediate context, to be sure, is "sacred history," and strictly speaking we are only exhorted to learn everything from this subject—but the examples leading up to the conclusion show that acoustical, arithmetical, and geometrical experimentation and astronomical observation are no less praiseworthy.

This same all-devouring and undistinguishing appetite is also obvious if we look at the list of translations undertaken by the single translators.[16] The interest was (if we restrict the investigation to mathematics) directed to *anything mathematical* at hand; no higher point of view (philosophical or other) beyond availability and a modicum of comprehensibility selected the material, which therefore turned out to constitute a rather eclectic heap by the early thirteenth century. Behind the total endeavor of translation lay, however, a philosophical formulation of the intellectual appetite: *the interest in the existing world*. As long as mathematics (and indeed *anything mathematical*) was understood as a necessary tool for this enterprise, eclecticism was in itself a consequence of the dominant philosophical point of view.

Referring to the title of the present essay we may therefore claim that philosophy was only implicitly expressed through the new character of twelfth-century mathematics, but on the other hand an essential background to this character and hence *not epiphenomenal*. On the other hand, however, the "philosophy" in question was to a large extent an intellectual attitude rather than an explicitly formulated structure of thought, especially in its relation to mathematics. Philosophy *stricto sensu* was therefore neither essential nor epiphenomenal in relation to the twelfth-century developments of mathematics: It was a *sleeping partner*. Moreover, the interest in mathematics was so unspecific, namely, an interest in what might serve as description of the existing world and in systematic thought, that *mathematics itself* could be declared an epiphenomenon: Mathematics was chosen because *it was traditional*, because it *promised to fulfill urgent intellectual needs*, and because it happened to be at hand.[17]

## 2. The Thirteenth Century

The thirteenth century is rightly known in the history of education and universities as the century of Aristotelization. Of course, Aristotle could not displace *everything* else, and from critical sermons we know that not only Aristotle but also geometry and mathematical astronomy could keep students from the pure spring of theology.[18] Still, both ecclesiastical trials, surviving university curricula, and the sources in general confirm that Aristotelian learning displaced every competitor to the position as the main intellectual challenge, tool, and stimulus.

What happened to mathematics and mathematical interests in the scholarly environment under these conditions? I shall try to approach this question from a variety of specific viewpoints before giving a synthetic answer.

First of all, it should be emphasized that the general mathematical level among arts students was apparently raised from (say) 1180 to 1280. The enthusiasm of translators and their immediate followers should not make us believe that normal students (even when sharing the enthusiasm) had digested much of the meal of translations. During the thirteenth century, however, elementary introductions to the art of algorism and to elementary spherical geometry became widespread at the universities, and *computus* remained a living subject, treated several times by Robert Grosseteste (c. 1175–1253) and even by as fine a mathematician as Campanus de Novara (c. 1220–1296).[19] Thanks especially to Alexandre de Villedieu's (d. c. 1240) and Sacrobosco's (fl. 1220–1244) pedagogical successes these topics were certainly better mastered by many scholars by 1280 than a century before (cf. Beaujouan 1954 and Evans 1977).

Interactions with philosophy are, however, not to be expected (nor, in fact, to be found) at the level of compendia and elementary treatises. At most they show us that the eclectic temper of the twelfth century had not vanished. What, then, about the *Elements,* probably the best occasion for metamathematical reflection that could be imagined?

One side of that question is the problem of diffusion: How much was generally taught? Hardly three or four propositions, if we are to believe Roger Bacon (c. 1219–c. 1292)[20]; 15 books, according to a collection of *quaestiones* from Paris.[21] A commentary probably written by Albert the Great (c. 1200–1280), dealing (not always very correctly) in full with Book I and briefly with Books II–IV (discussed in Tummers 1980) is probably the best hint we can get of the range of normal teaching at the highest level. If the usual discrepancy between teaching and learning is taken into account, we may safely assume that few scholars, and few active philosophers, were in possession of a knowledge of theoretical mathematics (or applied metamathematics) that could seriously challenge their philosophical tranquility.

The other way round, there is more reason to expect an influence, since everybody writing on mathematics in the thirteenth-century university would be well versed in Aristotelian philosophy and in the traditional metamathematical theory of the Latin quadrivium, and presumably more disposed to accept his upbringing than those rebels who had left "the cradle of [Latin] philosophy" for Toledo or Sicily a hundred years earlier.

The obvious place to look is the Campanus edition of the *Elements* itself.[22] Its character has been discussed on several occasions by John Murdoch (1968; and "Euclid: Transmission of the Elements," 446 f.), for which reason I shall be very brief. In a certain sense the hypothesis is confirmed: Especially in Book V, we find references to Plato's *Timaios,* to Boethius' *De musica,* and in particular to Aristotle's conceptual gunnery, needed for the discussion of quantity versus number, degrees of abstraction, and of the necessity that all four quantities in a proportion be of the same nature in the *permutatim* mode.

But only in a certain, restricted sense is Aristotelian philosophy an active molding factor. It contributes, no doubt, to that greater stringency which distinguishes Campanus's work from mathematical writings from the tenth through twelfth centuries. Campanus would never regard the abacus as a geometrical subject just because it makes use of a ruled board. But there is little specifically Aristotelian about the stringency, which is rather a stringent application of mathematical sources. This is revealed even in small details—for example, that Campanus speaks of *communes animi conceptiones,* a traditional Latin translation of Euclid's *koinaì énnoiai,* instead of using *dignitates,* the term used in current translations for Aristotle's *axioms.* What might look superficially as thorough orientation after Aristotelian modes of thought is rather a didactical dressing of the subject matter, connecting it to familiar patterns of thought without really determining the choice of subject[23] or approach.

In so deeply a didactically determined tradition as that of scholasticism (down to the name, we observe!) superficial correlations with other parts of the curriculum should cause no amazement, and they can be found everywhere in the medieval mathematical sources, which (like medieval learning in general) were somewhat at odds with the Aristotelian compartmentalization of knowledge.[24] A field such as mathematics might from its internal epistemological impetus have a tendency to be governed by its own laws and rules, in agreement with Aristotelian ideals[25]; but scholars moving in their teaching from one *artes*-subject to another, teaching students who followed a broad range of courses, would rather act against the inherent tendencies of the subject than let their avowed philosophy strengthen it.

A characteristic instance of such purely external Aristotelization of a mathematical subject is found in Petrus Philomena de Dacia's (fl. 1290–1300) commentary (written 1291/2) to Sacrobosco's *Algorismus vulgaris.* Already Sacrobosco had quoted Boethius' *Arithmetica* to the effect that the art of number be a prerequisite for knowledge of anything, and given the Aristotelian epithets *materialiter* and *formaliter* to Boethius' two different definitions of *number.*[26] In his commentary to this, Petrus Philomena takes the opportunity to speak broadly about *the four Aristotelian causes* of the art of algorism.[27] Neither in Sacrobosco nor in Petrus Philomena's commentary is there, however, any influence of Aristotelization in what Petrus identifies as the *pars executiva.*

If we go back in time from Petrus Philomena and Campanus to Jordanus de Nemore (fl. somewhere between 1220 and 1250) we shall find an author more governed in his whole mathematical activity by a philosophical stance.[28] To some extent this stance was Aristotelian. First, Jordanus appears to have taken the Aristotelian distinction between different sciences more in earnest than contemporary mathematicians. So, when writing mathematics (the only subject on which he wrote) he would not involve the usual array of didactical cross-references to other Liberal Arts, nor begin discussing the obvious astronomical applications of a theory of the stereographic projection. Second, one of his works, the geometrical *Liber philotegni* (which appears to have grown out of a series of university lectures, themselves known from a student's *reportatio*, the *Liber Jordani de triangulis*[29]) starts by a set of very Aristotelian definitions of *continuitas, punctus, continuitas simplex, duplex,* and *triplex, continuitas recta* and *curva, angulus*, and *figura*[30]. Third and finally, most of Jordanus' works were labelled *demonstrationes* in their own time, and probably by the Master himself, i.e., they were understood as faithful to the ideals set forth in the *Analytica posteriora*, in contrast to the *experimenta* of algebra and algorism.[31]

The second feature looks like expressions of explicit philosophical commitment; it is, however, superficial, and as irrelevant to the subject matter as the Aristotelian causes to the *pars executiva* in the algorism. It can safely be seen either as didactically motivated philosophical lip service or as joking flirt (indeed, the *reportatio* mentioned suggests that Jordanus's lectures were interspersed with irony). On the other hand, the first and the third feature touch the very essence of the Jordanian *opus*, and are truly exceptional in the thirteenth century. Since Jordanus was obviously a pure mathematician by inclination, there is no reason to believe that he needed Aristotle's permission to be one[32]; on the other hand, his personal inclinations, obviously inspired by the ancient pure mathematicians, made him agree much better with ideals formulated by Aristotle in the environment of these same authors than did those who tried to understand Aristotle on the conditions of the thirteenth century.

There is a certain parallel between Jordanus's Aristotelianism and that of the Averroists. In the *De eternitate mundi* Boethius de Dacia (fl. 1277) had distinguished *veritas naturalis*, the truth of natural philosophy, from *veritas christianae fidei et etiam veritas simpliciter*, "Christian, that is genuine, truth" (ed. Sajo 1964: 46; similar formulations *passim*). As I read the treatise there is no doubt that Boethius was sincere in admitting the ultimate truth of Faith; still, being a philosopher by profession, by training, and by inclination he claimed the right (and claimed it an obligation) to investigate that *natural truth* which was set into operation at God's creation. Boethius' position was only the extreme consequence of an otherwise accepted philosophy (and so indeed an appropriate expression of the inherent rationale of the Thomistic synthesis); but being extreme it revealed that the thirteenth century was not disposed to draw the full consequences of the Aristotelian division of the world into separate and semi-autonomous levels (nor that autonomy of single social groups which was its parallel), and Boethius was condemned in 1277.

Jordanus, too, was condemned—not by any bishop hostile to mathematics but by those closest to his enterprise. There seems, indeed, to have existed in Paris a whole Jordanian circle in or around the 1240s, embracing among others Campanus of Novara and in some way even Roger Bacon; but apart from the otherwise unknown Gérard de Bruxelles, author of a *Liber de motu* in Jordanian style, none of his disciples or associates cared to continue or defend the specific character of Jordanian pure mathematics. On the contrary: those who edited his treatise on the stereographic projection hurried to put in all those references to celestial circles, stars, and astrolabe which Jordanus had discarded.[33] Jordanus, the only mathematician in the Latin thirteenth century doing mathematics in reasonable agreement with Aristotelian precepts, was *eo ipso* unacceptable to his contemporaries.

Essentially, this agrees with the evidence offered by Albert the Great and Thomas Aquinas (1225–1274). As mentioned, Albert appears to have written a commentary on *Elements* I–IV. According to Tummers' analysis it is not very original, drawing heavily on al-Nayrīzī's commentary and other available material; the philosophical introduction is apparently "more in keeping with the Platonic-Pythagorean tradition than with the Aristotelian" (Tummers 1980: 483). In the Aristotelian paraphrases Albert is more philosophically stringent and clearly Aristotelian—but he demonstrates no striking mathematical competence, nor were discussions of infinity and continuity mathematically productive (at most they were counterproductive, since Albert's basic point was to "maximize the gap between mathematics and the natural world"[34] and to concentrate interest on the latter). St. Thomas's treatment of metamathematical questions (in the *Commentary* to Boethius' *De trinitate*[35]) is philosophically more original, more interesting, and much more positive in its evaluation of the relevance of mathematics for understanding the real world, but it floats miles above the level of actual mathematical work.[36] In its consequences, it will have been no more effective than Albert's more diffuse and more distrusting attitude.

The place to look for genuinely philosophical inspiration of mathematical activity is rather *outside* the most stringently Aristotelian circles, *viz.*—commonplace as it is—in the quarters of neoplatonic inspiration. Thomas himself is not fully a stranger to neoplatonic ideas, but dissociates himself from those, "for example, the Pythagoreans and the Platonists," who asserted "that the objects of mathematics and universals exist separate from sensible things" (*Commentary on De trinitate*, Question V, article 2, trans. Maurer 1963:34).

This statement expresses Aristotle's interpretation of the Platonic view; closer to those inspired by Platonism in the thirteenth century would be a claim that mathematics was closer to *real*—divine—reality than are the sensible things. This point of view results (and resulted) easily when the twelfth-century confidence in the descriptive power of mathematics (as described previously) is taken to its philosophical consequence. In Roger Bacon's diffuse mind the two views are not easily separated. In other authors, a clearly Augustinian illuminationist stance is more obvious, though rarely in sole and supreme reign.

A first name to be mentioned is William of Moerbeke (b. c. 1230, d. before 1286), who translated not only Aristotle but also Archimedes, Eutocios, Proclos, Alexander of Aphrodisias, Philoponos, and others directly from the Greek.[37] His neoplatonic convictions are visible in his choice of authors to translate; in a *Geomantia* probably from his own hand; and especially through a dedicatory letter written by his friend Witelo in the latter's *Perspectiva* (ed. Risner 1572: II, 1–2; reproduced in Clagett 1976: 8f., note 30). As is made clear through the testimony offered by Albert and Thomas, a decision to translate the full Archimedes was far from inevitable even for a dedicated translator. The translation itself (ed. Clagett 1976) gives no clues for Moerbeke's motives, but it is a fair guess that his neoplatonism played a major role, presumably together with an incipient, philosophically supported friendship with Witelo the mathematician (Witelo arrived at the papal court in Viterbo in 1268, at which occasion he met Moerbeke [see Lindberg 1971: 72], and Moerbeke's mathematical translations are from 1269).

The connection to Witelo leads us to a whole cluster of names, *viz.*, Grosseteste, Roger Bacon, Peckham (c. 1230–1292), Witelo (b.c. 1230, d. after c. 1275), and (as a partial contrast) Dietrich von Freiberg (c. 1250–c. 1310), and to *optics*, one of the two "new" mathematical disciplines of the Latin thirteenth century (Jordanus's statics being the other).

Grosseteste's role in this connection is mainly that of giving inspiration. Like every philosopher of his century, he was of course inspired by Aristotle—but *The Philosopher* was only one of several authorities to Grosseteste, who was definitely no Aristotelian as regards his position on the relation of mathematics to other subjects. In a small and probably very early treatise *De artibus liberalibus* (ed. Baur 1912: 1–7) he tells how informative these are for natural as well as moral philosophy. The discussion is not profound—astronomy, for instance, is important (in its astrological appearance) because it tells the right moment to act. Later works, however, show a fair acquaintance with astronomy, calendar construction, and the fundamentals of optics[38]; when seen together with his philosophical works this mathematical competence makes his influence in his own and later centuries understandable. Important in the present context is his illuminationist coupling of optics with epistemology and with theologically tainted metaphysics (the theory of "multiplication of species").

One of those to be impressed was Roger Bacon, whose grandiloquent confidence in his own mathematical competence has made later times accept it. So much truth is contained in the claim that Bacon was familiar with lots of mathematical authorities (and authorities of any scholarly discipline!), and that he was able to construct simple but relevant geometrical arguments pertinent to many optical observations and informal experiments. He combines sense for physical reality with a belief in the potency of mathematics, which seems often more phantasmagoric than just neoplatonic. This would certainly have had more appeal a century or two later, but even in the thirteenth century it might have aroused an appreciable echo, had Bacon not been kept imprisoned and the circulation of his writings restricted for reasons which are only indirectly connected to his mathematical philosophy (if at all). His influence in broader circles was therefore modest, and passed mainly through whatever Baconian material was adopted by Peckham and Witelo.[39] Peckham himself appears to have belonged to the environment inspired by Grosseteste (in any case, he was a Franciscan and one of the founding fathers of neo-Augustinianism—cf. van Steenberghen 1955:98–104) and wrote on "mystical" numerology in combination with Boethian arithmetic;[40] even in his *Perspectiva communis* he reveals himself as a neoplatonist of independent rather than Baconian inspiration—not only in his characterization of the Lord as *lux omnium* in the preface (which might be nothing but a poetical metaphor), but also, for example, in a discussion in prop. I.6[41] of Moses Maimonides's claim that there is an "influence of a particular star directed to each particular species" in this universe that is "like one organic body." Probably, Peckham was therefore primarily interested in optics because of Grossetestian inspiration and neoplatonic inclinations; Bacon supplied him with factual material only.

Witelo seems to present us with an analogous case. As he tells in the dedicatory letter mentioned earlier, Moerbeke made him commence the work as a means to know "how the influence of divine powers (*virtutes*) affects lower bodily things through higher bodily powers"—and according to a remark in prop. X.42, his first interest in the matter had been aroused by observations of intriguing physical phenomena.[42] It appears that Bacon's optical manuscripts simply happened to be present and available (through Moerbeke's influence?) in Viterbo when they were needed.

It is hence probable that *all* important writers on optics in the thirteenth century were inspired from neoplatonic philosophy (or at least neoplatonically tainted philosophy), and that most of them sustained a neoplatonic belief in the explaining power of mathematics. The case of the first fourteenth-century writer on the subject, Dietrich of Freiberg, *may* be different.

According to Wallace ("Dietrich von Freiberg," 92), he rejected the Grossetestian "metaphysics of light," as well as the belief in the mathematical structure of Nature. But he *was* interested in neoplatonic doctrines. All in all, he must probably be taken as evidence that a mechanical coupling between specific philosophical doctrines and the interest in optics was less important than more fundamental levels in the neoplatonic orientation. Nothing in the material suggests a mathematical autonomization of the subject, which by c. 1300 would have cut it off from philosophical inspiration. The interest in optics is hence a link backward to the twelfth-century protophilosophical enthusiasm for mathematics.

In this respect, the interest in optics is a close parallel to that in astronomy and astrology—and as we have just seen, the former is often interwoven with the latter. This role for astronomy is no marvel. If any field confirmed the twelfth-century conception of mathematics as a way to true knowledge of Nature's phenomena it was certainly astronomy, from *computus* to Ptolemean planetary theory—and astrology was then (as we have seen in Grosseteste) the way to make mathematics a way to knowledge of almost any kind. Certainly, *music*, through the mathematical theory of harmony, can be claimed to be equally well described by mathematics. The theory of harmony, however, would only ask for the use of a fairly simple arithmetic of proportions, and its relation to actual sound was limited. Planetary astronomy, on the other hand, dealt with the real celestial bodies, and could make use of almost any available level of mathematical sophistication. So, wherever you were on the level of mathematical learning, you might see *your own* mathematics as an efficient tool.

Astrology was similarly accessible on many levels. It could be justified through sophisticated neoplatonic philosophy, as we have seen in Maimonides and the perspectivists; but it could also be exerted as a complicated but aphilosophical technique of prediction and warning, and even be grasped as such by an illiterate public. No wonder, all in all, that the complex astronomy + astrology came to be regarded by many as the ultimate purpose of mathematics (its *final cause*, as stated by Petrus Philomena).[43] This scale of values was institutionally confirmed in the University of Padua, where quadrivial teaching was given in the common *artes* and medical faculty by physicians as a tool for astrologico-medical prognostication.[44] In other universities and centers of learning, where ecclesiastical, Thomistic, or Albertian skepticism might be expected to have a greater influence, no institutional fixation occurred, but on the level of scholarly interests the difference was faint. The irony of history even led to the result that Albert's fame in later centuries connected him mainly with astrological and other occult subjects, of which the list of spurious Albertian works contains an impressive number.[45] Even though astronomy led to no significant development of *new* mathematical results or mathematical creativity,[46] there is thus no doubt that the enthusiasm for astrology (and hence the quasi-philosophical attitudes giving rise to this enthusiasm) was the main incentive behind the spread of basic mathematical competence in the scholarly environment.[47]

So, if we expected the century of medieval Aristotelianism *par excellence* to be the century where Aristotelian philosophy shaped mathematics, we will be disappointed. If we expect (for example, because we know the wave of mathematical creativity that in the Islamic Middle Ages followed upon the translations) that the translations and the enthusiasm of the twelfth century should have led to intense mathematical activity during the thirteenth, we will be equally disappointed. How are we, then, to sum up the trend of scholarly mathematical development in the thirteenth century, and how are we to explain it?

First, there is the absence of an explosion of mathematical creativity and what might look like a tendency to trivialization. Before considering this as enigmatic we should, however, compare the "trivial" level of thirteenth-century scholars not with that of the translations but with that of twelfth-century schoolmen—in which case we shall see that the apparent trivialization is an illusion, caused by wider dissemination of mathematical competence at an intermediate level, i.e., at a level well beyond that of normal twelfth-century scholars (whose enthusiasm did not always entail competence, as observed). If we cannot speak of a blast of creativity, we can at least compare the situation to a blaze of wildfire. Because of its own "epiphenomenal" character, the twelfth-century enthusiasm for mathematics could hardly have produced anything more.

Next comes the question of Aristotelian influence on the development of thirteenth-century mathematics. As observed, it is generally superficial, one among several traces left on the material by the social (in particular the educational) context of learning. Only as far as the progress in stringency and the relative regress of omnivorous eclecticism is concerned was *Aristotelianism regarded as philosophy* of possible importance for the development of mathematics (and, as we have seen, the agreement of Jordanus's mathematical style with strict Aristotelian principles was no success). What can be seen in the sources is the imprint of *Aristotelianism regarded as a way to organize teaching*. The "four causes" turn up in Petrus Philomena's Sacrobosco commentary (and in numerous other works) not because of any relation to the *pars executiva* of the algorism but because *this was the way to speak when in the Arts Faculty*. The *falsigrafus* (cf. note 32) is there not because he represents an important point in Aristotle but because a *word* was found in the *Topica* that, when sufficiently twisted, could be identified with an opponent in a university disputation.

In the formulation of this I have grouped works like the Albert(?) commentary to the *Elements* with philosophy rather than mathematics. My reason for doing so is that they were apparently not mathematically productive, i.e., that no further work in the field was inspired by them (at the moment). If they represent a mathematical genre then (in the thirteenth century) it was only a mathematical dead end.

In the terms of the heading of the essay, Aristotelian philosophy was then no real cause (not even a contributory cause) for the direction taken by the development of thirteenth-century mathematics—at most an epiphenomenon and one of the forces holding back for a while that "neoplatonic" mood that saw in mathematics the principal way to real insight. If any philosophy contributed to the development of mathematics in the thirteenth century, it was, indeed, neoplatonism—or, better, the neoplatonic aspect of various eclectic philosophies. This was obvious when the perspectivists' motivations were investigated; as a mood rather than an explicit philosophy it will also turn up if we try to formulate the attitude behind the interest in astrology. It is one aspect of a *medico-astrological naturalism* that had already been active in the first import of the doctrines of Aristotelian natural philosophy, and which in later centuries (thanks not least to the influence of Paduan Averroism) led to the observation that "where three physicians meet two atheists will be present." In the thirteenth century things had not developed thus far. Already then, however, when we should still speak of a tendency rather than a current, this tendency would often imply a bent toward heterodoxy, and be at variance with Thomistic and Albertian rationality (until it succeeded in turning Albert upside down). In mathematics, it would care little for proof and demonstrative structure; Richard de Fournival, physician and "well versed in mathematics" according to his own words (*Biblionomia*, the introductory passage—ed. Delisle 1868: II, 520), and intimately familiar with Jordanus's works and ideals (cf. note 31), would himself prefer *experimenta* and astrology to mathematical *apodixeis*. He was no exception (and Roger Bacon, who shared in the same dispositions if only in stronger form, is hence only a caricature, but no stranger to his century[48]). In the centuries to come this combination of a neoplatonic confidence in the potency of mathematics with emphasis on immediate practical insight and utility[49] would lead to continued reliance on the mathematical compendia written by thirteenth-century scholars, which would thus form the mathematical culture of average scholars until well into the sixteenth century.

### 3. The Fourteenth Century

Whereas the stereotype of the thirteenth century is that of classical Aristotelianism and Thomism, the stereotypical picture of the fourteenth century displays the *via moderna* and Ockhamism. And whereas the thirteenth century saw the transformation of the schools of Paris, Oxford, etc. into "universities", the fourteenth century brought the spread of the universitarian idea into German land and the creation of a multitude of new *studia generalia*.

The fourteenth-century transformation of university learning can be discussed and described at many equally valid levels. On the level closest to the facts of daily teaching, the stabilization of the institution, of its teaching methods and of its curriculum, and the concomitant "professionalization" of the community of Masters of Arts[50] offer valuable explanations of the development. The social consolidation, the acquisition of a set of firm professional values and a teaching method based on intense and critical discussion allowed a systematic cumulation of scholarly insights—and since Aristotelian learning had been accepted as a common foundation, the cumulation resulted in cumulative development of Aristotelian doctrines in confrontation with new problems (development that at times brought them far away from Aristotle himself).

Some of these problems were of a mathematical nature, for example, the "quantification of qualities." And so, the fourteenth century produced that Aristotelian mathematics which had defaulted in the thirteenth. I shall not cover the current in depth nor mention all important authors or works,[51] but only discuss a few select aspects pertinent to my subject.

Oxford appears to have been the place where the new philosophy was first developed (see Weisheipl 1966). It might be alluring to see this as a consequence of an old bent toward mathematics aroused by Grosseteste and Roger Bacon; the difference between the inspired neoplatonism of *their* mathematization of nature and the stringent intellectual style of fourteenth-century scholars such as Thomas Bradwardine (c. 1290–1349), Richard Swineshead (fl. c. 1340–1355), and William Heytesbury (fl. c. 1335) is, however, too great to make such a guess plausible. Furthermore, even though optical models were often used by the scholars of Merton College, the key abode of the new philosophy, the "metaphysics of light" appears to have played no role for them.[52]

The problem that first comes to mind in relation to the Mertonians is the above-mentioned *quantification of qualities*, which has made many see the Merton school as a first step toward Galileo and Newton. A sense can of course be found that is vague enough to make such a conception plausible (especially if we content ourselves with the fact that Galileo was taught and learned from material going back to the quantifying schoolmen)—but in a strict sense it is definitely false: The Mertonian project was different, and had to be different, from that of the seventeenth century.[53]

The main reason to see the Mertonians as proto-Galileans is that part of their work was concerned with the mathematical analysis of *motion*, in crit-

ical continuation of Aristotle's *Physica*. This, for instance, is the theme of
Bradwardine's important *Tractatus proportionum seu de proportionibus velocitatum in motibus* (ed., trans. Crosby 1955). But other qualities were also
discussed from the quantitative point of view, both those of compound
medicine[54] and those of the alchemical "primary qualities" (see Skabelund
and Thomas 1969), even though it must be admitted that the highest mathematical development was reached in connection with the treatment of motion.
As is evident from Bradwardine's title as just quoted, the mathematical theory of "proportions" ("ratios" in modern language) and proportionality was
a central domain.

On the Continent, the most famous continuation of the Merton quan-
tifications were Nicole Oresme's (c. 1320–1382) works. His *Tractatus de
configurationibus qualitatum et motuum* (ed., trans. Clagett 1968), which is
perhaps the most direct continuation of the Merton discussions, is famous for
its introduction of a geometrical representation similar to a modern coordi-
nate system of intensities of qualities subject to change,[55] in which connection he also finds infinite sums geometrically (or rather, splits geometrical
quantities into infinite sums).[56]

Other works from Oresme's hand are original not only in contents but
also in aim. Both in the *Tractatus de commensurabilitate vel incommensurabilitate motuum celi* (ed., trans. Grant 1971) and elsewhere he uses arguments
from the theory of incommensurability against one of the pet tenets of astrology, *viz.*, the existence of exactly repeatable conjunctions. These arguments had been elaborated by himself in the *De proportionibus proportionum*
(ed., trans. Grant 1966). This is one of two treatise where Oresme makes pure
mathematics out of the problems of the theory of proportions as used to discuss questions of motion. The other treatise is the *Algorismus proportionum*,[57] which is much shorter but shares its main characteristics. As
indicated by the name, it is an algorism, albeit a very special one. Like other
algorisms, it describes basic rules for computation—though not computation
with Hindu numerals nor at all with numbers, but with ratios. The work is a
beautiful piece of mathematical generalization, making use of the fact that
the composition of ratios can be regarded as an addition—as the "addition"
of a musical fifth and a musical major third gives an octavo, whence (3:2)
"+" (4:3) = (2:1). This allows Oresme to define addition and subtraction of
ratios, and to multiply and divide a ratio by a positive integer (whence even
to multiply it by any rational number).[58] Mutual multiplication and division
of ratios cannot, however, be defined meaningfully, as observed by Oresme
(I.xi, ed. Curtze 1868: 19).

If we identify the ratio ($a : b$) with the real number $^a/_b$, "addition" of
ratios becomes multiplication of real numbers, and the "multiplication" of

a ratio ($a : b$) with the rational number $^p/_q$ becomes the power $(a/b)^{(p/q)}$. It has therefore been customary to interpret Oresme's theory as a theory for powers with rational exponents. In reality, however, Oresme's scheme is much closer to the spirit of modern abstract group theory; in the language of abstract algebra his algorism is indeed, as he argues, a group allowing root extraction but not expandable into a ring or a field through techniques at Oresme's disposal. (Similarly, the longer *De proportionibus proportionum* applies the structure abstracted from Euclidean arithmetic to the "addition" and "multiplication" of ratios and their inverse operations).

I have dealt at some length with this particular work because of its prototypical character for much of the new learning of the age. It continued a traditional subject (*in casu* one belonging to the rather elementary quadrivial level), but took it as nothing but a stepping stone for a discussion of pure principles (a second and third part then demonstrate "the very great usefulness" of these principles, *viz.*, as structuring tools mainly in geometry, showing that abstraction was no goal *per se* to Oresme but rather the style of his whole intellectual context[59]). In the present case the result was a piece of pure structural mathematics, the underlying principles of which could not be understood before the twentieth century (and even then they have normally been misunderstood by translators and commentators); in other cases the discussions led to semantic or logical theories with a similar fate. Generally, the *via moderna* led scholars into a highly sophisticated but also highly and narrowly specialized style of thought, particular results of which might at times be used in the following centuries, but whose outcome in the form of coherent structures was unable to find an echo in the early Modern age.

Up to now the discussion was concentrated on the highest level of fourteenth-century philosophical mathematics. As it is to be expected, this high forest was surrounded by a humbler underbrush of similar orientations. In order to get an impression of this broader environment we may look at a list of *quaestiones* from Paris, datable to c. 1330 (since the terminology is in many places too technical to allow meaningful translation without an extensive commentary, I leave it untranslated):

1. Utrum entia mathematica sint abstracta a sensibilibus qualitatibus.
2. Utrum mathematica abstrahant a motu.
3. Utrum mathematica sint coniuncta in esse cum qualitatibus sensibilibus.
4. Utrum mathematica sint priora qualitatibus sensibilibus.
5. Utrum de mathematicis sit scientia.
6. Utrum de substantiis possit esse scientia mathematica.
7. Utrum de qualitate sensibili possit esse scientia mathematica.

8.   Utrum omnia entia mathematica et omnia sensibilia communia sint per se sensibilia.
9.   Utrum scientie mathematice habeant aliquam communem materiam.
10.  Utrum sint tantum quattuor mathematice.
11.  Utrum geometria sit prior arithmetica.
12.  Utrum mathematice scientie sint certissime.
13.  Utrum scientie medie [like optics] sint magis naturalis vel mathematice.
14.  Utrum entia mathematica diffiniatur per materiam intelligibilem.
15.  Utrum quantitas sit per se divisibilis.
16.  Utrum de numero sit scientia.
17.  Utrum numerus sit ens reale extra animam.
18.  Utrum unum et multa opponantur.
19.  Utrum numerus componatur ex unitatibus.
20.  Utrum numeri differant specie.
21.  Utrum diffinitio numeri sit bene data.
22.  Utrum continuum indivisum sit principium numeri.
23.  Utrum subiectum in geometria sit magnitudo vel aliquid aliud.
24.  Utrum quantitas precedat formam substantialem in materia.
25.  Utrum sit dare dimensionem terminatam et indeterminatam.
26.  Utrum dimensio terminata et indeterminata sint generales et corporales.

27.  Utrum punctus diffiniatur.
28.  De diffinitione linee.
29.  Utrum linea componatur ex punctis.
30.  Utrum punctus sit aliquid vel nihil.[60]

Many of these questions point back to discussions in St. Thomas and Albert (and the discussions refer, indeed, to both). Others remind one of the introductory definitions from Jordanus's *Liber philotegni*; their real connection, however, is not to these (which, as we remember, had been completely external to Jordanus's *pars executiva*), but to active fourteenth-century discussions of the nature of the continuum and continuity, of the point, and of atomism; Clagett speaks of a general ''current of scholasticizing geometry by techniques followed in natural philosophy,''[61] and points out that it was felt even more widely: so in general theoretical geometry, as illustrated by Bradwardine's *Geometria speculativa*.[62] While the Albertian commentary to the *Elements* had not been mathematically productive in its own century, a similar orientation *did* hence lead to the creation of an integration of mathematics, metamathematics and Aristotelian philosophy in the early fourteenth century.

It goes almost without saying that the tendency toward integration of mathematics and Aristotelian philosophy (be it forest or underbrush) did not dominate the landscape completely—and far from that. In universities many habits from the thirteenth century were continued (as revealed by the continuous use of the compendia from good old times); the trend toward astronomical preponderance was continued. Nothing in this situation calls for discussion beyond what was given already. Finally, however, it must be mentioned that some university scholars began taking active interest in the higher levels of practitioners' mathematics. A first instance of this, *viz.*, an anonymous treatise *De regulis generalibus algorismi ad solvendum omnes questiones propositas* (ed., trans. Hughes 1980) may even go back to the very late thirteenth century. It contains rules for solving problems of the first degree familiar from many older problem collections inspired by practitioners' methods—for example, find the length of a (broken) lance when a third of it is embedded in the bottom of a pond, a fourth is in the water, a fifth is lying on the water, and 26 feet remain outside the pond'' (Hughes p. 270). The rules for solving the problems (rules given without any proof) are, however, *not* those current among the practitioners (in the case of the broken lance the very flexible ''single false position'') but derived from quadrivial arithmetic.

This short treatise is a first step toward integration between quadrivial and ''commercial'' mathematics. Another, much longer step is taken by Jean du Murs (fl. 1317–1345) in his *Quadripartitum numerorum* from 1343.[63] This treatise deals successively with the traditional quadrivial arithmetic (Boethian as well as Euclidean); with the art of algorism (concentrated on the treatment of fractions); with al-Khwārizmīan second-degree algebra; and with a richly varied material drawn from Leonardo Fibonacci's *Liber abaci* and *Flos*. The final part, on applications of arithmetic, presents *inter alia* Archimedean statics.

Another work written in part and brought to completion by Jean de Murs is the *De arte mensurandi*.[64] Thanks to Jean, a traditional mensuration treatise is integrated with much material drawn from Moerbeke's Archimedes and from *Elements* X; in many cases, proofs are given for theorems that are given, and in others it is told where material for a proof may be found (mainly in Euclid and Archimedes).

Two things should be observed concerning Jean de Murs and the two works just mentioned. First, Jean's approach to the material reminds one of that of the better mathematicians of the following century, as we shall see; Regiomontanus, indeed, planned to make an edition of the *Quadripartitum*[65] and possessed a copy of a treatise abbreviated from the *De arte mensurandi* (Busard 1974: 151). Second, Jean differed completely from, for instance, Oresme in his attitude to astrological prediction, and believed firmly in his own predictions[66]; in this respect he was no different from those of his contemporary mathematicians who, like him, fell outside the current of "philosophical mathematics."

What was said up to this point on fourteenth-century mathematics was concerned with mathematical currents associated with the university sphere. Universities, however, were not the only focus of mathematical development. Another one was the Italian abacus school (cf. note 11). Its early history is unclear, but around the end of the thirteenth century it seems to have reached maturity. The basic curriculum does not look particularly advanced, but it gave occasion for interest in more advanced extensions of the same fields. In this connection it is characteristic that only two of the manuscripts of the *Liber abaci* listed by Boncompagni (1851: 31–59) date from the thirteenth century (and both of these from the later part of the century); four were written in the fourteenth century, and four date from the fifteenth. From the fourteenth century a number of shorter abacus treatises are also known, dealing with the full curriculum as described in note 11 and containing a wide range of problems belonging to these fields, together with some practical geometry and occasionally some second-degree algebra.[67] Problem collections of a similar sort, but concentrated on the "pure," recreational outgrowth of practical arithmetic, are known from a number of fourteenth-century monastic manuscripts.[68] Neither abacus treatises nor monastic problem collections contain the slightest hint of philosophical ideas or attitudes.

The principles and main trends of the fourteenth century development are, then, easily summed up. Around the creative Aristotelianism and the professionalization of the Master of Arts a highly sophisticated type of mathematics emerged, which in itself was also an important tool for the new philosophical style. Here, for the first time, Aristotelian philosophy was a contributing cause for the evolution of a new kind of mathematics, acting together with the social organization of knowledge.

Concomitantly, the traditional mathematical style was still in existence, connected especially with astronomy and astrology. It was apparently not bound up philosophically, neither with Aristotelianism nor with neoplatonic currents,[69] but continues thirteenth-century medico-astrological naturalism. The utilitarian inclination of this current reflects itself in Jean de Murs' integration of practitioners' mathematics with theoretical and even high-level mathematics.

In Italy, finally, the social basis for a genuine development of calculating mathematics arose around the turn of the century; for the time being, however, no philosophical (or merely protophilosophical) formulations are found in this connection.

## 4. The Early Renaissance

To make "The Early Renaissance" follow upon "The Fourteenth Century" seems to presuppose a sharp but arbitrary boundary line at 1400 A.D. This is not quite the case. A closer look at the examples from the previous chapter will reveal that nothing significant took place after c. 1375. (Indeed, it seems that nobody entering a university for 50 years after the Plague contributed anything but the most faithful continuation of existing trends to the history of mathematics.) Nor was anything important going to happen to the development of mathematics during the early decades of the following century. There is thus no reason to locate the break more precisely than "somewhere between 1375 and 1425." During this half-century, however, a significant break *did* take place.

This is not to say that old habits and traditions were fully abandoned. To the contrary: until well into the sixteenth century most mathematics teaching, be it at the universities or in the merchant schools, was virtually unchanged—which not only tells something about what the students had to learn but is also informative about the knowledge and style of their teachers. This continuity holds on all levels. Elementary university curricula from the fifteenth and sixteenth century refer to subjects and compendia familiar from the thirteenth century (Book I of Witelo's *Perspectiva*, Campanus's *Theorica planetarum*, Sacrobosco's *De sphaera* and *Algorismus vulgaris*). But even the "philosophical mathematics" from the fourteenth century was transmitted and cultivated: Bradwardine's *Geometrica speculativa* was read, and so was Albertus Saxonus' (c. 1316–1390) *Tractatus proportionum,* an introduction to the ideas of Bradwardine's *Tractatus proportionum seu de proportionibus velocitatum in motibus.* On the whole, the list of scientific *incunabula* is dominated by traditional works—mainly by works written during the Middle Ages.[70]

The teaching of practitioners' mathematics was grossly unaltered, too. Until the mid-sixteenth century the only important change was geographical—*viz.,* the spread of the Italian *practica* to Germany.[71,72] (On the boundary between university mathematics and practitioners' mathematics a brief treatment of the rule of three in a number of algorisms can be observed.[73]) The novel trends in fifteenth-century mathematics are hence not to be traced quantitatively; only future events were to demonstrate that these trends *were* indeed the future.[74]

A modest illustration of the new tendencies is offered by Alberti's (1404–1472) *Ludi rerum mathematicarum* (ed. Rinaldi 1980; also in Grayson 1973: 131–73) from c. 1450. Its most striking novelty is perhaps found in a small phrase in the dedicatory letter, referring itself "all 'umanità e facilità vostra." This can be related on one hand to the dedication of the Latin version of Alberti's *De pictura* (ed. Grayson 1973: 9), which supposes that the Prince of Mantua will, on account of his *humanitas* and his interest in the *studia litterarum,* read, understand, and relish the book in his leisure; and on the other to the dedication of the Italian version of the same work to Filippo Brunelleschi, which tells its purpose to be the resurrection of one of those *nobilissime e maravigliosi* arts of antiquity *quasi in tutto perdute:* painting, sculpture, architecture, music, geometry, rhetoric, and augury (ed. Grayson: 7). As the title of the work suggests, the *Ludi* are meant recreationally; the various dedications show, however, that this recreation was meant as noble leisure, connected to the ideas of humanism and to the resurrection of ancient splendor. One almost starts wondering whether real history can fit the stereotypes of conventional periodizations so precisely. But since it does we may conclude that Alberti's *Mathematical Diversions* represents the mathematical version of archetypal humanist ideals.

How, then, does Alberti's "*humanist mathematics*" look? First of all, it contains no references at all to any philosophy or philosopher, be it Aristotle, Plato, neoplatonism, or anything else. Mathematics is *in itself* a representative of antiquity and humanity, and needs in Alberti's eyes no further philosophical justification.

The mathematical *contents* of the treatise marks no watershed in the history of mathematics. Much space is occupied by practical geometry, in particular by the measurement of heights and distances which is in touch with Alberti's conception of vision and hence related to his interest in the theory of the central perspective, but also fully traditional. Practical geometry is also represented by area measurement (referring in particular to Columella and Savasorda (!) among the ancients, and to Leonardo Fibonacci among the moderns—ed. Rinaldi 1980: 50), involving triangulation and the use of lunules. Finally, the treatise contains some statics, describes "Hero's bottle" and his hodometer, and explains how to find by systematic trial and error the elevation and the correct quantity of gunpowder to use for a bombardment.

According to this Albertian treatise, mathematics is hence *applied mathematics.* That does not change if one goes to Alberti's works on perspective, which are concerned precisely with a novel and ingenious application of (fairly unsophisticated) mathematics. In another respect, however, we should be careful not to judge Alberti's mathematical ideals on the basis of *Ludi* alone. In one work he shows that his pretensions go beyond those of Vitruvius and Columnella: The *Elementa picturae* constitute, indeed, at attempt to combine the practical aim with Euclidean systematics and structure, but in a way that is adapted to the subject and not expressed in the words of Euclid.[75]

Similar orientations are found in the works of some famous teachers of applied arithmetic. I think of Piero della Francesca (c. 1410–1492) and Luca Pacioli (c. 1445–1517).[76] Piero's *Trattato d'abaco* (ed. Arrighi 1970), as well as the works of both on the "golden section" and on regular polyhedra (ed. Mancini 1916 and Winterberg 1896) bring together methods of algebra and practical geometry with interest in art and references to antiquity.[77] Luca Pacioli's treatise is particularly interesting because of its copious extra-mathematical observations and commentaries, from which a totally eclectic conception of ancient philosophy is obvious. Mathematics is primarily an architectural (i.e., engineering) science, and Archimedes the supreme mathematician hence also an "ingenious geometer and supreme architect." Luca does not relate his mathematics to any philosophy; on the contrary, he claims priority for mathematics over philosophy. His work is indeed *necessary* to "everybody wanting to study philosophy, perspective, painting, sculpture, architecture, music, and other most pleasant, subtle and admirable doctrines," and he concludes from his discussions that

> the mathematical sciences of which I speak are the fundament for the ladder by which one arrives at knowledge of any other science, because they possess the first degree of certitude, as the philosopher says when claiming that "the mathematical sciences are in the first degree of certitude, and the natural sciences follow next to them." As stated, the mathematical sciences and disciplines are in the first degree of certitude, and all the natural sciences follow from them. And without knowing them it is impossible to understand any other well. In *Solomon's Wisdom* it is also written that "everything consists in number, weight and measure," that is, everything which is found in the inferior or the superior universe is by this necessity submitted to number, weight and measure. And Aurelius Augustine says in *De civitate Dei* that the supreme artisan should be supremely praised because "in them he made exist that which was not."[78]

It will be observed that a quotation from Aristotle (called by his medieval pseudonym *philosopho*) is twisted to bring home a point of view which is anything but Aristotelian (*follow* being understood as "follow by logical derivation and in rank," not merely as coming next in exactness), and that the familiar words from Wisd. 11:21 (see note 13) are still quoted—new and still fermentative wine in disparate second-hand bottles.

In his classification of mathematical disciplines, Luca presents us with similar eclectic innovations. His personal preference seems to be a scheme of three genuinely mathematical disciplines: *arithmetic*, *geometry*, and *proportion*. Then he runs into the traditional quadrivial scheme, based on distinguished authorities such as Plato, Aristotle, Isidore, and Boethius, but opposes to these eminent philosophers his own judgment (though *imbecille e basso*) that if *music* is included *perspective* must be, too,[79] leaving us either with a different set of three or with a set of five disciplines.[80]

In their way to bring together different ideas these men, from Alberti to Luca Pacioli, represent something original. Their ideas *inside* and *on* mathematics are, however, not very original except when it comes to the classification of disciplines. Who looks for real mathematical originality in Alberti's time should go to Nicholas de Cusa (c. 1401–1464).

This originality has several sides. First, there are the ideas that he brings to mathematics. Inasfar as they go beyond the limits of his own background (practical geometry and Bradwardine's *Geometria speculativa*) they are indeed so original that they have neither precursors nor followers, for the simple reason that the arguments are wrong—in a strictly mathematical sense often trivially wrong, because Cusanus takes his philosophical axioms for mathematical facts. Only thanks to Cusanus's philosophical importance and political role are his many rectifications of the circle available in modern print, and even in German translation with competent commentary (Hofmann 1952).

Another aspect of his originality lies in his *confidence in his own originality*. Whereas Alberti, and so many others with him, believed that all there was to do was to resurrect the forgotten knowledge of classical antiquity (another stereotype, but confirmed in the preceding quotation), Cusanus knew better in the introduction to *De geometricis transmutationibus* (German trans. Hofmann 1952: 3f.):

> It is true that the ancients, gifted with a strong inquisitive spirit, have tried with untiring industry, for the benefit of their own and future generations, to bring much to the light which was hidden until then; and it is true that they have worked with success in most of the high and the beautiful arts; yet in some of the higher branches of knowledge they have not achieved everything they aimed at. The supreme protector has indeed decided thus, in order that the divine force of comprehension should not be dulled in us but instead be guided with increasing vivacity toward that which is still hidden but none the less knowable [ . . . ]. Among the tasks which until now barred the way to geometrical speculation, one in particular remained unsolved even by all those of whose strength of mind the surviving books bear reliable witness, that is: to bring about the equality of a straight and a curved line or to convert one into the other. After toil beyond measure it therefore seemed to most, nay nearly all of those who had dedicated themselves to this inquiry

that the way to insight into this matter was unaccessible to us, namely because the enterprise was impossible since the nature of congruence opposes itself to such contradiction. I think, however, that the difficulty of the undertaking is rather due to deficient understanding, to insufficient care and to the lack of that unrelenting attention which is required by a totally unsolved problem [ . . . ].

In one sense, of course, this belief in one's own progress over all predecessors is of necessity common to all circle-squarers. It is, however, coupled to a general belief in the possibility of a continued cognitive progress guaranteed by the Lord. This brings us to the third aspect of Cusanus's originality, his metaphorical use of mathematics as a guide in philosophy.

Metaphorical use of mathematics in philosophy has a neoplatonic ring. If neoplatonism is used as an easy catchword, neoplatonic inclinations can easily be read into Cusanus's writings. Catchwords, however, are of little use if one wants to understand Cusanus's unique use of mathematics. In *De docta ignorantia* II,xiii, it is true, Cusanus quotes the invariable *in numero pondere et mensura,* and explains how the Creator made good use of all four quadrivial disciplines for his creation, "whence it comes that the machine of the world cannot perish."[81] This, of course, is fairly traditional (apart from the "machine of the world"), and not very different from the conventional neoplatonic spirit of various medieval writings. But already his metaphors are different from the uncommitted imagery of earlier times,[82] and a very definite epistemological role is assigned to them, the investigation through symbols *quasi in speculo et in enigmate* of that which cannot be reached through rational discourse.[83] In this use of symbols, Cusanus's thought is related to the Joachimite, alchemical, and cabalistic tendencies of surrounding centuries—but even this comparison doesn't do him justice. None of these currents went beyond numerology in its use of mathematics; Cusanus went astray, but so precisely because he went far, into what we might call a "dynamic approach to infinity."[84] The strict discipline of fourteenth-century Oxford scholasticism might have transformed this mixture of mathematical and theologico-philosophical inspiration into something more rigorous; a Renaissance politician-philosopher and mathematical amateur submitted only to discrete critical questions from friends did not possess these opportunities, even when he listened to the questions.[85]

We may contrast Cusanus to the academic mathematicians of the day. Most of these, it is true, were already discussed anonymously, as carrying on fourteenth-century traditions into the sixteenth century. They lectured on the basis of texts that had once been related to one or the other philosophy or protophilosophical attitude. But they did so indiscriminately, and hence apparently from institutional inertia rather than through any living philosophical commitment. They may have been Aristotelians when asked questions on epistemology or natural philosophy; but when asked about the importance of mathematics they would quote Wisd. 11:21 (*in measure, number, and weight*) and Boethius' *Arithmetica* (everybody did, as we have seen); under astrological examination the immense majority would have committed themselves to "medico-astrological naturalism." In German, one might speak of this mixture as *Gewohnheitsaristotelismus*—which has almost as much to do with serious philosophy as *Gewohnheitserotik* has to do with the passions of love.[86]

We shall meet an extreme form of this superficial "Aristotelianism by habit and convention" presently, but for the moment remember that some fifteenth-century academic mathematicians were more than mere transmitters. I think first of all of Peurbach (1423–1461) and Regiomontanus (1436–1476), who, like Alberti and others, though a small minority, represented the incipient transformations of mathematics.

Both also represent a new, autonomous professionalization of scholarly mathematics. In Vienna, their common home university, this had already begun in the early fifteenth century when Johann von Gmunden (ca. 1380–1442) became the first specialized professor in mathematics and astronomy; Peurbach, by becoming a court astrologer, represents another aspect of the new professionalization. Johann von Gmunden had combined the new specialization with a fairly traditional attitude to the contents of the subject[87]; his specialization was first of all an occasion for prolific work, interest in instruments, and care for the tools (bibliographic as well as instrumental) of the discipline; a skeptical opinion on astrology as more than a way to earn a living *may* also depend on that intimate familiarity with astronomical technicalities which resulted from specialization.[88]

Through writings and disciples, Johann was the indirect teacher of both Peurbach and the latter's student Regiomontanus. In their generation, specialization coupled to relations with the Italian humanist environment came to fruition, and for the first time Latin mathematics began catching up with its ancient and Islamic precursors *in one of their own fields* (the philosophical mathematics of the fourteenth century being original not only in contents but also as a field of study). The field was (of course, one might say) *mathematical astronomy* and *the mathematics of astronomy* (which includes trigonometry). Peurbach, who was intimately familiar with the *Almagest,* wrote a *Theoricae novae planetarum* intended to replace Gherardo di Sabbioneta's unsatisfactory but still popular thirteenth-century compendium and began working on an abridged version of the *Almagest* itself. Regiomontanus finished the latter after Peurbach's death; fully his own is a devastating critique of Gherardo's compendium.[89] The term *devastating* is to be read literally, in the sense that Regiomonatus's oft-printed attack undermined its popularity (see Schmeidler 1972: xix). Both scholars were hence actively engaged in the onslaught upon what could be seen as mathematically sloppy scholastic astronomy on behalf of ancient (i.e., Ptolemean) standards, and hence in full right as part of a humanist spring cleaning in the discipline. This interpretation of their common endeavor is corroborated by their biographies, showing close personal relations to Italian humanism (not least to Cardinal Bessarion); Peurbach, furthermore, used his university chair to propagate humanistic classical studies,[90] and Regiomontanus's literary style is clearly that of the humanistic scholar (be it the dedicatory letters, whole works in dialogue form, or even the terse definitions of the *De triangulis*).

It seems, however, that both were "humanism's good servants—but astronomy's first" (to paraphrase Thomas More on King and God). This becomes obvious in the plans that Regiomontanus had for the use of his printing establishment in Nürnberg[91]; besides his own works and the great works of antiquity (Ptolemy, Euclid, Theon, Proclos, Firmicus Maternus, Archimedes, Menelaos, Theodosios, Apollonios, Hero, and Hyginus), the list includes Witelo's *Perspectiva* (designated "an enormous and noble work"), Jordanus's *Arithmetica* and *Data* (a work on theoretical algebra), and Jean de Murs' *Quadripartitum numerorum* ("a work gushing with subtleties"). Regiomontanus was neither an enthusiastic amateur nor a mere ideologue, and would never claim, like Alberti, that the mathematical arts had been "almost lost" since antiquity; his own eminence as a mathematician made him recognize sophisticated mathematics even in scholastic garb.

One type of sophisticated work, however, is lacking from his circular: Bradwardine, Swineshead, and Oresme are all conspicuously absent. What they had made fell outside the canon defining Regiomontanus's enterprise. It cannot be because it was not astronomical—for Oresme *had* written on astronomical problems, and Jordanus's two treatises are on the other hand definitely non-astronomical; nor can it be because their works had been forgotten—as we have already seen, more traditionally minded printing houses *did* print them. It will rather be their primarily philosophical involvement and the purely hypothetical nature of their investigations *secundum imaginationem* that made them uninteresting—that same characteristic that had separated them from the astronomical naturalism of a Jean de Murs. Regiomontanus's interest in mathematics was indeed, though on a high theoretical level, an interest in a scientific *tool* to be used in the description of nature.

This is seen very clearly in the *De triangulis*. Not only is the whole work written to procure a mathematical underpinning for Ptolemean astronomy; this aim reflects itself even in the initial definitions, which build on actual *measurement*[92] while being metatheoretically problematic. The quasi-philosophical attitude expressed (directly as well as indirectly) by Regiomontanus combines the naturalism of previous centuries (and even medico-astrological naturalism, although he was somewhat less sanguinary on the direct applicability at least of existing astrology than, for instance, Jean de Murs[93]) with the conviction already pointed out in Luca Pacioli, that mathematics was *in itself* a way to ancient splendor as good as any philosophy.[94] No explicitly philosophical claims or professions of philosophical faith are to be found, only general expressions of reverence for, e.g., Plato, Aristotle, Plotinos, Anaxagoras, Democritos, John Scotus, and Thomas.

A final mathematician from this period to discuss at some depth is Cardano (1501–1576). His *Ars magna* from 1545 belongs (with Copernicus' *De revolutionibus* from 1543, *De humani corporis fabrica*, likewise from 1543, and a few other works) to the milestones demarcating the transition to the mature scientific Renaissance. Here, however, I shall concentrate on two of his earlier works.

Cardano was, like Peurbach and Regiomontanus, a professional scholar; but his profession was that of a physician, besides which he was by inclination a naturalist philosopher, maintaining that everything in the universe was living and animated. Mathematics was, as stated by his pupil Ludovico Ferrari, a subject he exerted "for enjoyment, to seize for himself some recreation and solace,"[95] but also to gain an income and to win fame (cf. Ore 1953: 10f.). His *Practica arithmeticae generalis* from 1539 and the *De numerorum proprietatibus* (both in Cardano 1663), extending one of its chapters, are thus products of a semi-professional mathematician of genius, addressing subjects that could be expected to have (and indeed had) resonance in his times.

The *Practica arithmeticae generalis* is, according to its title, a generalization of the *Trattato d'abaco* (cf. note 71), and it contains what *should* be contained in such a treatise: the numeration system, the basic arithmetical operations, the rule of three and related commercial arithmetic, basic algebra, and basic mensuration. Numeration and arithmetical operations are, however, discussed not only for integers and fractions, but also for surds and for powers of the algebraic unknown (which involves automatically some classification of irrationals). Also contained in the introductory chapter is a proposal for an (explicitly generalizable) designation of the first 10 algebraic powers. Further on, sexagesimal arithmetic, Boethian and Euclidean theoretical arithmetic, *computus* and astronomical regularities, magic squares coupled to "their" planets, biblical numerology, the conceptual instrumentarium from *Elements* X, the "rule of six" used in spherical geometry (with a reference to Regiomontanus and older astronomers), a chapter on games and one of the principle of *Data* (referring amply to Regiomontanus), and various geometric constructions (including Philon's extraction of a cube root by moving geometry[96])—truly an overwhelming work, and truly far from any standard of mathematical normality.

The *De numerorum proprietatibus* is an extended but purged version of the chapter "De proprietatibus numerorum mirificis" of the previous work. The astronomical regularities, the numerology, and the astrologically connected magic squares have disappeared; what remains is an account (with heuristic proofs) of the main concepts and results from the arithmetical Books VII–IX of the *Elements,* critically correlated with the Boethian theory of figurate numbers, and augmented with a number of observations on the proprieties of numbers (including the casting out of nines) and with an arithmetical translation of select results from *Elements* II and XIII. Here, as in the *Practica,* free use is made of current algebraic abbreviations for *plus, minus,* and *radix.*

In the first work Cardano behaves as the prototype of a "universal Renaissance genius," speaking of everything and respecting no customary disciplinary boundaries; in the second he shows that he *could* restrict himself to mathematics, but that this subject was on the other hand a single and undivided entity. Like Regiomontanus, though in a very different manner, Cardano brings together the practitioner's and the theoretician's view of mathematics, no practical question being too humble for theoretical elaboration, and no theory too elevated for application. Each in his own way achieved what had been foreshadowed by Jean de Murs, and what had been the implicit project of twelfth-century naturalism—impossible that early for lack of adequate practice, for insufficient understanding of theory, and for lack of adequate social structures able to carry on the project.

In neither of the two works does one find references to major philosophical systems. If we go to Cardano's encyclopedic *De subtilitate* (1550), which can be said to represent Cardano's own philosophy, we shall find that Book XVI, "De scientiis," is in fact mainly concerned with mathematics (some music and meteorology and a little medicine being included). The concluding list of key workers is headed supremely by Archimedes. Then come, in order of succession: Ptolemy; Aristotle (the naturalist); Euclid; Scotus (no further identification is given, but certainly Duns); Swineshead; Apollonios; Archytas; Eutocios; al-Khwārizmī; al-Kindī; "Heber Hispanus"[97]; Galen; and Vitruvius. The philosophical context can be (briefly!) characterized as naturalism bent toward occultism and influenced by neoplatonism; but the attitude toward mathematics seems to be a very open-minded "Archimedism'— Archimedes being both the ingenious geometer, the calculator of the apparently incalculable, and the great engineer.

The undomesticated style of the *Practica* corresponds well to the approach of *De subtilitate,* and can hence be said to be a consequence of Cardano's philosophy. It was, if we regard the detailed contents, idiosyncratically Cardano's own. The general tendency of both works is, however, close to what can be found in Michael Stifel (ca. 1487–1567), both his principal work, the *Arithmetica integra* (Stifel 1544) and the *Deutsche Arithmetica* (Stifel 1545).[98] The former, presenting "all that was then known about arithmetic and algebra, supplemented by important original contributions" (Vogel, "Stifel," 59), included also magic squares and extensions of the theory of irrationals, along with the rule of alligation and similar topics from practitioner's arithmetic, whereas the latter, popular book tried to propagate more high-level methods in the field of the *practica,* arguing that everything which could be done by false position could be done more easily by algebra ("coβ").

Stifel did not share Cardano's specific philosophy; whereas the latter was an astrologically minded, heterodox Catholic physician, Stifel was a cabalist and a Lutheran pastor (whose biblical numerology was only spared the heterodox epithet because of Luther's personal intervention). In both, then, a general occultist orientation, a deep interest in the secret forces of nature or number, can be found, and in both it was coupled to their unification of *all levels of* and *all approaches to* mathematics. Because both were extraordinarily gifted mathematicians, this unification (and hence ultimately this common orientation) brought the inherent tendency of early Renaissance mathematicians to a culmination.

It is well known that many contemporary figures shared the occultist orientation of our two eminent mathematicians, mostly without sharing their giftedness.[99] *They* did not understand the meaning of those traditional disciplinary delimitations which they disrespected, which then led only to confusion.

Up to this point, the early Renaissance was dealt with from the perspective of active mathematicians (including non-professionals such as Alberti). Other groups too, however, took part in the shaping of the new mathematics, moved by philosophical or quasi-philosophical ideas.

Previously, a number of *mathematicians with humanistic affinities* were discussed. Moving from them toward the center of the humanistic movement we meet a number of *humanists with mathematical affinities.*[100] We meet them in increasing numbers toward the second half of the period and further on in the mature scientific Renaissance, as translators and diffusors of Greek mathematics and as patrons of translation and diffusion. Early examples are Cardinal Bessarion (1403–1472), protector of Peurbach and Regiomontanus, whose desire to spread the full gospel of Greek learning was the motive force behind their common abridgment of the *Almagest* (see Rosen, "Regiomontanus," 348); and Pope Nicholas V (1377–1455), who, along with Homer, Herodotos, and the Greek Fathers, had Archimedes translated once again from the Greek (see Clagett 1978: 297f. and 321ff.). Later names are Giorgio

Valla (c. 1447–1500), physician, translator of Aristotle, and author of a bulky encyclopedic work[101] fusing mathematics (the regent discipline), medicine, and natural philosophy with the *studia humanitatis* and encompassing much Euclidean and Archimedean material; and his student Bartolomeo Zamberti (b. 1473, d. after 1539), aggressively anti-medieval translator of the *Elements* from the Greek. Continuators of the tradition into the next period were Maurolico (1494–1575), commentator on Archimedes and Apollonios, translator of Autolycos, Theodosios, and Menelaos; Commandino (1502–1575), translator of Ptolemy, Archimedes, Apollonios, Euclid, Aristarchos, Pappos, and Hero; and Baldi (1553–1615), friend of Commandino and author of more than 200 *Lives* of mathematicians.[102]

Detailed accounts of the activities of these humanists would lead much too far. I shall therefore just note that they confirm the impression gained from the humanist mathematicians discussed previously: Mathematics is *in itself* a part of and a path to ancient splendor; interest in mathematics is not in itself coupled to Platonic (or Aristotelian) predilections, but rather to that mood which saw both philosophers as great men who could easily go in company (a conception that in itself is of course a break with scholastic Aristotelianism, and which is hence often considered as "Platonism"[103]). Their common evaluation of the relative merit of ancient mathematicians is expressed by Baldi in the last paragraph of his extensive biography of Archimedes: "Archimedes was the Prince of mathematicians; whence Commandino said quite rightly that *he* can hardly call himself a mathematician who hasn't studied Archimedes' works with diligence."[104]

If we go to Northern humanists with mathematical affinities but not themselves mathematicians, *their* level would (at most) permit them to see *Euclid* as the "prince of mathematicians." An illustrative example is Melanchton, who generously lent his name to many a mathematical book. His preface to the Basel Euclid (*Euclidis Megarensis . . . Elementorum geometricorum libri XV*) shows him to be much better versed in Aristotle and Plato and their discussions of arithmetically versus geometrically proportionate justice than in the subject matter of the particular object of his praise; in another book to which he granted his favors, Vögelin's *Elementale geometricum ex euclidis geometria,* the author himself claims that his excerpts from *Elements* I–IV (the very books once commented upon by Albert) were "almost sufficient to lead to the summit of learning."[105] "Miserable summit, but still more miserable the frugality of a time which reprinted Vögelin's brief excerpt time and again and made preferential use of it in the most diverse institutions. Geometry, we have to observe once again, was not the strong point of German mathematicians in general," as Cantor (1900: 394f.) observes with affectionate irony.

As soon as we disregard the few creative mathematicians, we need not restrict ourselves to geometry nor to Germany alone. In France of the earliest sixteenth century, a small century of Renaissance innovations in mathematics had passed with as few traces as in Germany, even among those who were mathematically interested without being mathematicians themselves. As witnesses we can take Jacques Lefèvre d'Étaples (c. 1455–1536), the circle surrounding him, and the books published in this environment.

Lefèvre d'Étaples himself was responsible for a number of mathematical editions, including Sacrobosco's *De sphaera*, a common edition of Campanus's and Zamberti's *Elements*, and two editions of Jordanus's *Arithmetica*. He also wrote a number of mathematical manuals and compendia himself. All was done in an attempt to raise the mathematical level of the Parisian university environment, and from a broadly neoplatonic perspective[106]—and all was very traditional, though when evaluated on that background, of good quality. In spite of Italian acquaintances[107] and inspiration the neoplatonic paragon of French mathematical publishing around 1500 could do nothing better than resurrect the best quality of thirteenth-century mathematics.[108]

The low pretensions of Vögelin and the more brilliant but not much more innovative accomplishments of Lefèvre d'Étaples are not astonishing when seen in the light of the *real* success of the early sixteenth-century scholarly book market: The *Margarita philosophica* written in 1496 by Gregor Reisch (a Cartusian from Freiburg, and future collaborator in Lefèvre d'Étaples' edition of Cusanus's writings; c. 1470–1525), printed first in 1504 and reprinted in France, Switzerland, and Germany numerous times during the following decennia.[109] If anything is described by the concept of *Gewohnheitsaristotelismus*, this work is. For this reason; because its popularity demonstrated it to be a good exponent of the moods of its times; and because it was especially read as a mathematical textbook,[110] I shall describe it in some detail.

First of all comes a table of the division of philosophy. The subject falls in two parts, *theorica/speculativa* and *practica*. The former is divided into *realis* and *rationalis*, and the former of these into *metaphysicam* (including both theology and normal philosophical metaphysics), *mathematicam* (identical with the quadrivium), and *phisicam* (containing, title for title, the traditional curriculum of Aristotelian natural philosophy, and told also to embrace medicine). *Practical philosophy* is either *activa* (with subdivisions *ethica, politica, oeconomica* and *monastica*), or *factiva*, the subdivisions of which are nothing but those "mechanical arts" which were once listed by Hugue de Saint Victor in *Didascalicon* III,i (same sequence, same terms). Only the inclusion of theology as the main part of metaphysics and the specification of a moral science of monastic life is something new; the rest is a mix-up of twelfth- and thirteenth-century lore, humanistic only insofar as the strict scholastic conceptual organization of the whole has been dissolved into general benevolence toward everything revered and old. The inclusion of mechanical arts as "productive philosophy" is perhaps an expression of the Renaissance upgrading of applied knowledge; if so, the mere repetition of an almost 400-year-older list (which furthermore had already been bookish at birth) shows the upgrading to be totally empty.

A pictorial representation of the castle of philosophy shows the roof terrace to be occupied by Petrus Lombardus, representing *Theologia seu Metaphysica;* in the windows of the upper story we discover *Philosophus* (alias Aristotle, one must presume) representing *Physica,* and Seneca representing *Moralia;* the next lower floor is occupied by Pythagoras (*Musica*), Euclid (*Geometria*), and Ptolemy (*Astronomia*); third come Aristotle, this time under his own name, representing *Logica;* Tullius (Cicero) representing rhetoric and poetry; and Boethius (arithmetic). No more innovative, and no less medievally-eclectic, than the preceding description in words.

Book IV on arithmetic also opens with a woodcut, representing Boethius and Pythagoras, respectively, as a young, modish, apparently Italian merchant calculating with Arabic numerals, and as an elderly colleague in old-fashioned northern dress performing his computations on the ruled abacus board (traditional lore on the invention of the two notations). The description itself starts by praising the quadrivium, with unspecific references to Boethius and to Cusanus's *De docta ignorantia,* for being the key to many arcane places in the Scripture.

The discussion of speculative arithmetic is in the main taken over (directly or indirectly) from Isidore's *Etymologiae* III,ii–vii, but compares sometimes badly with this source.[111] Certain formulations are borrowed from Boethius, but even the modicum of theoretical reflection offered in the latter's *Arithmetica* is absent from the *Margarita.*

Practical arithmetic is mainly an algorism, dealing with both integers and fractions, "vulgar" as well as "physical" (sexagesimal), covering the subject up to the extraction of a cube root, and including calculation on the ruled abacus and a presentation of the rule of three with select examples. On this subject Reisch is hence as modern as the anonymous Vienna algorism mentioned in note 73.

The chapter on speculative geometry is liable to provoke as much bewilderment as the depreciated Isidore on arithmetic. Not that it is taken over from elsewhere. At least I have found no probable source for its peculiarities; I also doubt that any source can be found to maintain that no circle can be drawn through *two* or three points on a straight line, or to regard *diameter, axis, chorda, costa, latus, basis, cathetus, corauscus,*[112] *hypotenusa, diagonalis, perpendicularis, orthogonalis (et alia plura)* as different sorts of lines;

the author can have possessed little understanding of the terms he uses, and conveyed no more.

There is no reason to say more on the contents of this *Pearl of Philosophy*. It will probably be clear from the above that the work mixes up the traditional medieval authorities (some recognized, others unrecognized) in a totally uncritical way, misunderstanding, furthermore, much of what they have to say. Its immense success demonstrates that the mathematical *avant-garde* dealt with in the preceding pages (including Vögelin!) had left the main corps of the universitarian army of northern Europe far behind. Readers satisfied with a work like this cannot have taken its references to philosophical authorities as anything more than tokens of submission to revered old traditions—a queer sort of "poor scholar's humanism." Any finer philosophical points—including all such points that would make it meaningful to ask about philosophical influence of the discourse of mathematics—will have been beyond their horizon; so will, however, everything in mathematics that made it a fruitful field.

Hence, neither Northern humanism nor its reflection in writings such as the *Margarita* are engaged directly in the development of new kinds of mathematics, as sometimes seen in Italy. I shall, however, point briefly to more indirect influences contributing to later developments in the area: First, the idea that an educational system without mathematics was incomplete, which procured a chair at the Collège Royal for the "encyclopedic, elementary, and unoriginal" astronomer and mathematician Oronce Fine,[113] but even then contributed to broaden the basis for local mathematical activity and hence to make the general reception of new thinking on the subject possible. Second, the humanist interest in cosmography, which was coupled to map-making and mathematical geography and hence to mathematics in general; this had effects of a similar sort. Third, finally, the activity of certain printers, who worked systematically in favor of the new currents: Petreius has already been mentioned, who (in partial collaboration with Osiander) managed to print Copernicus' *De revolutionibus* in 1543, Stifel's *Arithmetica integra* in 1544, and Cardanos' *Ars magna* in 1545 (to be followed by the *De subtilitate* in 1550); another name that returns time and again is Ioh. Schöner (1477–1547) from Nürnberg, himself a teacher and writer on mathematics and astronomy,[114] who *inter alia* printed most of Regiomontanus's works (which an early and sudden death had prevented the printer-astronomer himself from publishing). Petreius's personal preface to *De subtilitate* shows him to share the philosophical attitudes of Cardano, whereas Schöner's presentation of his edition of the *Algoritmus demonstratus*[115] demonstrates him to be a true follower of Regiomontanus. Discussion of the scholar-printers will hence add no new dimensions to the foregoing. The only quasi-philosophical attitude to derive from the uncommitted goodwill toward mathematics exemplified by the Collège Royal and cosmography is also a repetition: the tendency to consider mathematical practice a legitimate entrance into theory, and theory (however understood) the best tool for practice. I shall therefore discuss none of these subjects any further.

In an overall discussion of this period, the first question will naturally be for the place of the prominent philosophers of the time. Where is Marsilio Ficino? Where is Pico della Mirandola? Nicoletto Vernia? Pomponazzi?

The truth seems to be that if we approach the question of relations between philosophy and mathematics from the mathematical side, all these major figures are absent. Among the characters from the period counted as philosophers, for example, in Randall's *Career of Philosophy from the Middle Ages to the Enlightenment* (1962), only Cusanus, Bessarion, Cardano, and Lefèvre d'Étaples are of any importance for mathematics—and among these only Cardano is of real importance, Cusanus being either regarded as a mathematical fool (by Regiomontanus) or used together with Isidore or Augustine as an advocate for the general importance of mathematics. Disregarding Cardano, who as a philosopher wholly of his own must count as a special case, philosophers of importance were simply not oriented in contemporary mathematics. An illustrative example is Pomponazzi, who *did* take up a mathematical discussion in 1514—but a discussion with Swineshead, belonging in the context of the early fourteenth century (see Wilson 1953).

Conversely, mathematicians were (Cardano being again by necessity an exception) not very well oriented in the actual philosophy of the day—either because occupational specialization separated the two, or because the precise discussions of contemporary philosophy were uninteresting to them (probably both, in varying balance and interplay).

Only as a very general background were the philosophical developments of the time of importance for mathematics. First, of course, most or all currents participated somehow in the humanist movement, which *was* of importance—this is a question to which we shall return in a moment. Second, the eclectic appreciation of everything in ancient philosophy and especially of Greek texts (at times regarded as "Platonism") would encourage work on all available ancient mathematical texts. Third, the breakdown of old philosophical fences between different ontological levels and categories—inherent in the prevailing eclecticism—contributed also to the breakdown of old barriers. Fourth, finally, *some* currents at least would not only break down barriers but also create connections. Thus, in *Theologia platonica* III,ii (quoted from Randall 1962: 60), Ficino declared about the soul, that she ascends to higher things and descends to lower. And when she ascends, she does not forsake the lower, and when she descends she does not leave the higher. For if she forsook either, she would fall into the other extreme, nor would she be the true link between both worlds.

Similarly, in his *Oration on the Dignity of Man,* Pico declared *Man* to be set "at the world's center," "neither mortal nor immortal," and "constrained by no limits," so that he may ordain for himself the limits of his nature, either degenerating "into the lower forms of life, which are brutish," or ascending "into the higher forms, which are divine," and for that reason "envied not only by brutes but even by the stars and by minds beyond this world" (trans. E. L. Forbes in Cassirer et al. (eds.) 1956: 223–25). Ficino as well as Pico represents a variant of neoplatonism, where the *Great Chain of Being* was no longer a unidirectional system channeling Divine emanations, and where *Man* (or the human soul) as an active being had the task to mediate the daily and the supreme levels of reality. If a mathematician would perform this task inside his profession he could do no better than regard practical applications as a way to theory and theory as the best tool for practice.[116] And the other way: Should a neoplatonic philosopher have translated an unspecific belief in the unity of all cosmic levels into the terms of *his* discipline, his course would be the one chosen by Ficino and Pico.

But the influence of philosophy remains on a very general level. Nowhere have I seen a mathematician argue directly for the unified treatment of theoretical and applied mathematics in philosophical terms. Thus, rather than really speaking of "philosophical influence" we have to do with an all-pervading and unspecific, quasi-philosophical conviction that *reality is One,*[117] which philosophers and mathematicians made specific, each party in its own terms. Ficino's philosophy will be no *cause* but rather a parallel that highlights a quasi-philosophical attitude shared by an increasing number of mathematicians.

Mathematicians' attitudes can be discussed under two broad headings: humanism and naturalism. Dependent on the context of humanism are a number of interrelated issues reflected in many mathematical works, including many of the works discussed already.

A recurrent theme is the creed that mathematics is in practice the real *first philosophy.* This is definitely not Aristotelianism; nor can it, however, be regarded as serious Platonism—after all, Plato had considered mathematics only a *prolegomenon,* a training and mental preparation by analogy to the real insight gained through dialectic. But the writings of both philosophers (and their commentators) abound in references to mathematics, and indirectly their writings would then carry the message that mathematics was fundamental.

Closely related to the appraisal of mathematics as the veritable first philosophy was hence the appraisal of the field as itself an expression of and a path to ancient splendor. In this connection it should not be forgotten that the appropriation of the mathematical heritage from antiquity had been an essential aspect of the High Middle Ages' reorganization of learning (and, as we have seen, that mathematics had rather been an *alternative to* than a genuine *part of* strict scholasticism). As Bessarion, already Gherardo di Cremona and the anonymous translator from the Greek had seen the *Almagest* as an indispensable work.

Closely related is also the character of humanism as a citizens' movement, a cultural current carried by people activity engaged in the higher and highest levels of civic life and emphasizing the public utility of their cultural skills. In many ways, mathematics had proved itself publicly useful since the late Middle Ages: as commercial arithmetic; in the administration of the city-states; in architecture; in "machines" used in military and architectural engineering as well as the clockwork "machines" referred to by Cusanus; in painting; in surveying and cartography; in other sorts of calculation combined with measurement (gunnery, *Visierkunst*, etc.); and—not to forget—as medical and courtly astrology (all of them fields that in some way or other turn up in the works discussed, and most of which are mentioned in Regiomontanus's Padua lecture—ed. Schmeidler 1972:49ff.). If utility in the cultural context of the Renaissance city-state was the yardstick, no wonder that mathematics came to be regarded, with rhetoric, *belles lettres*, and *beaux-arts*, as a major constituent of the ancient heritage[118]; and no wonder that the ancient philosophers' references to mathematics led not only Luca Pacioli but even Bessarion to treat mathematics as something more fundamental than philosophy.

However, if utility was the gauge, the traditional disrespect for applied mathematics as being of secondary rank would be an unbearable paradox; similarly, the quadrivial scheme was, in spite of its ancient legitimacy, an unacceptable straightjacket precisely for many of the new applications of mathematics. We remember Luca Pacioli's protest that the mathematics of painting was no less important than the mathematics of acoustical harmony. Even in the *Margarita philosophica* Reisch shows the quadrivium to be outdated by investing his real interest in algorism and practical geometry while filling the mandatory "theoretical" chapters with traditional lore and nonsense.

Traditionally, *music* had been regarded as the discipline of proportions and balance, and hence *of good life*.[119] Luca (and others with him, not least of course his friend and collaborator Leonardo da Vinci) had come to regard the theory of proportions as a theoretical discipline to which music, visual harmony, and so forth were subordinated. Melanchton (to mention only one example), in his Euclidean preface, shows that the old connection to moral and political balance was also present to Renaissance minds. Through the concept and theory of proportions mathematics was hence not only a theoretical and utilitarian but also a moral science, a *scientia activa*. Comparison of fourteenth- and fifteenth-century Italian painting gives an immediate impression that the political shift from communal to princely government in the Italian city states was reflected in (and not only accompanied by) a shift in interest in art from naturalistic truth and detail to balance, stability, and harmony ("realism" in the philosophical sense, as opposed to the "nominalist" acceptance of the phenomenal world). The arguments of mathematicians and their fellow-travellers on the importance of the mathematics of proportions, referring not only to the beauty of music and the visual arts but also to the moral and political sphere, thus appear not to be coincidental: mathematical interest in precisely that field seems to have been founded (at least in part— Boethian tradition and internal developments presumably played their roles too) on ideas deeply rooted in the political thinking and dispositions of the aristocratic and courtly environment. (Yet another point where the cultural conditions and attitudes influencing both philosophy and mathematics are easily mistaken for philosophical inspiration of mathematics).

Archimedes was known to have served his city and his king, biographical facts often referred to (for example, by Luca Pacioli). He was known as a fabulous military engineer; and his mathematical works would speak for themselves to anybody able to grasp their sophistication, making him the supreme representative of ancient mathematical splendor. From every aspect of humanist mathematics he was thus the paramount figure to be called upon as a witness. When personified, humanist mathematics therefore appears in the guise of *Archimedism*. On the other hand the Archimedes invoked was a protean figure, able to legitimate almost every kind of mathematical activity. Archimedism was hence no strict program or philosophy; it *was* apparently influential all the same, and no pure epiphenomenon, but only because invocation of one aspect would tend to call up the others as well: If architecture was supreme because represented by Archimedes, a really good architect had to be an eminent geometer too. Archimedism was hence the conceptual incarnation of the unity of courtly or civic science, utilitarianism, and theory at the highest level.

Whereas humanist attitudes were a specific Renaissance influence in mathematics, *naturalism* is an old acquaintance that will require fewer words, though it took on new forms in the Renaissance period.

Insofar as astrology is concerned these new forms were primarily social, *viz.*, the institutionalization of court astrology. Thereby astronomy was transformed into a lucrative career permitting specialization, permitting but also dictating a much higher level of sophistication, technicalization, and reflection on the subject. (Socially, the contact to the courtly environment was also a contact to humanist currents, which makes distinction between influences difficult.)

By being less established and thereby less governed by stable traditions than astrology, the broader naturalist and occult[120] currents are more interesting for our purpose. They are often neoplatonic, or at least tainted by neoplatonism, and as such they are of course related to the neoplatonic currents of precedent periods. Just as Ficino (a connecting figure, in his quality as an important philosopher who in his very philosophy represented the occult current) gave Man an active role as the mediator between the various levels of the Great Chain of Being, so occult Renaissance naturalism in general gave to him a central role as the executant of *natural magic*.

Renaissance occultism has been amply discussed since the early 1960s, and its influence in mathematics can easily be overrated. Direct reflection in the new trends of mathematical thought is no more visible than were direct reflections of philosophical stances. Once again, of course, Cardano is (with Stifel) a kind of exception, insofar as his undisciplined *Practica generalis* mixes up things that are unconnected from almost any viewpoint except his own hylozoic naturalism. But already in his *De numerorum proprietatibus* he had sorted things out; the *Ars magna* is totally free of them, and so is the reception of his work by other mathematicians. The effect of Cardano's distinctive naturalism on his mathematics was, in the end, only that of epistemological optimism—an observation that (with emphasis on the utility of knowledge) seems to hold throughout for the core of occult philosophy (in which Cardano is not to be counted) and its effects on mathematics. Archimedes, so important a figure in renascent mathematics, was more or less an unperson to the occultists.[121]

Furthermore, "epistemological optimism with emphasis on the utility of knowledge" was not a distinctive characteristic of the occult currents. I have already discussed it, as the shared attitude of humanist mathematicians. Naturalism, insofar as it influenced the development of mathematics, participates in broader currents; and occult naturalism, in its most specific form, was rarely shared by important mathematicians and had no influence on mathematics. Whereas the specialization made possible by the emergence of courtly astrology had given astrological naturalism the possibility to interact with the advances of mathematics and to enter into dialogue with the best of ancient astronomy, general neoplatonic naturalism, leaning increasingly toward occultism, lost the contact that had been fruitful in preceding centuries.

Equally devoid of specific influence was the northern universitarian tradition of "Aristotelianism by habit and convention." No doubt, the continued teaching of fundamental mathematical skills and knowledge (*Elements*

I–IV, algorism, *De sphaera*) supplied a certain basis of recruitment for professional astronomers and mathematicians and a living for a number of mathematics professors; but (for example) the incipient integration of the rule of three into algorisms (which was in harmony with the general unifying tendency in Renaissance mathematics) came to the universitarian environment from the outside, and was rather an influence of renascent mathematics on university teaching than the other way round. As illustrated by the *Margarita philosophica*, the whole environment was in general too philosophically flabby to be able to exert any influence on the increasingly vigorous field of mathematics.

## 5. Epilogue: Mathematics as Part of the Foundation of New Philosophies

We are approaching the end of the study, which I shall round off with some sketchy observations on the transformations of the structures discussed up to now during and immediately after the mature scientific Renaissance.

One astonishing development is a temporary union between Hermetic-occult-naturalist interests and mathematics. It was just argued that no such marriage took place before the mid-sixteenth century—but then it did. Three rather prominent figures can be mentioned as its representatives[122]: Bishop Foix de Candale (c. 1502–1594); John Dee (1527–1608), and Faulhaber (1580–1635). Candale, who made a new Greek edition of the *Corpus hermeticum* and was convinced of Hermes' sovereign wisdom (but did not accept Hermetic magic),[123] also made a large commentary on the *Elements* (description in Cantor 1900: 554). Dee, no less a universal genius that Cardano, was a cabalist, an astrologer, and a magician; he translated an Arabic version of Euclid's work on the partition of figures; he cooperated with Commandino, studied with Gemma Frisuis, and was friend of Pierre de la Ramée and Mercator; he took care of the first English translation of Euclid (which involved the writing of an extended and influential *Mathematicall Praeface* (ed. Debus 1975), of introductions to the single books, and of many commentaries and additional theorems), and was the proponent of educational reform, of English imperialism, and of the application of science.[124] Faulhaber, an extraordinary self-taught mathematician, engineer, and teacher of engineering mathematics, developed his own version of Cabala in which he was a firm believer.[125]

In the preceding chapter *Archimedism* was used as a keyword embodying the new developments in Renaissance mathematics. It is characteristic that all three figures just mentioned pass the Archimedist test one way or the other. Foix de Candale was regarded by contemporaries as "le grand Archimède de nostre age" (Jean Bodin, quoted in Westman 1977: 42); Dee refers to Archimedes time and again in the *Mathematicall Praeface*[126]; and Faulhaber translated Archimedes from Latin into German for his own use, along with Euclid, Apollonios, Regiomontanus, and Cardano.

In Foix de Candale the only connection between his Hermeticism and his Euclidean commentary appears to be a common search for pristine truth behind later errors and distortions.[127] For both Dee and Faulhaber, however, magic and applied mathematics were parts of a continuum: for Dee, as quoted in note 126, the squaring of the circle was arcane knowledge; Faulhaber, too, used similar words to speak of new geometrical instruments and of his new, wonderful cabala.[128] For both, the paramount emphasis placed on applied disciplines, as well as the explosion of the number of such disciplines, was directly dependent on their general utilitarian occultism (and *vice versa*).

The sudden occurrence of such major mathematicians involved in Hermeticism or occultism puts the absence of mathematically influential occultists in the previous period in a new light. It proves that early Renaissance mathematics was not at a technical or epistemological level that would make magical approaches impossible or necessarily unfruitful; the impotency must be found on the part of early Renaissance occultism, whose cultivators, far below the level of a Dee, were not even competent to grasp that of living fifteenth-century mathematics. It seems that the continuity of neoplatonic and related traditions (from Joachim of Fiore, Arnaldo di Villanova, and Meister Eckhart onward) was only philosophical and mystical. Mathematical results gained by one generation (as thirteenth-century optics) might be accepted as mathematics and then handed down through the continuity of the mathematical tradition (uneven as this continuity was); they were, however, not carried on by the philosophically and religiously heterodox current itself, and fifteenth-century *magi* had to start afresh with elementary number symbolism, and mystical geometry.

Above, Dee and Faulhaber were grouped together as equally technologically minded. Socially, however, they differ. Whereas Dee was an outstanding figure in a broad Hermeticist and occultist movement, Faulhaber was an eccentric in his occultism. In his generation, occultism was a *possible but strictly private* inspiration for a mathematician, and no longer something to be regarded as influence of an intellectual current on mathematics, regarded broadly. The Kepler-Fludd debate (see Westman 1984) provides another illustration of this: Fludd (1574–1637) the Rosicrucian had lost contact with serious mathematics; only Kepler's "Pythagorean" bent was still compatible with broad orientation and influence in the field[129]—but even that was on the wane.

Archimedism, the specific complex of admiration for ancient mathematics symbolized by supreme reverence for Archimedes the engineer and geometer, was another inherited structure. Much was already said about it; here I shall merely recall that the change of emphasis in the current interpretation of Archimedes (from the engineer to the subtle theoretician) was only a result of the mature Renaissance, to which both Baldi and Commandino belong (cf. above, text to note 104). As we have just seen, the involvement of occultism with mathematics at the professional level implicated its involvement with Archimedism, a fact that in itself shows that this attitude was still of fundamental (if not necessarily very specific) importance in mathematics.

A third structure of importance was the utilitarian orientation and persuasion of mathematicians. We met it in intensified form in Dee and Faulhaber, but it is a much more general characteristic. It found new support in the development of new or improved mathematical techniques, from *prosthaphairesis* (the use of trigonometric formulae and tables to transform multiplications into additive operations), logarithms, and the invention of new instruments to military architecture, cartography, and mathematical navigation.[130]

The conviction that genuine mathematics was, or was modelled upon, ancient mathematics, suggested the tools for two attempts to reform algebra. Both were French, and chronologically they were separated by only by three decades. The difference between their choice of ancient ideas and their corresponding fate is illustrative of the ways in which inspiration from quasi-philosophical attitudes can direct mathematical development, as well as of their frailty as determining forces.

One work is Pierre de la Ramée's (1515–1572) *Algebra*.[131] Its intended set-up is basically different from cossist or Italian algebra. Algebra is defined as "a part of arithmetic, which from imagined continued proportions establishes a certain form of counting of its own."[132] The imagined continued proportion is, of course, the sequence of algebraic powers *unitas* ($x^0$), *latus* ($x^1$), *quadratus* ($x^2$), etc. (An example using the powers of 2 goes to $2^{15}$). For all these, as for addition and subtraction, symbols are introduced. Part 1 is, then, a set of examples illustrating the use of schemes (borrowed without acknowledgment from Stifel's *Arithmetica integra*) for the calculation with algebraic expressions—for example, the scheme.

$$\frac{8q - 9 \qquad 8q}{64bq - 72q}$$

where $q$ means *quadratus* and $bq$ *biquadratus* ($x^4$), and the whole scheme $(8x^2 - 9) \cdot 8x^2 = 64x^4 - 8x^2$. The rules are given merely *as rules;* proofs are absent. Part 2 *in aequatione* falls into two subsections, the first of which deals with problems of the first degree and the second with second-degree equations. Here, at least, reasons are given for the standard algorithms used for the solution, *viz.*, geometrical representations and references to *Elements* II. 4–6. If we compare the work with al-Khwārizmī's classical treatise, the use of abbreviated names and of schemes for algebraic reductions are new; so are the Euclidean references and the use of the ancient concept of magnitudes in continued proportion. As far as mathematical substance is concerned, however, Ramus does not surpass the first Islamic treatise, nor its twelfth-century Latin translations, notwithstanding the possibilities offered by abbreviations and schemes; not only Cardano's *Ars magna* (which goes by itself) but also Luca Pacioli's short section on algebra in *Summa de Arithmetica* (1523: 144$^r$–150$^r$) bear witness of deeper insights. The idea to use the framework of ancient mathematics in order to put the old algebraic *experimenta* on a more solid base was good; but Ramus' choice was uncritical, depending on familiarity (what could be more familiar from ancient mathematics than the elementary concepts of proportion and the early books of the *Elements?*) and not on mathematical adequacy. Jordanus, when modelling *his* theoretical algebra on the analytic idea of the *Data* and on the mathematics of *Elements* VII–IX had chosen much better more than three centuries before.

Decisively better also was Viète (1540–1603) in his *In artem analyticam isagoge*,[133] published three decennia later, and in a number of other works. Viète considered existing algebra "so defiled and polluted by barbarians" that he found it necessary "to bring it into a completely new form." The enterprise was necessary because, as all mathematicians knew, "under their *Algebrâ* or *Almucabalâ,* which they extol and call the great art, incomparable gold is concealed, which however they cannot find at all."[134] The way was, once again, a return to the cleanliness of ancient mathematics—but for thorough reform, not just to dress up single concepts and procedures of the existing discipline in ancient garb, as done by Ramus. Everything had to be recast, and the mold was provided not by the works nearest at hand[135] but by Pappos's discussion of the concept of *analysis* (making possible the formulation of the theory and a definition of the metatheoretical status of algebra); by the stringent differentiation between quantities that differ in kind (contributing the principle of homogeneity); and even by a twisted borrowing from Aristotelian ontology, permitting Viète to conceive of the unknown not as something "imagined" but as pure "form."

Both representatives of mature French humanism thus show us the legacy of Archimedism: The way to correction of errors and to progress in mathematics was the return *ad fontes.*[136] Fruitful results, however, would evidently only follow when the ancient sources were used in critical integration with the actual state and urgent needs of mathematics as a living discipline: Archimedist ideology alone wouldn't do.

The breakthrough in algebra created by Viète and followed up by Descartes contributed to make the early Modern age conscious of its scientific advantage over antiquity, and to make it consider mathematics a major point of this advantage. Thus understood, even Viète's *algebrâ novâ* contributed to the foundation of the new philosophy. The contribution was, however, modest, indirect, and peripheral, insufficient to motivate the heading of the present chapter, and Viète is unlikely to be mentioned in even a broad-minded history of philosophy; on the other hand, Gilbert, Galileo, Kepler, Hooke, Boyle, and Newton can be expected to turn up, some of them as principal actors.[137]

I shall not venture into an investigation of these founders of Modern science. It would double the size of the present essay, which is already bulky. Instead I shall recall the unifying concept used in the seventeenth century as a common denominator for this somewhat uneven bunch (and used as a slogan by some of them): *Experimental philosophy*. Experimental philosophy was related to the *experimenta* of unorthodox medieval currents, and to the naturalism of the sixteenth century (including such figures as Cardano, Telesio, and della Porta). On the whole, however, sixteenth-century non-astronomical naturalism had not been mathematicized[138]; its use of mathematics had been symbolic and figurative; not descriptive and calculating. This was precisely what came to distinguish experimental philosophy from the precursor.[139] The mathematics of experimental philosophy was not necessarily of great sophistication[140]; but it was connected to actual and potential measurement, and was thus a modeling of sensible reality (as the mathematics of astronomy had been since antiquity). Often, furthermore, its modeling function came to require comprehensive mathematical deduction, i.e., direct involvement of description with the production of mathematical theory. This whole role of mathematics was something new, only anticipated by the creation of spherics and trigonometry as spin-offs from astronomy; it was fundamental for the new philosophy, and it was understood as such. Since Experimental philosophy was itself an essential constituent of early Modern philosophy, mathematics thus came to be a constituent part of its foundations.

It can easily be argued, texts in hand, that the new role for mathematics was understood by the proponents of experimental philosophy along Archimedist lines; Archimedism, once a mathematicians' private quasi philosophy, was dissolved inseparably into early Modern philosophy. But it was also adopted more directly and independently, as a general, paradigmatic "geometrical method."[141] It is perhaps no wonder that Galileo's *Discorsi* are organized in part in *theorems, problems, lemmata, corollaries,* and *scholia,* nor that Newton's *Principia* follow the same pattern with still greater consequence. But in Descartes' replies (1641) to the critics of his *Meditationes* we find that he was urged to propound his "arguments in a geometrical fashion, in order that the reader may perceive them as it were with a single glance"— which he then did perfectly, with definitions, axioms, and propositions in his *Arguments Demonstrating the Existence of God [ . . . ] Drawn Up in Geometrical fashion.*[142] Later in the century, both Spinoza's exposition of Descartes' philosophy from 1663 and his *Ethica* from 1675 (both ed., trans. Caillois, Francès, and Misrahi 1954) were *more geometrico demonstratae* as perfectly as anything, and in 1658 Pascal wrote an essay "De l'espirit géométrique et de l'art de persuader" setting forth the reasons why the geometrical method possessed this paradigmatic status—namely, that geometry complies with a method "consisting principally in two things, one of them to prove every proposition in particular, the other to dispose all propositions in the best order.[143] Evidently a perfect methodology, and according to many seventeenth-century thinkers, a goal within reasonable reach. After 500 years and many disruptions of continuous development, the phantasmagorically optimistic belief in Euclid, Ptolemy, and Nature that had once sent scholars to the outposts of Christian Europe and beyond had proved convincingly—if only temporarily—true.

# 6

## JORDANUS DE NEMORE: A CASE STUDY ON THIRTEENTH-CENTURY MATHEMATICAL INNOVATION AND FAILURE IN CULTURAL CONTEXT

### *Introduction*

*Whereas the previous essay had the character of an archaeological surface survey covering many square kilometers, this one can be assimilated to an archaeological pit dug at one place. Without changing the theoretical approach, it examines how a single thirteenth-century mathematical author appropriates and orients himself with regard to an intricate and contradictory set of ideas, aims, and scholarly norms, interpreting and recasting them in agreement with his own talents.*

*The essay as it stands has a complex prehistory, knowledge of which may facilitate the reading. A preliminary version of Sections 2–14 was presented to the XVI'th International Congress of the History of Science in 1981. These are the portions that directly discuss Jordanus de Nemore in the cultural context of his times. After this presentation I left the subject aside for some years until, after my participation in the International Symposium of Evolutionary Epistemology in Ghent in 1984, I was invited to prepare a contribution on a relevant theme for the proceedings volume. I found Jordanus's idiosyncratic reaction to his cultural background and the no less characteristic reception of his works and ideas to be a suitable topic, and worked the problem through once again. I used the occasion to address some questions raised by the appearance of a critical edition of essential sources; besides, because of the specific purpose I cut out as many of the historical details and the references to peripheral sources as possible. For the same reason, I added an introduction and an epilogue where the implicit epistemological issues are made explicit and related to the themes of evolutionary epistemology insofar as these are (i.e., were seen by me to be) relevant for the evolution of scientific knowledge systems.*

*In the end, Reidel Publishers found the volume as planned by the editors too costly and cut out some of the more bulky contributions, including mine. Instead, it was published in 1988 in a special issue of* Philosophica *(Ghent) concerned with philosophies of mathematics. This is, details apart, the form in which the study appears here.*

*Also in 1988, another version with reasonably complete historical underpinning was published in* Archive for History of Exact Sciences *(Høyrup 1988). Here, detailed discussions of and justifications for much that appears as rash assertions in the present version can be found, together with an extensive bibliographic apparatus. I shall add that several items appearing only in this full bibliography have been of fundamental importance for my understanding of the epoch and its scholars.*

*As to the relation between the following and Chapters 1 and 5, one additional remark should be made: The naturalist mood or current is mentioned directly at only one point, and its oblique appearances are few. This may astonish, given the importance of medico-astrological naturalism according to the earlier chapters. It seems to be a fact, however, that naturalism was completely ignored by Jordanus—not even implicitly recognized through deceit, veiled hints, and irony, as is the existence of the "illegitimate" mathematical disciplines. Only as a motive for the rejection of the Jordanian ideals does naturalism play a role.*

## 1. Preliminary Reflections on Principles

As a philosophically minded historian of science, I am particularly interested in the thought-style of whole periods; in the changes of such styles over time; and in the reasons that can be given for their existence and development and for the differences between the scientific thought-styles of different cultures. This tunes me sympathetically to the idea of an evolutionary epistemology. Comparison with other evolutionary phenomena, particularly the phenomena of biological evolution, may suggest analogies and models that can be tried as possible explanatory devices even for the development of socially organized and epistemologically coherent knowledge; it may also warn us by showing how many pitfalls (and which sorts of pitfalls) turn up when one tries to make causal explanations of evolutionary processes (even when "cause" is taken in the wide, quasi-Aristotelian sense of "explaining reason").

On the other hand, as an empirically inclined historian I am also suspicious of the adoption of the analogy as a full-fledged *model*—an adoption that tempts us to ask not, for example, *whether* the genotype-phenotype distinction offers something valuable to the understanding of epistemological evolution, but instead *what* are the analogies of these two levels. Furthermore, a tiny devil from my past as a physicist keeps whispering to me that when using water waves as an analogy for acoustic phenomena one should not look for foam in the microphones. It may be a sound metaphysical principle that reality is one, and that homomorphous structures can be expected to turn up at different levels; but its soundness presupposes awareness that different levels are also *specific,* and that the relation between shared and specific structures can only be known *a posteriori.*[1]

According to these reflections, the best thing I can offer to a discussion of evolutionary epistemology is probably an investigation of a concrete historical case, locating a specific innovation of knowledge and its fate in context—"internal" as well as "external," with regard to a dichotomy that I find misleading. On an earlier occasion (Høyrup 1980: 9) I have characterized the approach as "anthropology of mathematics" (since the investigation is in fact concerned with mathematics), using

> a term which suggested neither crushing of the socially and historically particular *nor* the oblivion of the search for possible more general structures: a term which neither implied that the history of mathematics was nothing but the gradual but unilinear discovery of ever-existing Platonic truths nor (which should perhaps be more emphasized in view of prevailing tendencies) a random walk between an infinity of possible systems of belief. A term, finally, which involved the importance of cross-cultural comparison.

The case study is concerned with one of the few great mathematicians of the Latin Middle Ages, Jordanus de Nemore. The choice of this figure has the advantage for our present purpose that he is completely unknown as a person; our only access to him is via his works. We shall therefore not be tempted to replace the search for explanations by the writing of biography.

At the same time, the choice of the *opus* of a medieval writer has its costs: The localization in the total directly relevant context requires that much material be brought together which is not otherwise found in one place or discipline (in order not to make things completely confusing I leave out the question of the indirectly relevant wider context, which would include the socio-economic background to the rise of the High Middle Ages' university, the overall social and epistemic organization of scholasticism, and the increasing "rationalization" of medieval urban culture in general from the eleventh century onward.[2] If the whole analysis is not to stand as a collection of gratuitous postulates, quite a few details must therefore be included in the exposition (and only because of the existence of a more complete presentation have I avoided making their number exorbitant). On the other hand, the clustering of details may serve as a reminder that evolutionary argumentation must not only come to grips with empirically established processes but even with the difficulties, to *establish* reliable empirical data pertinent to its inquiries.

The choice of a mathematical case involves advantages but also costs. Mathematics has always been considered a "hard case" for "epistemologically relevant sociology of science"—be it "internal" or "external." An investigation that exemplifies *which* features of mathematical cognition can be explained by reference to social context is thus inherently interesting. Since, however, these features turn out to concern the *organization* of mathematical knowledge rather than concrete algorithms and formulae, experimental verifications and the final test of technological applicability will be absent; this puts some constraints on possible processes of "selective retention." The case presented in the following can therefore only be one specific case, demonstrating possible but not necessary structures.

## 2. Jordanus de Nemore: Who, When, or What?

Since the mid-nineteenth century, Jordanus de Nemore has been recognized as one of the greatest mathematicians of the Latin Middle Ages. Thus, Siegmund Günther (1887: 156) characterized him as "second to only [Leonardo Fibonacci] and to no other aiming toward the same goal," whereas Moritz Cantor (1900: 53) considers the two together "the boundary posts of a new era for the mathematical sciences." No wonder that considerable amounts of effort and speculation have been dedicated to the identification of an author of such scientific merit.

Efforts expended in vain: Evidence believed to connect Jordanus to Toulouse University turns out to be a red herring, and the identification with Jordanus de Saxonia appears to build on loose rumors. No single source known to date informs us about Jordanus as a person. We are left with references to his works, and with these works themselves.

It is evident from the works that Jordanus wrote in the Latin scholarly environment; we know that he wrote before the mid-thirteenth century, when his major works turn up in the *Biblionomia*, a catalogue that Richard de Fournival, writer, physician, and chancellor at the Amiens Cathedral "well versed in mathematics," drew up of his extensive private library at some moment between 1243 and 1260. A similar time limit is furnished by a reference to Jordanus's *Arithmetica* and by extensive use of the same work in Campanus de Novara's version of the *Elements,* written presumably between 1255 and 1259.

As to the *terminus a quo,* Jordanus makes use of a wide variety of twelfth-century translations, some of which date from its later part; hence it is improbable that he can have worked before the early thirteenth century, when a reasonable diffusion of manuscripts had taken place, and impossible that he can have worked before c. 1175.

Jordanus thus worked during a critical period of the history of European learning: the period when Latin Europe had to digest the rich meal of translations made during the twelfth century, and to re-create its own system of learning, socially and intellectually, on a higher level. It is the period when the schools and guilds of *magistri* and *scholares* of Bologna, Paris, and Oxford developed into universities, and when the first universities created as such came into being in Cambridge, Padua, Toulouse, and various other places. It began with the spread of Aristotelian philosophies tainted by neo-platonism and Avicennism, and it ended with the accomplishments of Saint Thomas and Albert the Great. It was marked by a series of processes and prohibitions, beginning with the condemnation of Amaury of Bène and David of Dinant and of the Aristotelian *Libri naturales* in Paris in 1210. Later, in 1215 and 1231, the teaching of Aristotle's natural philosophy was again forbidden in Paris, and in 1229 Pope Gregory IX warned the Paris theologians not to adulterate the Divine Word by the admixture of profane philosophy.[3] In 1228, the statutes of the recently founded Dominican order—an order to which *study* was a central activity—forbade the reading of pagan philosophers and of secular arts in general, including the Liberal Arts, reviving the old Augustinian fear of secular intellectual curiosity.[4] And yet, already in 1231 the prohibition of the *libri naturales* was mitigated by a promise and an attempt to make a censored and "inoffensive" edition, and when in c. 1250 Albert the Great wrote the first part of his huge paraphrase of Aristotle, beginning with *Physica,* the "book on nature" *par excellence,* it happened according to his own words on the request of the "fratribus ordinis nostris"—i.e., on the request of Dominicans in need of a book from which they could understand Aristotle's natural philosophy.[5]

If we cannot know his father, his mother, his genealogy, or just his nationality, we may therefore approach Jordanus and the Jordanian question from another angle, more interesting, in fact, from the points of view of medieval cultural history and the dynamics of the evolution of knowledge. We may ask how Jordanus fitted into the world of learning inside which he lived, *insofar as we may judge from his works.*

### 3. The "Latin Quadrivium"

Jordanus was a mathematician, and so we shall have to take a closer look at the development of medieval mathematics up to the late twelfth century.

The early and central Middle Ages had inherited from antiquity the scheme of the seven Liberal Arts: grammar, rhetoric, dialectics (those three *artes sermonicales* constituting the *trivium*), and the four mathematical arts, arithmetic, geometry, music, and astronomy (the quadrivium). Take as witness that most cited authority of the Middle Ages next to the Bible: Isidore of Seville. Book I of his *Etymologies* deals with grammar, Book II with rhetoric and dialectics, Book III with the quadrivial arts.

The level of mathematical knowledge as presented by Isidore in the early seventh century is low—no lower, perhaps, than the mathematics of the average Roman gentleman who had been taught in *his* times the Liberal Arts, but significant in view of the fact that Isidore was probably the most learned man of his century in Latin Europe.

Isidore can hence hardly be considered a transmitter of mathematical *knowledge*. But he did transmit. He transmitted the conviction that mathematics was a legitimate and an important part of Christian culture. Thus *Etymologies* III,iv,1:

> Knowledge of numbers should not be held in contempt: In many passages of the Holy Scripture it elucidates the mysteries therein contained. Not in vain was it said in the praise of God: *You made everything in measure, in number, and in weight.*

And III,iv,3:

> By number, we are not confounded but instructed. Take away number from everything, and everything perishes. Deprive the world of computation, and it will be seized by total blind ignorance, nor can be distinguished from the other animals the one who does not know how to calculate.[6]

In the seventh century, "mathematics" was thus mainly a shell of recognition containing almost no substance. Numbers were less the subject of arithmetic than that of sacred numerology. Only Bede's writings on *computus,* i.e., sacred calendar reckoning, gave to the early eighth century the beginnings of a new tradition for elementary applied mathematics, which later met with great success.

The early "Carolingian Renaissance" meant little for filling the empty shell, except that Bede's writings were spread on the Continent. The shell itself, however, was strengthened—that is, the recognition of mathematics as an important subject was enhanced by the spread of Martianus Capella's *Marriage of Philology and Mercury* and of Cassiodorus's *Institutiones.* Due, however, to this same recognition, those mathematical works which were discovered were increasingly copied at the *scriptoria* of the reformed Carolingian monasteries in the ninth century. Thus, a number of copies of Boethius' *Arithmetic* began to circulate, and various geometrical writings were discovered and spread: excerpts from Boethius' translation of the *Elements,* and various extracts from Roman agrimensor writings (which were apparently used as a basis for Liberal Arts geometry rather than as surveyors' manuals). When monastic and cathedral school teaching expanded in the later tenth century, this came to serve.

For one thing, teaching of Boethian arithmetic by means of a board game, *rithmomachia,* was practiced at least from c. 970 onward; Boethian arithmetic must hence have been studied systematically in certain monastic schools from that time on.

Next, and more significant, a new abacus was invented, apparently under some inspiration from the Islamic world.[7] A number of treatises describing it date from the late tenth and the early eleventh century. It was combined with the agrimensor and "Boethian geometry" traditions, and so became part of quadrivial *geometry,* paradoxical as this may seem. It had intercourse with the *computus,* too.

Finally, we have direct testimony of an awakening mathematical interest by the turn of the century, both as concerns Gerbert's teaching in the cathedral school at Rheims (organized according to the quadrivial scheme) and a number of letters exchanged between various scholars.

The general growth of intellectual interests at the cathedral schools during the eleventh and early twelfth century had to a large extent the form of interest in the Liberal Arts. Theology, medicine, and canon law were estimated, it is true, but in practice they were carried by no educational program comparable to that of the basic Liberal Arts. Concomitantly, the flourishing of the "twelfth-century Renaissance" was seen by contemporaries as the apotheosis of these arts—as can be seen in all sorts of sources, from the Royal Portal of the Chartres Cathedral to humble treatises on the abacus.

By the early twelfth century, the Isidorean shell of recognition had thus been filled out by what we might call the *Latin quadrivium,* i.e., a quadrivium founded almost exclusively on the tradition of Latin Christian learning and on the innovations that had been introduced homeopathically inside the Latin Christian world: Boethian arithmetic; the "sub-Euclidean" agrimensor-geometry including Euclidean fragments, the abacus, and the figurate numbers; music, as dealt with in Boethius' *De musica;* and astronomy, including elementary theory of the celestial sphere, *computus,* some elementary treatises on the astrolabe, and some astrological *opuscula* that had made their way into Latin learning from the late tenth century onward, primarily through the contact in Lorraine. It is an open question how much of this was learnt by normal students—the quadrivium was not *à la mode,* as was dialects[8]; but much of it was *taught* in a number of schools, and some fragments and outlines were remembered by former students. In this way, the Latin quadrivium became an integrated part of the cultural traditions of the High Middle Ages' scholarly environment.

A few distinctive qualities of the Latin quadrivium deserve to be listed. First, *arithmetic* had the absolute primacy as regards prestige. This may owe much to the circumstance that Boethius' *Arithmetic* was the only really coherent treatise at hand, for all its lack of mathematical substance; it may also reflect a connection between the growing interest in mathematics and a slow and gradual but indubitable calculational rationalization of aspects of social practice[9] and the growing interest in mathematics. Second, the quadrivium was a not quite autonomous part of the global system of Liberal Arts, as it is well illustrated in the didactical introductions to Boethius' *Arithmetic:* They are stuffed with references to the verbal arts of the trivium, especially to dialectics.

A final characteristic of the Latin quadrivium was that is was recognized *not to be complete.* From its own basic authorities (Boethius, Cassiodorus, Martianus Capella, Isidore) it was known that a number of inaccessible Greek authors had laid its foundations. Pythagoras, Archytas, Euclid, Archimedes, Apollonios, Nicomachos, and Ptolemy are mentioned along with a number of others, and of these only Nicomachos was known through translations.

### 4. "Christian" Quadrivium and "Christian" Learning

That this flaw of the Latin quadrivium was really felt as one by those interested in quadrivial matters is seen in the heroic story of the twelfth-century wave of translations—a wave of which it was a motive force. I shall only quote a fourteenth-century biographical note on Gherardo of Cremona, the greatest twelfth-century translator: "Educated from the cradle in the bosom of philosophy," he became dissatisfied by the limits of Latin studies and "set out for Toledo" to get hold of the *Almagest:* he stayed there translating the Arabic treasures "until the end of life" (the whole impressive note is published in Boncompagni 1851a: 387f.).

By the mid-twelfth century, translations of the *Elements* began to circulate, and some treatises ascribed to Ptolemy were in use. When soon afterward Gherardo's translation of the *Almagest* became available, Latin Europe had at long last got a complete, "Christian" quadrivium—the term *Christian* taken without its religious connotations, as "belonging legitimately to Christian culture, to its background in antiquity, and to its school curriculum"—i.e., taken as an ethno-cultural characterization.

If we look at the non-mathematical branches of twelfth-century "Western" culture, there are good reasons to use the word in this sense (although, of course, men of the twelfth century never distinguished between a religious and a cultural Christianity). For one thing, pagan classical antiquity was the ever-present background in all branches of scholarly activity—and it was often more than background. No accident that Dante, though writing at a time when medieval scholarly culture was more autonomous than ever before or after, took Virgil as his guide through Inferno and placed so many of the grand legendary figures of antiquity in the noble castle of Limbo, that circle of Inferno which only lack of Hope distinguishes from Felicity, along with only Saladin, Avicenna, and Averroës from more recent times (*Inferno* IV).

Not only, however, were there reasons to see antiquity as belonging to one's own circle, "les gens d'ici". There was also a tendency to see Islam as a distinct, strange, and intimidating world—"les singes d'à côté." This did not regard only faith, culture, morals, and political and ecclesiastical power. Even the modern stereotypes of the Healthy West and the Unhealthy and Venomous East find their parallel in the twelfth century.

In general it is not easy to find such antagonism expressed directly in writings concerned with learning. The great twelfth-century translators seem to have been just as eager to transmit Islamic or medieval Jewish as Greek authors. Enthusiasts as well as antagonists of the new, imported learning were, so it looks, just as enthusiastic about genuine Islamic learning as about the full Greek heritage, or just as much against Galen and Euclid as against any modern pagan philosophy.

Still, at closer inspection much of the evidence suggests a hidden affinity for the Greek part of the new learning. This is revealed if we ask the questions, *who* among the many translated authorities became favorites, and *for which purpose* they were used.

Two things become clear as soon as these questions are posed. Those authors who belonged to the ancient heritage became the main authorities, and the main threat according to those suspicious of the new learning; further, Muslim authorities were most often used not independently but rather as support or commentary in connection with questions or works correctly or falsely ascribed to antiquity. There are also certain indications that ancient and Jewish authorities were considered either more reliable or less dangerous to cite than Muslims.

A comparison of the different styles of translations from the Greek and from the Arabic leads to similar conclusions. Arabic texts, even when translations from the Greek, were treated as *objects to be used*, and not always very formally. Greek texts, on the other hand, were sacred objects and handled as such; they were normally translated *de verbo ad verbum*, in spite of the impossible Latin constructions that would result. No wonder that the translations which came in fact to be used were initially those made from the Arabic.

All this rather diverse evidence appears to call for one plausible interpretation: Through the import of translated learning, Latin Europe procured for itself the fund of learning that it considered its own legitimate heritage, a fund of learning which can be characterized by the term *Christian* in the above-mentioned secular sense. The central authors were familiar by name from the encyclopediae of late antiquity and the early Middle Ages; along with their works are found others that could be considered commentaries, explanations, or continuations of the tradition.

Watt (1973: 150) goes a but further. Noticing that

il aurait été difficile, sinon impossible, de tourner le dos d'une façon purement négative aux musulmans, surtout au moment où l'on avait tant à apprendre d'eux, en science comme en philosophie,

he concludes that

il fallait trouver une contrepartie positive. Cette contrepartie fut con-stitué par le recours au passé classique, grec et romain, de l'Europe.

More specifically, he argues in connection with the spread of Aristotelianism that

les Européens étaient attirés par Aristote non pas simplement à cause de ses qualités propres, mais également parce qu'il s'apparentait, par certains côtés, à leurs traditions spécifiquement Européennes. En d'autres termes, le fait d'assigner à Aristote une place privilégié dans les domaines scientifique et philosophique peut être interprété comme une des façons qu'avait la chrétienté d'affirmer son autonomie par rap-port à l'Islam.

I shall not discuss the validity of this interpretation of the (conscious or subconscious) motives of the Aristotelians of the twelfth-thirteenth centuries. I shall argue, however, that it has a good deal to say about Jordanus and *his* motives.

### 5. The Wider Context of Mathematics

Before advancing directly toward this discussion, however, we shall need to return once more to the situation of mathematics as encountered by Jordanus.

The twelfth century had not only produced the translations needed to round off the quadrivium. Translators such as Gherardo of Cremona, Adelard of Bath, John of Sevilla, Robert of Chester, and others had translated works representing almost all branches of Islamic mathematics. Moreover, besides being known implicitly through these translations, a global picture of (Is-lamic) mathematics had also been transmitted explicitly to the West through translations of Islamic philosophy, not least through al-Fārābī's *De scientiis,* which was known from two translations (one prepared by Gundisalvo, the other by Gherardo[10]) and from Gundisalvo's syncretic *De divisione philoso-phiae.*

Al-Fārābī's scheme is a product of the old quadrivial scheme and of those developments which had taken place in later antiquity and during the earlier phase of Islamic mathematics. It divides the subject into seven sub-disciplines: *Arithmetica; scientia geometriae; scientia aspectuum* (optics/perspective); *scientia stellarum*, "science of stars"; *scientia musicae; scientia ponderosorum*, "science of weights"; *ingeniorum scientia*, "science of ingenuities" (Arabic *'ilm al-ḥiyal*, "science of artifices/stratagems/devices").

*Arithmetic* is divided in two: "Active" or practical arithmetic, dealing with concrete number, and used, for example, in commercial calculation; and "speculative" or theoretical arithmetic, dealing with abstract number—the Old "Pythagorean" arithmetic as found in Boethius, and the post-Pythagorean arithmetic of *Elements* VII–IX.

*Geometry* is also divided in an "active" and a "speculative" branch. Active geometry deals with the geometry of physical bodies, exemplified in the text by timber, iron objects, walls, and fields. Speculative geometry deals with abstract lines and surfaces; like speculative arithmetic, it is said to lead to the summit of science. It investigates lines, surfaces, and bodies, their quantity, equality or being-greater, position, order, points, angles, proportionality and disproportionality, "givens" and "not-givens" (cf. Euclid's *Data*), commensurables, incommensurables, rationals, surds, and the various species of the latter (cf. the classification of irrationals in *Elements* X).

The treatment of geometry is rounded off with the remark that arithmetic and geometry contain elements and roots, together with other matters derived from these roots, and that the roots are dealt with in the *Book of Elements* by "Euclid the Pythagorean."

*Perspective* "deals with the same things as geometry," but in lesser generality. The details are irrelevant to Jordanus.

The *science of stars* also consists of "two sciences": one dealing with the signification of the stars for future events, and one that is "the mathematical science of stars." In other words, predictive astrology is not counted among the mathematical sciences (and in fact, it is also said, only counted as a means for judgment and not as a science). The mathematical theory includes, apart from genuine mathematical astronomy, Ptolemean mathematical geography.

*Music* is, like perspective, irrelevant to Jordanus.

The *science of weights* is treated in only seven lines. One, basic, part is the theory of weighing and statics. The other considers the movement of weights, and the means to move them; it investigates the foundations of the instruments by which heavy things are lifted up and on which they are moved.

The *science of ingenuities* is the preparatory science for the application of the preceding disciplines "to natural bodies, their invention and their rest and their working." The term *ingenio* can hence be read in its double Latin sense, as "cleverness" and as "instrument" (in agreement also with the Arabic text)—we might speak of the discipline as "engineering" or as "applied theoretical mathematics." One subdiscipline is *ingenia* of numbers; of these there are several, including *algebra et almuchabala*, although "this science is common for number and geometry." Under the heading of algebra comes the whole theory of surds, "both those roots which Euclid gives in Book X of the *Elements* and that which is not dealt with there." Geometrical *ingenia* are several, too, the most important being construction of masonry. To this genus belongs even mensuration of bodies and "the art of elevating instruments, of musical instruments, and the instruments of still other practical arts, like bows and sorts of arms." Equally, "the optical *ingenium,* the art which directs our view to comprehend the truth of things seen at a distance, and the art of mirrors," further the science of burning mirrors; the *ingenium* of the art of very great weights; etc. [not specified]. In the end the contents of the discipline are summed up as the *principles of the civilian practical arts,* like those of masons and carpenters.

Most of the translated mathematical works can be classified under these headings. Often, of course, specific works belong to subdivisions that go unmentioned. So *algorisms,* treatises explaining computation with Hindu numerals (in certain cases expanded to include some commercial arithmetic or algebra); so also spherical geometry (including the theory of the astrolabe) and trigonometry, both belonging with mathematical astronomy.

Even though the substance of al-Fārābī's arithmetic, geometry, astronomy, and music is to be understood in much greater depth than the corresponding Latin disciplines, they were familiar as members of the traditional quadrivial family. Other parts of the import were known, from Aristotle or from attributions to ancient authors, to belong to the ancient heritage (perspective and its subdivisions; science of weights; and spherical geometry).

Some of the imports, however, had to be recognized as genuine non-"Christian" contributions: Algorism was, by the very words of the basic treatise introducing the subject, to be recognized as calculation with Indian numerals; in other treatises that kept silent about the Indians a foreign symbolism could not be covered up in spite of all philosophical commentaries and technical terms borrowed from the schools and the abacus tradition. And *algebra et almuchabala* was equally foreign, by name and by contents.

### 6. Jordanus and Illegitimacy. The Jordanian Corpus

In general, the mathematical writings of the twelfth and thirteenth centuries betray no worry about the existence of non-"Christian" mathematical disciplines. Neither is there any explicit expression of dismal feelings in Jordanus. Still, the total body of his works displays a structure that strongly suggests that he was not only aware of the state of things but also determined to do something about it.

This is a bold statement: First because the canon of Jordanian works is not definitively established; second because several of the definitely Jordanian treatises exist in two or more versions.

To overcome the first problem, I shall build my argumentation solely on such works as can be ascribed with certainty to Jordanus (for which I follow the list established by Ron B. Thomson [1976]). Problems arising from the existence of different versions I shall take up case by case. In order to locate Jordanus's work with respect to an advanced contemporary understanding of the composition of the science of mathematics, I shall classify the writings under al-Fārābī's headings.

### 7. Arithmetic

Like al-Fārābī, we can distinguish practical and theoretical arithmetic. Concerning the first of these subdivisions, one cannot maintain that Jordanus wrote really practical arithmetic. But he wrote some treatises on a subject belonging to the domain, *viz.* on algorism. There exist four treatises ascribed to Jordanus and dealing with algorism, two on integers and two on fractions.[11]

There can be no reasonable doubt that Jordanus wrote at least one treatise from each set (I shall argue below that he wrote all of them). There are important differences of style between the two sets, but they have one thing in common: Their aim is not to *teach the art of algorism,* as do other algorism treatises. Instead, they are to *demonstrate from mathematical principles that what is currently done in the art is in fact correct.* In the *Biblionomia* (N° 45), Richard de Fournival speaks of "Jordanus de Nemore's apodixis on the practice which is called algorism." In the following, where we are repeatedly to meet the same genre, I shall borrow Jordanus's own term and speak of *demonstrationes.*

Parts of the *De numeris datis* (see below) can also be regarded as *demonstrationes* of standard methods and problems belonging to commercial and "recreational" arithmetic (e.g., the rule of three and the "purchase of a horse").

Under the heading of theoretical arithmetic belongs one of Jordanus's major works, the *Arithmetica,* presented in the manuscripts as *Elements of the Art of Arithmetic* and on the title page of Lefèvre d'Étaples' printed edition from 1514 as *Arithmetica demonstrata.* Both titles are justified. The whole domain of theoretical arithmetic, Euclidean as well as Boethian, is collected and ordered in this work; the exposition is founded upon definitions, *petitiones* (postulates) and *communes animi conceptiones* (''common notions'' or axioms), and it is organized deductively in propositions.

All propositions are provided with demonstrations. In these, Jordanus makes (sparing) use of letters as representatives of unspecified numbers, instead of the exemplifying numbers used by Boethius. The flavor of the work will appear from proposition I.12:

> *If a number measures the whole and what is detracted, it also measures the remainder.*
>
> For instance, if $c$ measures $ab$[12] and $a$, it will also measure $b$. Indeed, it measures $ab$ as $de$ and $a$ as $d$. Now since $de$ times $c$ is as much as $d$ times $c$ and $e$ times $c$ according to proposition 9, $a$ together with $e$ times $c$ makes $ab$. By the [common] notion, $b$ will thus be as much as $e$ times $c$. Therefore, if $c$ measures one, it also measures the other.

The *Arithmetica* is hence a *demonstratio,* as claimed in the printed edition. But it is more than that: The ordering of the material in the 10 books, the start by definitions, etc., the coherence and the progressive construction of the presentation—everything shows that Jordanus has tried to write the *Elements* of arithmetic, and hence to elevate his subject, which in the demonstrationless Boethian version had formed the apex of the Latin quadrivium, to the level of Euclidean order (hence providing a basis for a ''legitimization'' of algebra; cf. section 12).

## 8. Geometry

One major work belongs to this field: The *Liber philotegni,* the *philotegnus* of which is probably to identify either author or user as a ''friend of the art'' (as *philosophus* is ''friend of wisdom'').

Before the genuinely geometrical contents of the work comes a metatheoretical introduction, defining (in a way related to the Aristotelian and not to the mathematical tradition) ''continuity'' ( = manifold): ''single,'' ''double,'' and ''triple'' continuity, ''right'' and ''curved'' continuity, and so on. The definitions are not very coherent, and they are not used in the main text, which instead draws heavily and extensively on Euclid and on a number of other works in the Archimedean tradition.

The work is concerned with advanced and occasionally somewhat specious problems, organized in a loose deductive scheme in series of propositions concerned with similar problems.

Whereas most of Jordanus's treatises contain no clues as to their institutional origin, a few clues are provided here, pointing to the environment of an arts faculty. First, of course, there is the "friend of the art" of the title—but in itself, this could hint simply at the "art of geometry." More decisive are the preliminary definitions. They have no bearing on the mathematical substance of the treatise, and their style is opposed to Jordanus's normal habits (and hence presumably to his personal inclinations). They may be present because he felt the need to clear away in advance some metamathematical questions on the status of those plane figures with which he is concerned in the following text. Such a need must probably be connected either to audience expectations, or to Jordanus's own expectations concerning the interests and background of his audience—and the foundations on which he builds the explanations are not related to the Euclidean commentary tradition (nor, for that matter, to the old Latin tradition), but instead to that Aristotelianism which was gradually rising to hegemony at the arts faculties during the first half of the thirteenth century.

Another hint is the repeated occurrence in the text of a peculiar figure, a *falsigraphus,* who has to be brought to silence. Originally, he descends from Aristotle's *pseudographos* (cf. Chapter 5, note 32), but he is re- (and mis-)interpreted here (and in other contemporary works) as the opponent who has to be reduced *ad absurdum*—a truly familiar person in the medieval university disputation.

A final piece of evidence is inherent in a *Liber de triangulis Jordani.* This has long been regarded as another name for a long version of the *Liber philotegni,* but Marshall Clagett's critical edition of both works (1984) showed them to be different, though related. Furthermore, his analysis demonstrates that the *Liber de triangulis* was not written by Jordanus.

At closer stylistic analysis of the work, a wealth of evidence turns up that it was not written on the basis of a manuscript of the *Liber philotegni* as it is known to us. Instead, it contains (aside from its last part) a student's report of a series of lectures (a *reportatio*) held over an earlier version of the *Liber philotegni.* The lecturer *need* not have been Jordanus himself—but *if* the two persons differ, the lecturer must have had access to Jordanus's own work before it reached its final form as we know it. He must, so it seems, have been a member of that "Jordanian circle" to which I shall return.

The oral style of the work has conserved some polemical remarks that have no counterpart in Jordanus's written works. The plausibility that the lecturer was either Jordanus himself or some student of his suggests that the polemics reflect Jordanus's own attitudes—and in fact they correspond well to those attitudes which are expressed indirectly in Jordanus' writings. Noteworthy are two references to the assistance of the Almighty; as they stand they can hardly be meant as anything but parody of the Muslim habit of invoking God not only in the beginning and the end of mathematical treatises but often even in the middle of the text. Taking over the Islamic level of mathematical rigor and proof, Jordanus and his circle stay culturally aloof.

## 9. Perspective

Nothing but a spurious ascription connects Jordanus's name to this discipline. The last part of the *Liber de triangulis Jordani* (which is not part of the *reportatio,* but consists perhaps of notes collected by the lecturer but not covered in the lectures) shows, however, that ibn al-Haytham's extensive work was known in the Jordanian environment.

## 10. Science of Stars

Nothing in Jordanus's writings reminds even slightly of a reference to astrology.

Even in astronomy, he is not interested in the movement of the celestial bodies. Still, he has contributed to the discipline—but once more his contribution turns out to be a *demonstratio.* The work in question is the *De plana sphaera,* a treatise on the design of the astrolabe, the favorite astronomical instrument of the time.

The relation of the treatise to the practice of the art is, *mutatis mutandis,* pinpointed by Joan Robinson's description (1964: 25) of the relation between economists and businessmen: "It is the business of the economists, not to tell us what to do, but to show why what we are doing anyway is in accord with proper principles." Since antiquity, indeed, the stereographic projection had been used in the astrolabe, and Ptolemy had stated that the equator, the tropics, and the ecliptic all projected as circles—but he had given no proof of this, nor had any proof reached the West. Jordanus's treatise teaches nothing different—but he shows Ptolemy's assertion to be truth "in accord with proper principles."

Furthermore, where Ptolemy's *Planispaerium,* the standard treatise on the astrolabe, deals with the location of particular celestial circles, Jordanus generalizes, never speaking of the celestial sphere, equator, tropics, meridians, or ecliptic. The problems of the astrolabe are used as the occasion (not to say the pretext) to construct a general theory of the stereographic projection.

This is a move back toward the style of ancient spherics. This discipline, too, had been a pure mathematical science, the applicability of which was in principle left to the reader to discover. Islamic workers in the field had normally been dissatisfied with this opaque purity and had made their own versions of the works, referring constantly to the material implications of the theorems. In view of his broad knowledge, Jordanus must be assumed to have been familiar with these Islamic departures from ancient style and ideals, and his own style can then be seen as a conscious reaction or correction to this deviation from ancient norms.

The *De plana sphaera* exists in three versions. In versions II and III, foreign hands have inserted explanations of the "astrolabic" implications of the theorems. The way back to ancient ideals accomplished by Jordanus was thus traversed immediately in opposite direction by his adaptors.

### 11. Science of Weights

This is the sole field where Jordanus's work gave rise to a genuine school. One work (*Elementa super demonstrationem ponderum*) can be ascribed with full certainty to Jordanus, and another (*De ratione ponderis*) with almost full certainty. The purpose of the *Elementa* was summarized by Joseph E. Brown (1967a, abbreviated from 1967: 3f.):

[ . . . ] the catalyst for the medieval "auctores de ponderibus" was a short treatise of Hellenistic origin on the steelyard, *De Canonio*. The Greek work appealed to other treatises, lost since antiquity, to justify its assumptions. Jordanus attempted to supply the missing justification in an exceptionable, nine proposition treatise that often precedes the *De Canonio* and comes to be joined to it. Its starting point was an Aristotelian, dynamic principle applied to the balance and its high point a mathematical statement of the principle of work to prove the inverse law of the lever. Jordanus seems to have worked hastily and loosely so that his treatise became a further catalyst to commentators.

The title of the work, including the terms *elements* and *demonstration,* betrays the usual Jordanian concerns and agrees well with Brown's account. It seems as if Jordanus did here exactly the same thing as he did by his *Arithmetica* for algebra (except that in the case of algebra he had to re-create that discipline in order to obtain agreement between the arithmetical "elements" of the field and their advanced, algebraic applications—cf. the next section).

The *De ratione ponderis* is a more rigorous extension of the *Elementa*.

### 12. Science of Ingenuities

Jordanus was no engineer. But he certainly wrote works belonging under the heading *ingeniorum scientia* as al-Fārābī understood it: not "practical" mathematics but "theoretical mathematics" especially prepared to serve applications if practitioners would want to apply it. Thus, the *De plana sphaera* could be classified as an astronomical *ingenium;* and in the *Liber philotegni* one finds a series of propositions on circular arcs (growing out from the trigonometry of chords) that comes to serve both in Jordanus's own *Elementa* and in the *Liber de motu* by one Gerard of Brussels, in spite of its apparently pure character.

Both of these cases stand, however, at the boundary between the disciplines where I have discussed them, and the *ingeniorum scientia.* I guess thirteenth-century readers of al-Fārābī would have classified them as I did. One of Jordanus's major works belongs, however, to a discipline that al-Fārābī locates unambiguously as an *ingenium:* the *De numeris datis,* which is not only a work on algebra but also a work written in veiled dialogue with the Islamic algebraic tradition.

To a first inspection it does not look so. Even the title, recurring in every proposition, places the work in the tradition of the Euclidean *Data.* The propositions are, with a few variations of form, of this sort: If for certain numbers $x$, $y$, . . . certain arithmetical combinations. $C_1(x, y, . . . )$, $C_2(x, y, . . .)$, . . . are equal to given numbers, then the numbers themselves are given. Everything appears to be concerned with generality, solvability, and uniqueness, not with solution for particular values of the given numbers. In many cases the discussion is even stopped at the point where the problem has been reduced to one previously dealt with. Only as illustrating *examples* are the generalizing propositions followed by numerical algebraic problems approaching those known from other algebra treatises from the time.

The same interest in solvability and generality is found occasionally in Abū Kāmil and Leonard Fibonacci, but only as exceptional deviations from the normal style. *Their* algebraic works, and all others that could be known to Jordanus, concentrate on the solution of particular problems, aiming at conveying thereby an *implicit* understanding of the general method.

In his methods, too, Jordanus follows other paths than those of the Islamic algebrists. Neither their rhetorical exposition nor their fixed algorithms for the solution of standard cases (nor, for that matter, the geometric methods used to justify these fixed algorithms) can be traced in his text; instead, Jordanus uses the letter symbolism and the propositions of his *Arithmetica.* Obviously, *his* work on given numbers bears the same relation to his arithmetical *Elements* as Euclid's *Data* to the *Elements* of geometry. It is no simple further step in the track of an already established algebra tradition.

Still, a closer investigation of the text turns much of this upside down. A fair number of problems and problem types, including rather specious problems and the specific values of the given numbers in the illustrating examples, coincide with well-known problems from the Islamic tradition. The possibility of random coincidence can be safely disregarded. Jordanus appears to have known not only al-Khwārizmī but also Abū Kāmil and probably even Leonard Fibonacci. His omission of all names (and even of the name of the discipline) is striking.[13]

At the same time, the fair number of full agreements should not make us forget the still larger number of propositions that have no antecedent. Jordanus does not merely offer an alternative formulation of a set of cherished problems, or a *demonstratio* of current algebra. As in so many other cases, he builds up a coherent structure of his own, in which his new formulations of old problems suggest analogies and extensions that had not been scrutinized before.

Summing up, we may characterize the *De numeris datis* as follows:

First, it is a *demonstratio* consolidating the current usages of a practice—just like the algorisms and the *De plana sphaera*. Whether the traditional methods are considered sufficiently supported by arguments may be a matter of taste and epoch; but there is no doubt that Jordanus's treatment of the subject is *more* rigorous.

At the same time, Jordanus's treatment is also a *transformation* of the subject. The ordering of the material in most precursors is rather confusing, when not chaotic. If not comparable to the strict progress of Euclid's *Elements*, the *De numeris datis* supports comparison with his *Data*. The *De numeris datis* thus transforms a mathematically dubious *ingenium* into a genuine piece of mathematical *theory*.

Finally, it is worth noticing which sort of theory is created. As already pointed out, algebra is brought under the sway of theoretical arithmetic, the discipline *par excellence* of the Latin quadrivium—whereas for al-Fārābī algebra was "common for number and geometry," as we remember. Algebra, a bastard by cultural as well as by disciplinary standards, had been legitimized (in such a way that only those who explicitly wanted to could trace the illegitimate origin through the numerical illustrations). In addition, theoretical arithmetic itself was brought still closer to classical ideals. Jordanus had already given it its *Elements;* now it was also provided with its *Data.* In spite of the new, shattering claims on mathematical rigor imported during the twelfth century, *arithmetica* could still claim to be second to none of her sisters.

## 13. Algorism Revisited

This finishes the list of indubitable Jordanian works. Before discussing the overall character of the Jordanian *corpus* we shall, however return for a moment to the algorism treatises, certain features of which are better discussed at the end of the list than at its beginning.

It will be remembered that two sets of algorism treatises exist, one short and one long. A content analysis shows beyond doubt that the short set is original and the long set a revision.

The curious thing is that it is the revised version which looks Jordanian. Most striking is the construction of the proofs. The structure of the proofs is the same in both sets—but whereas the revised recension makes use of Jordanus's normal letter symbolism, this technique is lacking (and wanted!) in the first recension, except in cases where, for example, an unidentified five-digit number is represented as *abgde*. It looks as if the original recension is a juvenile work, already containing in germ the ideals that are expressed in the mature *opus* (cf. below), but not yet in possession of the techniques permitting it to live up to these ideals. When the techniques had been created, Jordanus returned to the subject and rewrote the treatises as he was now able to do.

This interpretation is, of course, tenable only if the short treatises are indeed from Jordanus's hand; the best argument that they are is that they express *specific* Jordanian ideals (together, of course, with the ascriptions to Jordanus).

This can be seen at different levels. First, the basis for the proofs is, as in other works, Euclidean theoretical arithmetic. Second, the treatises are *demonstrationes*, but also more than that. The tendency to substitute for the mere demonstratio of a practice a mathematical generalization is already found in the first treatise on fractions. It does not argue mathematically for the usual manipulations of sexagesimal fractions, no more than the *De plana sphaera* argues explicitly about the heavenly sphere. Instead it invents a generalization of the sexagesimal fractions, replacing the powers of 60 as denominators by the powers of any integer ($>1$). This generalization, moreover, is made non-trivial by a further generalization opposing these "consimilar" fractions to "dissimilar" fractions, where the denominators are not the series of ascending powers but a gradually extended product. These are in fact nothing but the *ascending continued fractions* known from works translated from the Arabic ("two fifths and one third of one fifth" instead of "seven fifteenths"—in general numbers of the form $n_1/P + n_2/(PQ) + n_3/(PQR) + \ldots$). The author of the short treatise thus creates a theoretical framework encompassing not only the legitimate sexagesimal fractions but also the more dubious *partes de partibus* usage, thereby assimilating the latter into the "legitimate" tradition.

That legitimization is indeed aimed at is demonstrated beyond doubt by the revised recension. There, a large section of the introduction explains the concept, not by reference to the Islamic treaties where it belongs, but through reinterpretation of material from the Latin (Isidorean and *computus*) traditions, where it had never been in use.

A similar trick is used in the short recension in order to legitimize the whole dubious subject of algorism. The introduction to the short treatise on integers refers to the ancients and contains large passages in the Boethian tradition. "Algorismus" himself (i.e., al-Khwārizmī) is bypassed in silence, reappearing only later when he is given credit for a specific subtlety (as if only this trick were of gentile origin—a strategem that was also used in the *De numeris datis*).

In one way, several of the algorism treatises depart from normal Jordanian stylistic ideals, *viz.*, by possessing long, discursive introductions—the first part of the introduction to the short treatise on integers speaking, furthermore, directly to the audience as "you" in flourishing, classicizing Latin. This feature, like the lack of letter symbols in the short recension, suggests an early origin for the treatises. It would be no wonder if a young teacher in the beginning of his career were still impregnated with the stylistic ideals of this environment—and these were, like the introductions to the short treatises, discursive and heavy with Boethian phraseology, and classicizing (not always as successfully as here) in dedicatory and similar formal passages.

As in the case of the *De plana sphaera*, Jordanus's followers appear to have missed his specific points. One "Master Gernardus" wrote an *Algorismus demonstratus* that used Jordanus's technique of proofs, and which became quite popular. Significantly, however, he reduced the scope of the work to that of a *demonstratio:* He dropped the dissimilar fractions, and instead of consimilar fractions in general, he dealt with the specific sexagesimal fractions.

### 14. Jordanus's Achievement

We have now finished our tour through the Jordanian *corpus*. The stylistic landscape was not quite homogeneous: discursive and Boethian in the early algorism, slightly shaped by the philosophical context in the *Liber philotegni,* terse in most other works. Still, a coherent picture appears to emerge, a picture that would not change even if a few dubious items had been included in the touring plan.

The important general features of the picture are as follows.

1. Jordanus wrote *demonstrationes:* Works that, in Joan Robinson's words, give mathematical reasons why "what we are doing anyway is in accord with proper principles." Normally, however, a Jordanian *demonstratio* is generalizing and abstract, hinting little if at all at the specific applications—cf. feature 3.

2. Jordanus wrote *elements,* works constituting a coherent mathematical basis for a theoretical subject.

3.  Jordanus made pure mathematics—or, better, *theoretical* mathematics, in the ancient sense repeated by al-Fārābī: mathematics considering abstract principles, not concrete applications or sensible entities.

4.  Jordanus worked on subjects belonging under most of al-Fārābī's subdivisions, i.e., over the whole width of mathematics as understood in his days. But everywhere he would seek out what called for *demonstratio* and *elements*, and regardless of the starting point and the occasion the result would be theoretical mathematics.

5.  Jordanus followed the mutual ranking of the mathematical disciplines current in the Latin quadrivium, putting *arithmetic* at the top of the scale. Methodologically, he pursued the standard of Euclidean rigor and organization that had been set by the translations of the twelfth century.

6.  Combining features 1 to 5, Jordanus succeeded in subsuming the main mathematical currents at hand under the ideal conception of the Latin mathematical tradition, "legitimizing" thereby all dubious subjects.

7.  On the other hand, subsuming all important currents and raising arithmetic, that traditional *pièce de résistance* of the Latin quadrivium, to the best methodological standards, he made the structure of legitimate mathematics competitive on any intellectual market, in agreement with the motto "anything you can do, I can do better."

Motto apart, there is little doubt that Jordanus was conscious of most of this. He need not have *pursued* competitiveness—he may just as well have made mathematics as he thought mathematics should be. But in a scholarly environment more impregnated with metatheoretical discussions than with genuine mathematical innovation he must surely have known that he was pursuing a particular standard.

Thus, Watt's explanation of the attractiveness of Aristotle seems certainly relevant as an analogue when Jordanus's work is concerned, especially regarding feature 7.

Feature 6, on the other hand, reminds of another aspect of the reception of Aristotle: the papal order of 1231 to prepare an inoffensive edition of Aristotle's books on nature (cf. section 2):

[ . . . ] since, as we have learned, the books on nature which were prohibited at Paris [ . . . ] are said to contain both useful and useless matter, lest the useful be vitiated by the useless, we command [ . . . ] that, examining the same books as it is convenient subtly and prudently, you entirely exclude what you shall find there erroneous or likely to give scandal or offense to readers, so that, what are suspect being removed, the rest may be studied without delay and without offense. (trans. Thorndike 1944: 40)

The papal committee never set its pen to paper. The legitimization of Aristotle was only carried out by Albert the Great and Saint Thomas with a delay of some 20 years; anyhow, *their* achievement put a decisive mark on Latin Europe for centuries.

In mathematics, no books had ever been forbidden. Algebra and algorism had never given rise to learned heresies as had the reading of Aristotle. Still, *mutatis mutandis,* Jordanus produced the mathematical analogue of the Albertian paraphrases and the Thomist synthesis. Like theirs, his work is an exponent of the great cultural conflicts of the thirteenth century.

### 15. Failure and Its Reasons

And yet, in spite of this apparent harmony with the dominant mood of his times, Jordanus's influence was not only small compared with that of the Great Dominicans—it was virtually absent.

Admittedly, a "Jordanian circle" existed in Paris at some period during the first half of the thirteenth century. In all probability, the Master himself was at its center, and several well-known scholars must have been members or have had contacts with it[14]: Richard de Fournival, Campanus of Novara, and Roger Bacon; to these come Gerard of Brussels, author of *De motu;* presumably that "Master Gernardus" who wrote the *Algorismus demonstratus;* the student who wrote down the *Liber de triangulis Jordani;* and perhaps even the authors of some commentaries to Jordanus's works on statics and the adaptors of *De plana sphaera* II and III. Furthermore, Jordanus's works gave rise to a minor current that was never quite forgotten; the *Algorismus demonstratus* may have gained a certain popularity; and throughout the Middle Ages, the *science of weights* was marked by its Jordanian beginnings. Nonetheless, the Jordanian current remained insignificant: Even in Paris, the original domicile of the Jordanian circle and the place where Jordanus himself must in all probability have worked, the rumors reaching in the late thirteenth century the ears of an ordinary student reflect only his works on statics and, probably, the *Liber philotegni.* During the first 50 years of bookprinting, only one of his works was printed in a single edition (the *Arithmetica*); Boethius' *Arithmetic* was printed thrice, whereas Albert of Saxony is represented by 13 editions (10 of his *De proportionibus*), and Ptolemy by 12; outside mathematics and related subjects, Albert the Great is represented by 153 editions.[15]

Worse: Jordanus' rare followers missed his specific point. They accepted his theorems, the letter symbolism for numbers, and his rigor (I disregard Roger Bacon, who missed even this part of the message)—but the spread of versions II and III of the *De plana sphaera* and of the *Algorismus demonstratus* shows that they did not share his care for ancient purity and standards. Richard de Fournival and Campanus may have grasped Jordanus's project—but they demonstrate in their own works and interests that if they understood his ideals, they were not inspired by them.

In part, the insignificance of Jordanian mathematics is of course to be explained through a relative lack of interest in mathematics in medieval learning from the mid-thirteenth century onward. Even inside mathematics, however, Jordanus was a relatively lonely figure, and at that ill understood by his followers (as we have seen). How could such a situation arise?

If one reflects once more upon the style of Jordanian mathematics, it becomes obvious that Jordanus was lonely for good reasons. Those mathematical works which really gained popularity from the late twelfth century onward were discursive and metatheoretical—precisely in the vein of an introduction to the algorism on integers that Jordanus deleted when revising the treatise. Such is the character of the popular ("Adelard II" and Campanus) versions of the *Elements;* such is also the character with which the non-Jordanian revisions of the *De plana sphaera* tried to impregnate the work, in order to make it agree with current style and intellectual demands. The evidence can be multiplied *ad libitum.*

Those *Elements* which spread widely did so because they were *aimed at* and *fit for an actual social community,* a quasi profession the demands of which it met: the community of present and former Masters of arts. As it appears from his first algorism introductions, Jordanus had grown out of this community (from where else should he have come?), and he had been impregnated by its ideas and ideals. But growing, he had grown away from it. His mathematics soon stopped being didactical and philosophical; it became *pure mathematics* almost of the ancient brand—cf. even his references to the ancients in the early didactical introduction. His standards were defined by those ancient mathematical works to which he had access, reflecting hence the needs and the structure of a social community long since deceased: the circle of corresponding mathematicians connected to the Alexandria Museum and schools (cf. Høyrup 1980: 46). Seen from this angle, Jordanus stands out as a Don Quixote, fighting a fight in which nobody really needed his victory. But even a Don Quixote expresses the ambiguities of his time, though he seem an anachronism—and in his striving toward anachronistic ideals Jordanus expresses another aspect of Western medieval culture: the constant longing for a *renaissance* of ancient splendor and the ever-recurring tendency to produce "renaissances" by every revival at least from the seventh-century Visigothic prime onward: "Carolingian Renaissance," "Ottonian Renaissance," "twelfth-century Renaissance"—and finally the *real* Renaissance. Jordanus's loneliness expresses symptomatically that the cultural glow of the thirteenth century (and especially that of university learning) was less of a renaissance, i.e., less of a digest of ancient culture on the conditions of the day, than most medieval flourishings, and correspondingly more of its own (just the reason why the real Renaissance had to react so sharply against its vestiges)—whereas on the other hand Jordanus *was* of real renaissance character.

## 16. Epilogue: Epistemology

It is not customary that miners follow the iron ore to the forge in order to work it up, nor are smiths necessarily the ones who know best how to use the tools they produce. Similarly, I may not be the one who sees most implications of the preceding study for understanding epistemological development, or who can state them with greatest sharpness and clarity. Still, since the study was made as an investigation of the evolution and molding of knowledge as a socio-historical process, I may perhaps be permitted to point out two sets of implications that seem important to me.

The first has to do with the very character of our concept of knowledge. It can be stated in the aphorism that *the way we know is an integrated aspect of our knowledge.* This may seem a rather empty truism, but it is my claim that it is neglect of this truism which has led to the consideration of mathematics as "the hard case" for sociology of scientific knowledge.

To exemplify this, we may consider algebra. If we concentrate exclusively on the "contents" aspect, we will, for example, be led to collect all "homomorphous" second-degree problems into a common, featureless heap, whether they deal with concrete or abstract number, or with concrete or abstract geometric quantity; we will also be led to identify all treatments of such problems where only the solutions follow algorithms containing the same numerical steps. As long as we accept that the number of dots in the pattern

is the same as the number in the pattern

irrespective of time and culture (I do), this opens no place for sociological understanding of the evolution of knowledge; the process becomes one of *discovery* of discovery-independent truths, and only the cadence and the gross direction of voyages of discovery become problems open to sociological attack.

In Jordanus's case we see that the concentration on the aspect of contents would have made it impossible to discover the most fundamental differences between the *De numeris datis* and the Islamic algebras. It would have closed our eyes to the elevation of the subject to ancient, "Euclidean" rigor, and to the exclusive subsumption under arithmetic; these, however, were precisely the aspects that had general parallels in scholastic culture, and which hence called for explanations through sociocultural forces.

This could in itself be said to be of minor importance for a theory of the evolution of mathematical *knowledge*—unless the quest for coherence and system (itself a Euclidean and hence a "formal" feature) had led Jordanus to take up a number of problems that had not been investigated by his precursors. Thus, innovation in content is a consequence of the organization, shape, and metatheoretical ideals of mathematics, and hence *indirectly* of those forces which shape these; it is the search for direct explanations which turns out to be a hard case (and often impossibly hard).

This two-level approach to knowledge may be insufficient in the case of empirical and technologically testable sciences (and in definitely non-empirical sciences such as theology and in the pseudosciences—I shall not venture sharp distinctions in this delicate field). But it can be taken for granted, I assume, that a *one-level* approach to any system of socially organized and epistemologically coherent knowledge is as impossible as in the case of mathematical knowledge.

It can also be observed that the two-level description is an approximation that hardly deserves as much as the (amply misused) name of a "model." Here as elsewhere, an Aristotelian distinction between "form" and "contents" helps us remember to—distinguish; and here as elsewhere we must be aware that we should not stop at a distinction by simple dichotomy, not even dialectical dichotomy.

Before, I spoke of the explanation of epistemological evolution through unspecified "sociocultural forces." No study can claim to discover and describe all effective forces, nor can a single case be expected to display all generally important types of "force." What can be done through a case study is to pinpoint some forces that in the specific case appear to have been fundamental, and to investigate their interplay. This is, I hope, what I have done in my study of Jordanus. The presentation was kept close to phenomena, and therefore a short systematic summary may be useful. My second set of implications is simply that such forces *can* be important, and that matters are *at least* as complicated as this if we want to understand the evolution of scientific knowledge.

Jordanus was part of a general culture, and shared a number of its values. The longing for a "renaissance" and the awareness of an "ethnic" (and maybe religious) "Christian" identity appear to shape his whole project and to pop up in the few scattered non-mathematical remarks of the works. But he expresses the general values in specific form (everybody does!), and *some* widespread attitudes and values appear to have been explicitly discarded. Thus, in contrast with the tendency of his age to adore relics (secular as well as sacred relics) he did not regard (or at least did not treat) the works of ancient authors as sacred. *His* renaissance was to be a fresh birth, no resurrection.

Jordanus was also part of a scholarly culture, and must have been brought up in a particular professional environment—that of incipient universities. Here again, he shared some values and rejected others. His acceptance of the norms of the "Latin quadrivium" is an acceptance of traditional values, but also a rejection of contemporary tendencies to displace the quadrivial arts through Aristotelian philosophy as the summit of secular learning. But also his implicit definition of himself as a "pure" mathematician was a rejection, namely of the current tendency to integrate all Liberal Arts; his replacement of traditional Latin mathematics with works oriented toward ancient standards was an acceptance of the "new learning" and a rejection of scholarly traditionalism, whereas his suppression of all traces of illegitimacy goes against the normal habits of that same new learning (although hidden affinities of the same character were revealed in its general pattern). Whatever he did can hence in some sense be said to be "wrong" with respect to some forces in his professional environment and "correct" with respect to other forces. The way he oriented his work in this contradictory situation can be assumed to depend in part on his personal history; but it was certainly also dependent on his *access to and understanding of previous knowledge.*

Indeed, Jordanus was a great mathematician. He was able to read from the classics not only their theorems or specific techniques but also a global view, and to assimilate a way of thinking developed in response to a professional social situation that had disappeared many centuries ago. He was able to read the ancient works *as mathematics,* with all the implications of an ancient or modern understanding of that word—prominent among which is a refusal to read those same works as *authorities* to be accepted on faith.

This fascinating understanding of the possibilities of mathematics will certainly have been an important factor for Jordanus when he oriented himself, consciously and unconsciously, with relation to the contradictory forces and claims of his environment as to the nature of "good" knowledge and style of thought. On the other hand, Jordanus was no Archimedes dropped accidentally in a High Medieval cradle or arts faculty. Even when understanding ancient mathematics *as mathematics* he understood it in part on premises of his time. It will have been his medieval upbringing that fastened his thought so firmly to arithmetic that he was able to build up a Euclidean understanding of that field and to develop an adequate apodictic technique dealing with general, unspecified numbers directly and not through geometric mediation.

If we relate this observation to the biological metaphor of evolution we see that past knowledge cannot be given the status of a fixed gene pool. Knowledge cannot function in the evolutionary process as long as it stays up in Popper's Third World. It will have to *be known* by somebody, and nobody—be he as oriented toward ancient models as Jordanus—is able to know in complete abstraction from the conditions of his own upbringing, where values, connotations, and color were given to his conceptual world. The process can be said to possess neo-Lamarckian or teleological features. The requirements of the time are co-determinants of the interpretation and organization imposed upon transmitted knowledge (when this knowledge attains its effective form in human minds).

Even in another connection is it important to remember that knowledge is only effective when known by somebody. Much of the knowledge developed by Jordanus turned out to be ineffective because no one was able or willing to know it. Because of specific historical circumstances and personal aptitudes, Jordanus was able to develop a form of mathematical knowledge that was largely out of key with the predominant tune of the "tribe" of medieval university scholars.[16] This can, if one wants, be described as a "random" (if definitely not "blind") variation with respect to the predominant conditions of the environment in regard to possession and transmission of knowledge—as random, at least, as the reading of a die according to classical mechanics, which also results from a multitude of uncontrolled influences. The rejection of Jordanian mathematics by the same tribe is, on the other had, a paradigmatic example of "selective (non-)retention." No matter how "good" when measured by a presumed abstract scale was the mathematics Jordanus created, it remained out of key with the (sociologically well rooted) norms, ideals, and needs of contemporary scholars, who therefore did not care to invest the time and energy required to carry on with the singing. Whereas the statistics of small numbers had created the possibility that a single scholar could deviate strongly from the ways and needs of his fellow tribesmen in his reaction to the requirements and possibilities of the time, the statistics of large numbers assured that this deviation should remain without significant effect.

True enough, a small clan inside the tribe was convinced of the qualities of the Jordanian tune and went on singing it. To a restricted extent, Jordanus's manuscripts were continuously copied throughout the Middle Ages because his mathematics was known *in the abstract* to be good. We have also observed the existence of a small circle of followers and admirers; yet even they could not hold the key but switched to that of the larger tribe: Campanus took over some of Jordanus's axioms but produced an eminent *medieval* Euclid; Richard de Fournival had all his works copied but was in reality more interested in medicine, astrology, and alchemy (the pivots of naturalism); the *Algorismus demonstratus* took over Jordanus's apodictic technique but displaced the general theory of consimilar and dissimilar fractions through an explanation of the sexagesimal fraction system; and the adaptors of the *De plana sphaera* pushed Jordanus's chastity aside and spoke shamelessly of celestial circles and stars. All these, of course, are not instances of "selective non-retention"; instead, they parallel on a larger scale and in more radical form the teleological (mis-)understanding of transmitted knowledge already discussed on the occasion of Jordanus's "arithmetization" of the ideals of ancient mathematics.

# 7

# PLATONISM OR ARCHIMEDISM: ON THE IDEOLOGY AND SELF-IMPOSED MODEL OF RENAISSANCE MATHEMATICIANS (1400 TO 1600)

To Marshall Clagett
who restored to us the medieval Archimedes and all his kin

## Introduction

*Like Chapter 6, the following study takes up a specific theme from Chapter 5 and submits it to closer examination. In contrast to the investigation of Jordanus, however, which antedated the broad survey of Chapter 5, the following is crucially dependent on a pattern that I became aware of only when preparing what appears as Chapter 5 (viz., on the notion of "Archimedism"), opposing it to the widely accepted thesis that Renaissance mathematics sprang from Renaissance Platonism. Once again I try to trace, through an example,* the way *in which external ideas imprint the aims and standards of mathematical workers—the* manner *in which it impresses their actual* work, and *the* extent *to which it does so—the impact of the mathematicians'* own situation *on their reception of general ideas (leading in this case to a genuine* construction *of the figure of "Archimedes")—and* the level *on which these interactions take place.*

*A first version of the essay was presented at the symposium "Archimede, mito traditione scienza," Syracuse and Catania 1989, and appears in the proceedings of that meeting. In its present form, it was discussed at the seminar* Renaissance-Philosophie und neuzeitliches Denken III: Die klassischen Interpretationen im Lichte der modernen Forschung, *Dubrovnik 1990. Section 6, considering the implications of the inquiry for the general interpretation of the Renaissance and of Renaissance science, was added on that occasion.*

## 1. The Question of Platonism

Interpretations of the Renaissance abound—many of them in obvious mutual conflict. Less obvious are the implications of these conflicts. Do they concern the identification of the Renaissance phenomenon? Do they concern different aspects of a phenomenon that is understood globally in largely the same way by authors? Or do they really offer competing explanations of the same aspects of the same historical phenomenon?

It is easy to point to instances of all levels. Seeing "science" as an expression of his "Counter-Renaissance," Hiram Haydn (1950) would certainly have looked in vain for that "Scientific Renaissance" which is fundamental to the "Whig interpretation"[1] (and quite a few other interpretations) of the Renaissance, from Whewell to James Burke. A significant part of the difference between Robert Westman (1977) and Frances Yates (1964) hinges on the question whether certain *magi* are part of the "modern" movement—but another, no less significant part concerns genuine disagreements on the affiliation, motivation, inspiration, and impact of specific actors. And so forth.

Ideally, inquiry into the validity of interpretations of the Renaissance should thus be made at several levels at a time. In practice, such multifaceted exploration may easily lead once again to overstating theses which the inquirer has derived from unacknowledged delimitations of the phenomenon or to mistaking in new ways, depending on the scholar's own interests, *partem pro toto*. An alternative, legitimate component of the evaluation of traditional theses will therefore be application to specific segments of Renaissance culture—not in order to verify or falsify, which cannot be done as long as the range of theses remains unspecified or open-ended, but with the aim of specifying ranges and differentiating meanings. This is what I intend to do in the following, questioning the general *Platonist* thesis and concentrating upon *mathematics* and *mathematicians*.

The following set of stereotypes is familiar and central to the theme of "Platonism in Renaissance mathematics":

- Scholastic natural philosophy was Aristotelian and therefore non-mathematical.
- The Renaissance and the humanist movement rediscovered Plato, and were fundamentally and strongly Platonic.
- Plato held a mathematical view of the world.
- The Renaissance produced a fresh start in mathematics and in mathematicized natural philosophy.

It is easy to array arguments in favor of these stereotypes, and even if they represent violent simplifications (especially in their undifferentiated view of "Scholasticism" and "Renaissance") none of them is totally mistaken. It is therefore attractive to combine them into a stereotyped conclusion, *viz.*, that

- Renaissance mathematics and mathematicized science were expressions and consequences of Renaissance Platonism.

In the following I intend to demonstrate that this conclusion[2] is far less true than the partially true premises from which it is drawn (no unusual situation within the polyvalent logic so characteristic of historiography turning explanatory). Next, I propose to show that another intellectual giant of antiquity can be argued with much better reason to have served as an example and as a navigation mark to Renaissance mathematicians, and that *Archimedism* provided them with an ideology almost as protean as the Platonism of miscellaneous philosophical currents, yet serviceable as inspiration and guidepost and for professional identity formation. Finally, I shall reflect upon the implications of these observations regarding the overall role of Platonism for the Renaissance movement, and upon their compatibility with certain competing interpretations.

First to the question of Platonism. It would be easy to point to large amounts of definitely non-Platonic mathematical activity made during the Renaissance era (say, 1400 to 1600). The whole quadrivial tradition as it lingered on at many universities until well into the sixteenth century could be used for that purpose; most German *Rechenmeister* are also pretty unphilosophical, as is much of their Italian model, the tradition of the *Trattati d' abbaco*. The existence of these traditions, however, is a trivial and rather uninteresting point. The real question is whether the *renewal* in Renaissance mathematics—*renascent* mathematics in the following—can in some fruitful sense be described as Platonic.

The delimitation of the *new* tendencies is in itself a problem, which I shall answer—or, it may be thought, avoid—by discussing a selection of outstanding examples. But at first, any claim about "Platonism" requires that we clarify the meaning of this term.

The most obvious sense in which somebody could be considered a Platonist is if he adheres to (some interpretation of) the complete philosophical system. But this sense has little meaning in the present connection, since the above-quoted stereotype "Plato held a mathematical view of the world" is only a half-truth. The *Timaeus,* admittedly, proposes a cosmology founded upon geometrical atomism. In other, more central, works, however, mathematics is only considered a *first, approximate* model for true, dialectical knowledge, a means for "the awakening of thought" (*Republic* 523A). This Platonism does not turn up in Renaissance mathematical writings, although traces of it can be found in certain *philosophical* writings, for example, in Ficino's *Liber de vita*[3] and Nicholas de Cusa's *De docta ignorantia,* which states mathematical objects to be mirrors and mysteries through which the Creator can be apprehended symbolically by his creatures.[4]

The general neoplatonism of Ficino's circle is no more relevant. Its interest in Hermes Trismegistos and in the *Prisca theologia* is shared by a few mathematical writers (among them Lefèvre d'Étaples and Foix de Candale), but it remained peripheral, being usually combined with cabala and numerology and not with genuine mathematics.

Mathematicians could also be considered Platonists if they were or believed themselves to be inspired by a particular part of the Platonic corpus, either to do mathematics at all or in the choice of a particular approach to the subject. The latter possibility can be excluded immediately: No part of the corpus is precise enough in its description of mathematics to convey a mathematically specific message to a generation that had understood its Euclid. The former possibility is open in principle but closed *de facto*. Admittedly, many writers refer to Platonic passages in order to legitimate the pursuit of mathematics. But Plato is never alone in the works that I have looked at, and never of outstanding importance.

Neither in the former (strong) nor in the latter (diluted) sense is it thus possible to verify the claim that Renaissance mathematics was Platonist. Why, then, is the claim so persistent?

The real intention of the claim seems to be a contrast. Thirteenth- to fourteenth-century mathematics may not always have been Aristotelian in the strong sense[5]; but in the diluted sense it was—*vide,* for example, the recurrent lip service offered to the four Aristotelian Causes in the most unlikely connections. To claim that Renaissance mathematics was Platonist should read as a claim that mathematics was *no longer* (*diluted*) *Aristotelian.*

The reason to identify non-Aristotelianism with Platonism is not exclusively endorsement of the dichotomizing topos which knows only these two directions in philosophy (even though this fallacy certainly plays a role). To see what more there is to the claim for Platonist Renaissance mathematics we may look at the material presented by A. C. Crombie in an article (1977) on sixteenth-century "Mathematics and Platonism. . . ". Renazzi's early nineteenth-century history of the University of Rome is quoted for the claim that a number of Roman sixteenth-century mathematicians were Platonists; most vivid is the statement that Giambattista Raimondi (c. 1536–1614) "led the way through his lectures in toppling Aristotle from the philosophical throne and replacing him by Plato" (trans. Crombie 1977: 63). In a corresponding note, on the other hand, a work by Girolamo Lunadoro from 1635 is quoted for the information that Raimondi had "beautiful thoughts about the doctrine of Plato, and that of Aristotle, as he was very well versed in both of these authors."[6] Quotations from the writings of Clavius (1537–1612) and Possevino (1533/34–1611) carry the same message. In his discussion of "The Way in Which the Mathematical Disciplines Could Be Promoted in the Schools of the [Jesuit] Society" Clavius refers to "the infinite examples in Aristotle, Plato and their more celebrated commentators, which can by no means be understood without a moderate understanding of the mathematical sciences," whence "teachers in philosophy should be skilled in mathematical disciplines, at least moderately" (trans. Crombie 1977: 65f.). In the Jesuit *Ratio studiorum* from 1586, strongly influenced by Clavius, mathematics is said to assist the *Analytics* by providing "examples of solid demonstrations," and to be useful for metaphysics by telling "the number of spheres and intelligences" (trans. Crombie 1977: 66)—and for poets, for historians, for politicians, for physics, for theologians, for law and ecclesiastical custom, and for the state, all for various more or less plausible reasons. Possevino, finally, in the chapter on mathematics (which Clavius had co-authored) of his encyclopedic *Bibliotheca selecta* from 1587–91, cites Plato's *Timaeus* and Aristotle's *Physics* as "very great proof of how much light mathematics itself sheds on philosophy" (trans. Crombie 1977: 70).

"Platonism" it thus not unspecific "non-Aristotelianism"; it reduces to that contrast to strict (Averroist and similar) Aristotelianism which regarded Plato and Aristotle as nothing but reconcilable exponents of *Ancient Wisdom*. In this sense, many Renaissance mathematicians were evidently just as much Platonists as the majority of Renaissance thinkers in general. But this is a "Platonism" where the specific character of Plato's own philosophy had largely disappeared; its chief argument for the philosophical importance of mathematics, moreover, is the necessity of fundamental mathematical knowledge for anybody who wants to enter the world of philosophical authors (no wonder that the favorite quotation from Plato on mathematics was the legendary "Let nobody unskilled in geometry enter"). In practice, mathematics was thus a *first philosophy* common to Plato, Aristotle, Ciceronian Stoicism, and so on, and its necessity was not to be derived from any particular system.[7]

It seems to be a legitimate conclusion that the "Platonism" of Renaissance mathematicians only diverts attention from the real issue. It has little to do with any serious philosophical Platonism (not even the specific Platonisms of the Renaissance); it was only one ingredient in a more general reverence for ancient wisdom *in toto;* and it only influenced the way mathematics was done quite superficially, namely, by legitimating the mathematical enterprise as a whole.

Yet: what *is* the real issue?

## 2. Renascent Mathematics

Not all mathematical activity of the Renaissance era was involved in renewal and in break with the medieval past. As already mentioned, much fifteenth-century university mathematics and much of what went on in the vicinity of the abacus school would only blur the picture if we look for the characteristics of "renascent mathematics." Only by concentrating on the latter current (insofar as it is permissible to speak about *one* current) will we be able to distinguish its particular traits.

*Renascent mathematics* is *mathematics conscious of its own rejuvenating role*. In the actual historical context this meant mathematics somehow connected with the humanist movement[8]; an appropriate first example will thus be Leone Battista Alberti (1404–1472), who was not only a mathematician but also a major humanist.

One of his mathematical works is the *Ludi rerum mathematicarum* (ed. Grayson 1973: 131–73), a recreational opuscule mainly on practical geometry and dedicated "all'umanità e facilità" of the princely recipient. It thus unites civic utility[9] with noble leisure; or, to be more precise, presents dilettante interest in a subject of public utility as a noble way to fill out leisure—in full harmony with the tendency of fifteenth-century Italian humanism.

Aristotle is not mentioned in the work, and neither is Plato nor Euclid.[10] Only Archimedes, "uomo suttilissimo," turns up (p. 172), *viz.*, because of his exposure of the fraud committed by Hieron's goldsmith.

In the dedicatory letter of the Italian version of the *De pictura*, Alberti tells more about his conception of the mathematical sciences: They are among those "elevated and divine arts and sciences" which had flourished in antiquity but now were "missing and almost completely lost": painting, sculpture, architecture, music, geometry, rhetoric, augury, and similar noble and wonderful undertakings.[11]

No doubt, then, that for Alberti the resurrection of mathematics was in itself a legitimate resurrection of ancient splendor. He did not need, and made no appeal to, any philosophical legitimation for his efforts.

Another architect-mathematician of somewhat later date is Luca Pacioli (c. 1445–1517). He was more loquacious than most mathematicians, and his *De divina proportione* (written 1496–97, published 1509; ed., trans. Winterberg 1896) contains an introduction to his views of mathematics extending over several chapters. The relation of the subject to philosophy is made clear by the claim that his work is necessary to "everybody wanting to study philosophy, perspective, painting, sculpture, architecture, music, and other most pleasant, subtle and admirable doctrines," and no less by the statement that

the mathematical sciences of which I speak are the fundament for and the ladder by which one arrives at knowledge of any other science, because they possess the first degree of certitude, as the philosopher says when claiming that "the mathematical sciences are in the first degree of certitude, and the natural sciences follow next to them." As stated, the mathematical sciences and disciplines are in the first degree of certitude, and all the natural sciences follow from them. And without knowing them it is impossible to understand any other well. In *So-*

*lomon's Wisdom* it is also written that "everything consists in number, weight and measure," that it, everything which is found in the inferior or the superior universe is by this necessity submitted to number, weight and measure. And Aurelius Augustine says in *De civitate Dei* that the supreme artisan should be supremely praised because "in them he made exist that which was not."[12]

For Luca, then, mathematics is as much the real *first philosophy* as it was to be for Clavius and his associates; to drive home the point he even enlists Aristotle, twisting his "being secondary to mathematics in exactness" to mean "follow by logical derivations from mathematics." But mathematics is more than that, and in the following passage Luca describes the utility of mathematics in warfare—no doubt more interesting to Ludovico Sforza of Milano, for whom the work is written. Here Archimedes is brought into the argument, "el nobile ingegnoso geometra e dignissimo architetto" and "gran geometra." He is the man who provides the connection between mathematics and its civic uses through his legendary defense of Syracuse, which also corresponds to

the daily experience of your Ducal Highness [ . . . ] that the defence of large and small republics, called by another name the military art, is impossible unless the knowledge of Geometry, Arithmetic and Proportion can be applied eminently and with honour and utility.[13]

Alberti and Luca represent the level of distinguished but not eminent fifteenth-century humanist mathematics. A representative of the truly eminent level (and probably the most eminent of all) is Regiomontanus (1436–1476), whose opinions on mathematics are perhaps expounded most clearly in a lecture "explaining briefly the mathematical sciences and their utility," held in Padua in 1463/1464 (printed by Schöner in 1537, facsimile in Schmeidler 1972). Following upon a general introduction comes a history of the subject, in which it is said, after the presentation of Euclid and his medieval translators, that this "father of all geometers" was

followed by Archimedes, citizen of Syracuse, and by Apollonios Pergaeos, customarily called the Divine because of the height of his genius, of whom it is not easy to say whether one is to be preferred to the other. While namely Apollonios described the elements of conics in eight books, which have never been put into Latin, the first rank appears to be given to Archimedes the Sicilian by the variety of publications, which under Pope Nicholas V were rendered in Latin by a certain Jacobus of Cremona; he composed two books on the Sphere and the

Cylinder, two on Conoids and Sphaeroids, and as many on Equilibrium; he also wrote about Spiral Lines, where he undertakes to designate a straight line equal to the circumference of the circle, thus permitting the squaring of the circle, which on their part several very ancient philosophers had sought for but nobody had found out until the time of Aristotle, and for the glory of which several most distinguished men of our own age stand ready [apparently a hint at Nicholas de Cusa]. From Archimedes we have, moreover, the Measurement of the Circle, the Quadrature of the Parabola, and the Sandreckoner. Some would claim that he has written a work on Mechanics, where he collects select devices for various uses, on Weights, on Aqueducts, and several others which until now it has not been possible to see.[14]

Later in the lecture Regiomontanus comes to the utility of mathematics for philosophy, and to an ensuing ranking of the two. Once again mathematics comes in as the authentic first philosophy while being itself above the judgment of philosophies. As with Luca, mathematics possesses "the first degree of certitude," and nobody will understand Aristotle who does not know the liberal arts of the quadrivium, neither *De caelo* or *Meteora,* nor *Physica* or *Metaphysica.* Philosophy is split nowadays into warring schools, and victory comes from mastery of sophisms. On *mathematics,* on the other hand, nobody but the insane would dare say something similar, and

neither time nor human customs can detract the least from its validity. Euclid's theorems have the same certitude today as a thousand years ago. Archimedes' discoveries will inspire no less admiration in the men to come after a thousand generations than pleasure in us when we read about them.[15]

It is therefore a triumph of the present day that so many theologians and other eminent men, from Bessarion and Nicholas de Cusa to Alberti, are interested in the subject.[16]

In spite of all differences of level and orientation, Alberti, Luca Pacioli and Regiomontanus thus agree on several important points. Firstly, mathematics does not derive its justification from any particular philosophy; but part of its justification comes from its *utility for philosophy in general.* Apart from that, its civic utility and its root in antiquity makes it a sublime subject.

One figure is mentioned by all of them as a token: Archimedes. To Alberti and Luca he is first of all a magnificent engineer and architect, to be extolled for his service to his king and for his ingenious mathematical machines. Regiomontanus, the most illustrious mathematician of the century, praises him as a theoretical mathematician and dismisses his engineering feats as undocumented in extant works (and, less directly expressed, as less interesting than his mathematical achievements).[17]

As amply demonstrated by the first volumes of Marshall Clagett's *Archimedes in the Middle Ages,* Regiomontanus and his contemporaries were not the first European mathematicians after the classical era to acknowledge the importance and eminence of Archimedes' mathematics. But his scholastic admirers were not interested in the historical person, and did not endow him with the status of a symbol. That idea seems to have come from the biographical interest of the Roman epoch and its revival in early humanism. Already in antiquity the subtlety of Archimedes had become proverbial (see Simms 1989). Basing himself on Cicero, Livy, Firmicus Maternus, and other Latin authors, Petrarch (1304–1374) had written several biographical notices on Archimedes (quoted in Clagett 1978: 1336–40). They describe how Archimedes' absorption in geometrical thought caused his death; the wonderful machines he made for the defense of his city; and his mechanical model of the heavenly system. Further, they speak about his interest in the heavens and his preeminence as an astrologer (claimed by Firmicus Maternus); and do not mention a single mathematical work, not even that *Measurement of the Circle* which had gone into the biographic notices written by Vincent of Beauvais and other scholastic authors.

The reason that humanist mathematicians might adopt Archimedes as the supreme representative of their art was thus that humanist culture was from its beginnings aware of his importance, though not of its foundation. An initial reason that neither Euclid nor Apollonios could be adopted in a similar way may be the lack of adequate biographical substance, and thus their absence from the humanists' picture of antiquity (later on, the respective levels of Euclid and Archimedes would of course ensure that Euclid could not take over the role as the paragon of mathematics).

The general status of Archimedes even outside the mathematical environment is illustrated by the pet name given by the citizens of early fifteenth-century Siena to their foremost engineer Mariano Taccola (1382 to c. 1453). If he was called "the Archimedes of Siena" and felt that appellation to be deserved (Prager and Scaglia 1972: 17 and *passim*) it was certainly not because of mathematical proficiency—he was no mathematician and never mentions either Archimedes or Euclid in his writings (Prager and Scaglia: 156). But he was a skillful designer of military and other engines and a fine artist, which was sufficient to justify the illustrious title.

In spite of the preponderant interpretation of Archimedes as a supreme engineer, the veneration of his name also affected the development of fifteenth-century mathematics proper. If Archimedes was the great geometer of antiquity and his works were still extant, it would, first, be important to possess a Latin translation unpolluted by the stylistic errors of Scholasticism.[18] Second, it would be important for those who took up work in his field (or who wanted an all-encompassing familiarity with ancient culture) to know him. Though mathematically incompetent in its beginning, humanist Archimedism was thus a spur to translations and manuscript collection, concerned not only with Archimedes himself but (by way of the general justification of mathematics it implied) also of other ancient mathematicians. Finally, it was an incentive for working mathematicians to take up as much of Archimedes as they were able to—who knows whether Regiomontanus the astronomer would have set himself the task[19] of publishing Archimedes and Apollonios if he had not been influenced by Bessarion's circle? Helmuth Gröβing (1980: 74f.), at least, suggests that this encounter changed both his mathematical and his stylistic ideals; indeed, the contrast between his actual mathematical interests and his enthusiasm for Archimedes suggests that the latter was imposed by more general ideals.

### 3. The Mature Italian Renaissance

I shall not go into the details of these processes of translation, manuscript collection, and the like, which have been expounded by Rose (1975: 26–56, *passim;* 1973). Instead, I shall look at the way Archimedes is seen by some of the mathematicians and the mathematical *dilettanti* of the sixteenth century. First, however, I shall point to a change undergone by the topos "the Archimedes of . . ." during the Renaissance period and its immediate sequel. In the earlier fifteenth century, we remember, it had been used of (and, in pride, by) a skillful but mathematically rather uninformed engineer. In the later sixteenth century it was used by Jean Bodin of François Foix de Candale because of his contributions to pure geometry.[20] Galileo, finally, uses it twice in the *Discorsi*[21] of Luca Valerio, author of a work on centers of gravity based on the Archimedean finitist method of limits.[22]

The changing use of the topos illustrates the gradual shift undergone by the Archimedist idea during the late Renaissance, which will be more clearly visible if we take a closer look at the attitude to Archimedes as expressed by some significant mathematical authors. For this purpose, Cardano will provide us with an adequate starting point. His was a profuse mind,[23] and we shall not wonder that he does not list Archimedes alone. Yet Archimedes receives special treatment. One work of interest is the brief *Encomium geometriae*, which was read in the Academia Platina in Milano in 1535.[24] It makes use of the "catalogue of geometers" from Proclos's commentary on *Elements* I "which has just been printed together with the Greek Euclid" (namely by Grynaeus in 1533), but inserts explanations and commentaries.[25] Basing himself on Plato's *Phaedo* (where the philosopher Euclid from

Megara is mentioned) he inserts Euclid as Plato's contemporary, which gives him occasion to bring in the whole progression of commentators and translators, from Theon, Proclos, Pappos, Marinos, and Hypsicles to Campanus and Bartolomeo Zamberti. To Proclos's list he further adds Aristarch, Porus (probably Sporos, as Richard Lorch has pointed out to me), Nicomedes, and Menelaos,

> all of them Greeks, and writing in Greek, whose writings will have been lost, but testimony of whom by others is alive. But they are, indeed, all defeated by Archimedes of Syracuse, almost all of whose findings we possess. A man of the highest genius, and who will have shown the circumference of the circle pretty closely, and taught by solid geometry how to interpose two lines between two others in continuous proportion. But that has been lost.[26]

In the *De subtilitate*[27] from 1550, Book XVI ("De scientiis") contains a list of the ten foremost authors,

> of whom Archimedes is the first, not only because of his works which have now been published [by Tartaglia in 1543] but also because of his mechanics which, as Plutarch relates, shattered the Roman troops time and again [ . . . ].[28]

Archimedes is then followed by Aristotle; Euclid; John [Duns] Scotus; John [*sic*] Swineshead, the author of the *Liber calculationum;* Apollonios "almost from the same epoch as Archimedes"[29]; Archytas; Muhammed ibn Mūsā [al-Khwārizmī], and Jābir [ibn Aflaḥ] of Spain. Number 11 as to subtlety is Galen—and as number 12 comes Vitruvius, "who could have been counted among the first if only he had described his own findings and not those of others." As a commentary to the list it is explained that each is excellent in his own way, Aristotle in genius (*ingenio*), Archimedes in genius and imagination (*imaginatione*), Swineshead in imagination, and so on.

Cardano was a tolerant eclectic to whom "every truth is divine"[30]—yet in view of his own very un-Archimedean approach to mathematics (not to speak of his philosophical and scientific orientation in general and his recent clash with Tartaglia the publisher of Moerbeke's Archimedean translations) it is striking that Archimedes comes in first, both *within* geometry and in a discussion of the sciences in general (in a list dominated by mathematiclans, it is true, but where Cardano's own arithmetical and algebraic interests are given as modest standing as Galenic medicine). Cardano's Archimedes, even more clearly than Regiomontanus's, was clearly not a figure molded freely to fit and justify Cardano's own doings, but an archetype representing a general idea of the nature and role of mathematics.

That those who translated Archimedes (Commandino) or purported to have translated him (Tartaglia) had a high opinion of him is hardly astonishing, and proves less than the very fact that they undertook to translate and/or to publish, and I shall discuss neither in detail. Yet before we leave sixteenth-century Italy it will be appropriate to look at Baldi's 201 *Vite di matematici*. The most extensive of these is the *Life of Pythagoras*[31] (64 pages in the autograph), which, however, leads off with a series of excuses for including Pythagoras (since Thales was included, who inaugurated philosophy in Greece, the *princeps* of Italian philosophy should be included too; and so on). Obviously, the length of the biography does not reflect Pythagoras's status as a mathematician in the opinion of Baldi's "model reader." Second in length is the biography of Archimedes, with 51 pages in the autograph (more than double any other). It begins in a totally different vein:

> In all domains there have been some who, having arrived at the peak of excellence, have demonstrated how far the human intellect could advance in that direction. Without doubt Archimedes was such a man in mathematics, since the first place is due for good reasons to him. That I shall write about him therefore distresses me for several reasons. Namely that my talent is not proportionate to the topic; that the high age of the subject does not allow me to know everything concerning him and worth telling; furthermore the lack of books, and the place where I sojourn, far not only from the famous libraries but from the most tiny ones,[32]

after which he discusses all aspects of Archimedes' life and accomplishment that a critical mind could read about in a famous library, including that which is found in Archimedes' own works, that which is told by other mathematicians, and those anecdotes about his engineering triumphs which are told by Plutarch and similar writers. In the final paragraph he concludes that

> Archimedes has been the Prince of mathematicians; whence Commandino said with good reason that the one who has not studied Archimedes' works with diligence can hardly call himself a mathematician.[33]

These lines were written August 25, 1595, some 250 years after Petrarch and a few contemporaries had instituted humanist Archimedism in Italy. They, and Baldi's biography in general, demonstrate to what extent humanist scholarship on ancient mathematics and mathematical work on Archimedean foundations had grown together while fecundating each other—in Italy.

## 4. Northern Humanism

—in Italy—but what had happened elsewhere? The first observation to make in order to answer this question is that the humanist movement itself was originally an Italian movement. Petrarch's contemporaries in Paris and Oxford were precisely those producers of *sophismata* and *calculationes* who were covered with scorn by fifteenth- and sixteenth-century literary humanists. There were also those such as Jean de Murs, who could be recognized as a good mathematician by Regiomontanus and who anticipated the Renaissance integration of theoretical and practical knowledge, as well as the civic utilitarianism of the following century, but who no more than the Italian abacus school mathematicians of the fourteenth century belongs to the Renaissance movement proper.

Then, from the mid-fifteenth century onward there is a definite revival of mathematical interest in the Provençal-French area somehow related to Italian abacus school mathematics and culminating with Chuquet's works.[34] As far as I know from the surviving writings (primarily Chuquet's works), however, the movement partakes no more in the Archimedist current (or in the assimilation of Greek mathematical writings) than does its Italian counterpart before Luca Pacioli. The best-selling *Margarita philosophica*, written by Gregor Reisch in 1496 and printed time and again in French and German territories between 1504 and 1517 (facsimile from Reisch's 1517 edition in Geldsetzer 1973), is equally irrelevant to the *new* tendencies in mathematics (in several respects Reisch manages to get below the level of Isidore of Seville's *Etymologies!*); to humanism (if any characterization can be given to its messy philosophical eclecticism it is *diluted medieval Aristotelianism*); and to Archimedism (Archimedes is a non-person).[35]

With a few exceptions (the most important of which are Lefèvre d'Étaples and Melanchton, to both of whom I shall return) Northern humanists excluded ancient mathematics and technology from their field of interest until c. 1550. Mathematics, it is true, or at least the science of numbers, was not unknown to them. Reuchlin's and Agrippa's mathematical interests, however, were cabalistic and magical. Even in connections where an Italian humanist would normally have referred to Archimedes (*viz.*, in connection with combined mirrors, automata, and similar astonishing technologies) Agrippa's *De occulta philosophia* (ed. Nowotny 1967: 111–13) leaves him unmentioned. Thomas More refers to the excellency of the Utopians in "musica [ . . . ] ac numerandi et metiendi scientia" (ed. Surtz and Hexter 1965: 158) but omits both Euclid and Archimedes from his list of important Greek books (pp. 180–82) in a way that stuns his commentators (p. 435) but which is in fact characteristic: More was a good Renaissance utilitarian, as documented in many ways in his *Utopia*, and he recognized the importance of basic mathematics ("counting and measuring . . ."). But since mathematics was not part of his picture of *ancient* culture his mathematics remained at this utterly elementary level.[36] In this respect he did not differ from his friend Erasmus nor from most other Northern humanists.[37]

The first exception of some importance was Jacques Lefèvre d'Étaples (c. 1455–1536). He was a sincere neoplatonist and had been to Italy; there he had been in touch with Ficino, Ermolao Barbaro, and Pico della Mirandola, and he was inspired by his knowledge of Italian philosophy to elevate the mathematical level of French learning.[38] Lefèvre undertook a number of mathematical publications, among which were two editions of Jordanus's *Arithmetica* (1494 and 1514); yet precisely this acme shows the "medievally solid" character of the venture—the most modern piece being perhaps the least solid, *viz.*, Charles Bouvelle's squaring of the circle etc. (1503). The paragon of French mathematical publishing around 1500 had learned of the importance of mathematics through his acquaintance with Italian humanism; but his idea of mathematics (or the idea that could be expected to gain acceptance in Paris) was traditional, far from any renascence of ancient mathematical writings or style. Far, indeed, from anything reminding of Archimedism.

The other main exception is Luther's follower Melanchton (1497–1560). Whereas Lefèvre had written at least an introduction to Boethius' *Arithmetic* and a few similar things, Melanchton was no mathematical author at all. But he lent his name and fame to many a mathematical book as a writer of prefaces. One such preface (written in 1537) is found in the Basel edition of Campanus's and Zamberti's translations of the *Elements* (*Euclidis megarensis* . . . 1546, fol. [+]2–[+]4). Here Melanchton shows that he is very well versed in what Plato and Aristotle write about mathematics—but especially in their opinions on the moral implications of mathematics (the central point being the identification of Justice with geometrical proportionality). These moral teachings are *alpha* as well as *omega* to Melanchton. In a curious way, mathematics is *first moral philosophy*. But this role is, it goes without saying, totally alien to all those ideas which could be attached to the figure of Archimedes: the wondrous technician, the master of civic utility, the prince of pure geometry. Melanchton, together with Lefèvre, represented a new strain in Northern humanism, *viz.*, the opening up of humanist thought to mathematics. But because Northern humanism had another perspective on ancient thought and culture (and, we might say, a more narrow, literary, and moralizing perspective—"Northern antiquity" was dug out from libraries and written on parchment, whereas "Italian antiquity" was full-bodied and everywhere present, though in ruins), its way toward this opening to mathematics was different from that of Italian humanism. *Archimedism* was beyond its horizon; the best it could aim at for the moment was *Euclideanism*.

Eventually, however, Archimedism did penetrate Northern humanism. In contrast to the development in Italy, where glorification had preceded understanding of the mathematical works, Northern humanism only assigned paradigmatic status to Archimedes when it had grasped at least vaguely what he stood for *in mathematics*.

The beginning was made in France. In the generation after Lefèvre, the paramount French mathematician was Oronce Fine (1494–1555), who was certainly not of a mathematical stature to grasp it however vaguely. In those of his works which I have looked at there is, correspondingly, no real trace of Archimedism,[39] nor is any hint in this direction to be found in his vast bibliography (Hillard and Poulle 1971; R. P. Ross 1974).

Even Fine's student Petrus Ramus (1515–1572) was certainly no mathematical genius. But it was Ramus who, as the first Northern humanists to do so, knew both classical literature and mathematics well enough to inaugurate the era of French Archimedism—not least, perhaps, because he was also enough of an ideologue to disregard disturbing facts.

All this is clearly borne out by the *Scholarum mathematicarum libri unus et triginta* (1569). Book I (pp. 1–40) deals with "the first inventors and authors of mathematics" (p. 41), from Adam to Theon and Proclos. Already the table of contents (p. β3) is striking: no other author is dealt with on more than two consecutive pages; Archimedes occupies seven (pp. 26–33). A look at the text shows, first, that Ramus was just as erudite as Baldi, though certainly less modest; second, that Ramus impressed his own utilitarianism on the material, censuring Plato and Euclid because of their unmistakable alignment of mathematics with philosophy or pure knowledge and weighing the evidence on Archimedes according to its agreement with his own predilections; third, that Archimedes was, to Ramus no less than to Baldi, the Prince of mathematicians. The account begins with almost the same topos:

> God has decided that there should be in each art something like a unique idea which everybody studying the discipline would propose to himself as a model—as in eloquence, Demosthenes and Cicero, and in medicine Hippocrates and Galen: thus Archimedes in mathematics.[40]

Ramus then tells about Archimedes' excellence in all mathematical disciplines, which is sufficient to outweigh the "obscurity of his method and whole way to present the matter," and which even makes Ramus accept and praise his infinitesimal investigations (p. 26; otherwise hardly an interest in agreement with Ramus's utilitarian canon). On the testimony of Plutarch it is admitted that Archimedes may have been imbued with the "error of Plato," but from other biographical anecdotes it is argued that he must at least have been much less so than Euclid (p. 28)—and most of the treatment from then on deals with Archimedes' technical feats and the sources ascribing technical writings to the great man, interrupted (p. 30) by another statement that Plato's "blind ambition" not only spoiled the applicability of geometry but virtually the science itself, causing the subsequent 1500 years to bring nothing new to geometry.

In the final page of his history of ancient mathematics Ramus concludes that

> If the composition and arrangement of mathematical teaching be looked for, Hippocrates, Leon, Theudios, Hermotimos, Euclid and Theon will carry off the first fruits of praise as authors of *Elements;* if the nobility and breadth of mathematical leisurely studies be assessed, Pythagoras, Plato and Aristotle will deserve authority in mathematics. But if not only the scholastic truth and proofs from books but also its public use and its utility be appraised, which indeed is the most valuable, Archytas, Eudoxos, Eratosthenes, but, greatest and towering over everybody, Archimedes alone is to be elevated to the skies.[41]

More lavish praise could hardly be imagined; formulated by an author whose natural bent was toward the engineer but whose competence was sufficient to see the real greatness of Archimedes in geometry, it shows that not only veneration for Archimedes but full-fledged *Archimedism* had now entered French mathematics—namely, with the acceptance of Archimedes as "an archetype representing a general idea of the nature and role of mathematics," as Cardano's attitude was formulated. Now as always, it is true, this archetype was of necessity identical with the understanding of the historical Archimedes that was held at the time, and no transhistorical entity. But even if historically conditioned it could not be twisted at will; commitment to the Archimedist ideal was a constraint, and the commitment to ancient mathematics inherent in the ideal will have been important in blinding Ramus to the agreement between his utilitarian ideals and the organization of much of the Islamic mathematical heritage.

In itself this blindness was no gain for Ramus's mathematics, as demonstrated by his anonymously published *Algebra* (1560). It is typically Ramist in its way to set up schemes for calculation with algebraic expressions, and tries to assimilate the subject to ancient mathematics through a definition of *algebra* as "a part of arithmetic, which from imagined continued proportions establishes a certain form of counting of its own"[42]; i.e., it assimilates the sequence of algebraic powers *unitas* ($x^0$), *latus* ($x^1$), *quadratus* ($x^2$), etc., with a continued proportion—a concept that was familiar from elementary Euclidean mathematics but which offered no really new insights into the nature of algebra. But apart from the not very useful schemes (presented as mere rules without proof) and the use of abbreviations (*u* for *unitas*, *l* for *latus*, *q* for *quadratum*, *c* for *cubum*, *bq* for *biquadratum*, etc.) nothing surpassing the level of al-Khwārizmī's original treatise or its twelfth-century Latin translations will be found. In this particular field, the constraints presented by Ramus's Archimedism and his love of schemes made him change the formal dress of the subject—but they did not help in creating a breakthrough.

That was not the fault of the basic Archimedist idea, and another French mathematician—less outspoken ideologically but of greater mathematical perspicacity—was able to bring the idea to fruition. François Viète (1540–1603), indeed, was no less sure than Ramus that what had happened to mathematics since the end of the ancient era was better not counted at all.[43] In the dedicatory letter of the *In artem analyticam isagoge* he explained that existing algebra was "so defiled and polluted by barbarians" that he found it necessary "to bring it into, and invent, a completely new form."[44] This new form, inspired by Pappos's discussion of the concept of *analysis* (making possible the formulation of the theory and of the metatheoretical status of algebra) and the stringent distinction between quantities of different kind (required by Aristotelian philosophy and entailing the principle of homogeneity) was, as we know, the starting point of Modern algebra (or, as some authors would have it, the starting point of *algebra* as distinct from a mere *algebraic approach*—see Mahoney 1971: 372).

## 5. The New Philosophies

Further illustrations of a sudden but diverse impact of Archimedism in late sixteenth-century France and England would follow if, for example, Foix de Candale and John Dee were scrutinized. I shall abstain from that, and close my investigation by suggesting how the Archimedist ideology was transformed and absorbed into the hegemonic ideas of mid-seventeenth-century philosophy.

This transformation can be illustrated, for example, by Marin Getaldić's *Promotus Archimedes seu De variis corporum generibus grauitate et magnitudine comparatis* from 1603 (facsimile edition in Dadić 1968: 1–80). The work, dealing with compared specific gravities, is composed "in the geometrical manner," as it would be called later on, from propositions (some of them "theorems," others "problems") with enunciation and proof and interspersed examples. Contemporary thinking tends to identify this style as "Euclidean." But it was Archimedes, not Euclid, who was known to have used it to describe physical problems, and the references to Archimedes in the title of the work is therefore most apt. Still, the construction *more geometrico* is only one aspect of the work; another aspect quite as important is the *experimental foundation* provided by actual measurement.

*Experimenta* was no new idea in late medieval and Renaissance philosophy. Originally, however, the term was seen in opposition to stringent demonstration,[45] and even in the sixteenth century experimentation was normally not quantified nor mathematicized (optics, one of the classical "intermediate" sciences, being an exception). The introduction of quantified experiment by Getaldić and his contemporaries is, significantly, referred to Archimedes and is, indeed, a result of the synthesis of the two aspects of the traditional, dichotomized Archimedes: the stringent and ingenious geometer and the wondrous mathematical engineer.

Getaldić's work is thus, first, a portent of the spread of the "geometrical method" to a great many branches of seventeenth-century philosophy, and evidence that the geometer inspiring that method is Archimedes rather than Euclid (even though this distinction will have been blurred when we arrive at the epoch of a Spinoza). Investigation of the Galilean corpus would point in the same direction.

Second, it points to the importance of the Archimedist ideology for the formation of seventeenth-century "experimental philosophy." For a movement inspired by utilitarians and sceptical empiricists like Bacon and Ramus it lay near at hand to dismiss mathematical exactitude as irrelevant (as indeed often done, for example, in chemistry until Lavoisier). But as long as Archimedes served as a "culture hero" it lay equally near at hand for those inclined toward mathematization and exactitude to make appeal to his authority, whereas it was correspondingly difficult for others to dismiss the appeal. This appeal was made time and again by Galileo. On the other hand, the appeal to Archimedes could also be used to legitimate the use of empirical data against any a-prioristic opposition (be it traditionalistic-Aristotelian or rationalistic-metaphysical). This was done no less often by Galileo, but also, for instance, by Descartes in a letter to Mersenne:

> You ask me whether I think what I have written about refraction is a demonstration. I think it is, at least as far as it is possible, without having proved the principles of physics previously by metaphysics [ . . . ]. But to demand that I should give geometrical demonstrations of matters which depend on physics is to demand that I should do the impossible. If you restrict the use of "demonstration" to geometrical proofs only, you will be obliged to say that Archimedes demonstrated nothing in mechanics, nor Vitellio in optics, nor Ptolemy in astronomy, etc.[46]

The historical Archimedes was, no doubt, an extraordinary mathematician and probably an extraordinary technician, too. Still, he could not avoid being a man of his times, answering the questions of this time, even formulating them, but still formulating questions that had meaning in their own historical context. The "Archimedes" of the Renaissance, from Petrarch to Getaldić, Galileo, and Descartes, was a man of *his* times, i.e., of the fourteenth to the seventeenth century. He changed with these, in particular with the advances in mathematical and classical scholarship. But because of the advances in scholarship, *Archimedism*, that mathematicians' ideology which brandished the "Archimedes" of the time as "something like a unique idea which everybody studying the discipline would propose to himself as a model" (to quote Ramus), carried decisive messages from antiquity to Renaissance and early Modern Europe. Here they went into the synthesis that gave rise to Modern science and at least to a certain moment of Modern philosophy—and, mediated by Hume's and Kant's and the Romanticists' reactions, indeed to any Modern philosophy. In this way, Archimedes as well as "Archimedes" may be more with us today than we usually notice. No doubt Plato is so, too—but *not* because of influence in the formation of Modern mathematics.

## 6. Generalizations

The preceding investigation was restricted in scope. It did not say much on Renaissance culture in general; it did not even speak of mathematics broadly but of "renascent mathematics," "mathematics conscious of its own rejuvenating role," leaving out the *maestri d' abbaco* and *Rechenmeister* as well as astronomers who were not (like Regiomontanus) also mathematicians in a more general sense.

Within this limited field, the stereotyped belief in the importance of Platonic inspiration turned out to be mistaken. Plato was never more than one among several equally revered representatives of ancient splendor. Mathematicians, instead, took their inspiration from one of their own kind, the very *incarnation of the idea* of mathematics, as Baldi and Ramus tell us: from Archimedes or, rather, from "Archimedes."

To groups with other interests, *Archimedes* was evidently only "one among several equally revered representatives of ancient splendor." Rejecting the general "Platonic" thesis in favor of an equally all-encompassing "Archimedist" thesis would certainly be an ill-advised step. Already, if we go to Copernicus' *De revolutionibus*, Archimedes' star is below the horizon; in Kepler's historical *Apologia pro Tychone contra Ursum* (ed., trans. N. Jardine 1984), it is true, Archimedes is mentioned time and again. But he is only used as the source of a statement on Aristarchos—and in the sole passage where he is more than mentioned (pp. 117/181f.), Kepler takes the idol of the mathematicians to task for distorting somewhat Aristarchos' words.[47]

The only aspect of the "Archimedist thesis" that can be generalized is the choice of a model which is specific to the profession. The particularly strong Archimedist affinities of the "Urbino school"—Commandino, Baldi, Guidobaldo—are in perfect harmony with Mario Biagioli's hypothesis that its "humanistic concern for restoration of Greek mathematics is a direct reflection of the 'courtier gestalt' of its members. The ennoblement of mathematics through the rediscovery and celebration of its famous classical ancestors was necessary to present their professional identity as fitting the upper crust of court society" (1989: 60); but since Archimedism was pervasive, it must also have served ennoblement of the field on somewhat less prestigious levels.[48]

Other groups, be it professions or *dilettanti* with particular interests, will have had to look for other particular idols. Alberti's and other architects' relation to Vitruvius is a well-known example; we may also remember the way Alberti lists Plato and Socrates as "illustrious in painting" in *De pictura*, along with a number of authentic painters, and ask whether Erasmus and others opted for Augustine and Jerome because of their epoch or because of their theology. All this is well-tilled land. But similar phenomena might perhaps be found in other places. It seems plausible, for instance, that the shift from general *occultism* to the characteristic *Hermeticism* of the Renaissance should be explained in the same manner, Hermes being to Ficino's circle an adequate archetype in a field whose medieval paragons—"Geber," "Albumasar," and so on—had ceased to be *comme il faut*. Even Ficino's Platonism—the Platonism *par excellence* of the Platonist thesis—might then turn out to be to some extent secondary, derived from a neoplatonism that itself had come to the fore as a legitimation and an underpinning of Hermeticism.[49] At the same time, Plato's whole style made *him* the perfect field-specific archetype for humanist philosopher-*literati*: the same reasons that caused mathematicians to be "Archimedists," however far their own ideals and techniques were from Archimedes', would cause philosophers to be "Platonists" regardless of their own philosophical opinions.

So much about the possible transfer value of the "Archimedist thesis." But we may also ask for the implications of this *particular* study for other *general* theses.

Located as it is on the level of ideas and ideology, it does not discriminate directly between "materialist" and "idealist" interpretations of the period. But it suggests that general humanist culture was the central determinant, and philosophies only secondary and "opportunistic" expressions of this culture in the context of professional sociocultural needs and opportunities (or a way to express professional needs, and such, in the context of general culture).[50]

The role of Archimedism in the emergence of the new philosophies may also carry a message for the "Zilsel thesis," according to which the scientific method of the seventeenth century was born through a synthesis between the experimental and quantitative attitude of "superior manual laborers" and the "methodical intellectual training" of university scholars and humanists (Zilsel 1942; quotations from the abstract). It is certainly not correct, as claimed by Zilsel, that humanist culture was unambiguously contemptuous of manual techniques[51]: for a long time, as we have seen, humanists admired Archimedes not as a pure mathematician but as an ingenious engineer, as the supreme higher artisan. Indeed, there was always a strong utilitarian strain in humanist thinking, in full agreement with its connection to the government of city republics and princely courts—these were not the places where ivory towers could be quietly built. But there was also a tendency to *accept practice as it was exerted by practitioners,* or at least to let its improvement result from purely empirical methods. This was so when Taccola was honored by his Archimedean title, and Ramus as well as Bacon was still inclined to think in the same way in an era when traditional humanism was giving way and utilitarianism grew heavier under the impact of emerging capitalism. From Alberti to Galileo it seems to have been the merit of the Archimedean "culture hero" to have brought about Zilsel's real synthesis instead of a mere change of balance,[52] by embodying the claim for optimal mathematical rigor *and* fantasy in the description of physical phenomena while at the same time legitimating mathematical argumentation based on empirical foundations—as discussed in Section 5. Archimedism appears to have been essentially involved as a cultural catalyst in the phase shift postulated by Zilsel; it can, in another chemical metaphor, be assimilated to that kind of optimal (and fateful) glue which binds by dissolving the surfaces of the items which it connects and by entering in chemical composition with both, obliterating thus their distinct identities forever.

# 8

## ON MATHEMATICS AND WAR: AN ESSAY ON THE IMPLICATIONS, PAST AND PRESENT, OF THE MILITARY INVOLVEMENT OF THE MATHEMATICAL SCIENCES FOR THEIR DEVELOPMENT AND POTENTIALS

### WRITTEN IN COLLABORATION WITH BERNHELM BOOß-BAVNBEK

Dedicated to
Horst-Eckart Gross
Mathematician, champion of comprehensive understanding,
Berufsverbot in the Federal Republic of Germany
and to
John Lamperti,
Statistician, "counter expert"

### Introduction

*For several decades, the involvement of the sciences with war has been a key theme in radical debates about the nature of science and the role of the scientific community and establishment—next, perhaps, only to the discussion of the responsibility of the sciences for engendering the ecological crisis.*

*The following essay belongs within this tradition, as illustrated by its history. A first, incomplete version was presented in 1982 at a symposium on "Military Influences on the Sciences and Military Uses of Their Results" held in Oldenburg, Western Germany, in memory of Carl von Ossietzky (co-founder of War Resisters International in 1922, and awarded the Nobel Prize for Peace in 1936 while a concentration camp prisoner). A full German version was published in 1984 by Bund demokratischer Wissenschaftler (West German branch of the World Federation of Scientific Workers, and thus a post-Hiroshima organization).*

*Apart from links to the war-resisting aftermath of the two World Wars, the article was thus connected to the debate on the Euromissiles, which was particularly intense in Western Germany during the early 1980s, and further to the discussions of the first-strike strategy of the Reagan strategists (the memorable "decapitation of the Soviet hen").*

*The essay takes it for granted that the sciences, and not least mathematics, are important for the waging of a modern war. This is a point that once needed to be made, but which has been made amply since the 1960s, and which, moreover, has no clear implications as to what should be done. Instead, questions are taken up that have such implications for various subgroups within the mathematical community, even though they may seem academic for others. Some mathematicians, indeed, have deserted their science in the conviction that it was corrupt through and through, socially as well as epistemologically; we ask whether this is true, trying to sort out the various senses in which it may be true or false. Others suppose that their science can only thrive (for epistemological or for trite financial reasons) if it is intimately connected with the military sector, and are willing to take the necessary risks in the interest of their field and profession; so, as a matter of fact, are quite a few members of any profession, but what we can do in the present context is to analyze whether the assumed necessary condition is really necessary for the prosperity and well-being of this specific field—trying, once again, to sort out meanings in order to make possible the formulation of precise answers. Still others simply strive to build up insights that are deep and precise enough to allow fruitful debate with colleagues and efficient political action within and by means of their science and profession.*

*The essay has two authors (thus the "we" of the preceding paragraph). One (Bernhelm Booß-Bavnbek) is a mathematician with strong interest in the relation between mathematics and its uses, including the philosophical aspect of this question; the other (Jens Høyrup, author of this introduction) is a historian of mathematics with an outspoken philosophical and sociological bent. Both authors share political commitment to the question, and the same fundamental stance; this did not prevent us from heated discussions over the content and formulation of at least every third paragraph; no single phrase was written, however, before we had come to an agreement on the point involved.*

*The English revised translation was prepared by me in 1988 as a contribution to the Sixth International Congress of Mathematics Education, Special Day on Mathematics, Education and Society. On that occasion, I attempted to weigh our original conclusions against the central events of the intermediary four years. I made no effort to update our statistical information, which had been meant only to create a basis for our theoretical discussions; the aim of the revision was, rather, to increase the precision of these.*

*Over the years, Bernhelm Booß-Bavnbek and I had collected and discussed material pertinent to this partial revision. However, during the phase when the translation was formulated we had very little opportunity to collaborate, and the text as it was then formulated was thus entirely my responsibility, though not my merit; it may not at all points have corresponded to what my co-author would have said.*

*Since 1988 another four years have passed, and the political geography of our world has changed. A proposal (made in Section 6) to "consider the implications of [the military employment of mathematicians], concentrating upon the situation of mathematicians of our own block—those of the socialist countries [being] in a better position to assess their own dilemmas" is outdated, since the "other" block and the*

*Soviet Union have lost the Cold War and do not exist any longer. Trying to camouflage the original context of the essay through removal of such phrases would be dishonest, I felt. Attempting to update statistical information would be without purpose, and replacing the original references to a specific political condition with others of equal precision would be impossible, given the instability of our actual situation.*[1]

If questions of war in general *had been obsolete, and not only those raised by* this particular *(cold) war, republishing the study would have been an expression of grotesque nostalgia. But they seem not to be. During a surgical operation involving carpet bombing and other blunt instruments, some hundred to three hundred thousand persons were killed in Iraq in early 1991. On March 8, 1992, the* New York Times *leaked a Pentagon plan to conserve world hegemony and to avoid the emergence of any competing global or regional power; even though the paper in its actual formulation may have been meant as an electioneering maneuver, part of its substance is certainly honest.*

*Bellicose* ideologies, *moreover, appear to be stronger than ever since the sixties. How else should we explain, for example, this advertisement from the* Evening Standard *(October 9, 1990, p. 32) directed not at potential members of the British Volunteer Reserve Forces but at their employers:*

*LEADERS AREN'T BORN EVERY DAY, JUST AT THE WEEKEND.*
*Before you can become a leader, you need to experience responsibility.*
*In the Volunteer Reserve Forces, men and women are given the opportunity to take control of challenging situations early in their careers.*
*It's an essential part of their training.*
*As their confidence grows, they grow as leaders. And this will show in their civilian jobs. As an employer you will certainly notice.*

*[etc.]*

*Or how explain that Time-Life Books makes a special offer just after the Iraq War, allowing the recipient of their colorful envelope to sign the statement "YES! I want to explore the new frontiers of modern warfare. Please send me* Special Forces and Missions *[etc.]"?*

*Everybody looking around will be able to find both preparation for actual war and evidence for militarization of dominant media and business culture. Which wars, and which reflections, may depend on the date; but nothing suggests that the main trends will change on short notice.*

*Nor is there any reason to believe that scientists and mathematicians are not, or will not be, involved in the development of new weapons and new tactics and strategies.*

*These observations are the background for the republication of a study whose concerns appear, unfortunately, to be less outmoded than some of its actual formulations. Since useful updating seems not to be possible, I have preserved the text as it was formulated in the translation of 1988. It is left to the reader (and it may be a sound exercise in times like these) to reflect critically upon the text, and to disentangle whatever still-valid theoretical understanding it may contain from formulations referring to a world that has now disappeared.*

---

**I.  . . . that the modern military business consumes every nourishment . . .**

Missile control, e.g., consumes systems theory and control theory, i.e., group 93 of the *Mathematical Reviews.* In civilian life, these theories are also used for process control in chemical plants, breweries, rolling mills, in the construction of industrial robots, and for optimizing measuring instruments.

Looking through the keywords used in this group during one year (1981), one encounters the following concepts recurrently: approximation, autoregulation, boundary condition, characteristic, convergence, correction, correlation, curvature, difference equation, differential equation, dimension, distribution, eigenfunction, error, filter, frequency, graph, harmonic linearization, information, nonlinearity, order, parameter, perturbation, phase space, probability, process, randomness, singularity, stability, state, steering symmetry, time lag, transition process, turbulence, variable.

An array, indeed, that points to an abundance of consumed disciplines.

---

The following examines the relation between mathematics and the trade of war. This question has a double interpretation. One may ask for the importance of mathematics for armament and war. Or one may explore the importance of armament and war for mathematics.

The first problem is only dealt with as a subordinate theme in the following. We all know that the modern military business consumes every nourishment within its reach: science, money, people, etc. We also know that modern warfare and its preparation build intensively on advanced technology, and thus, under the conditions of the ongoing scientific-technological revolution, on scientific knowledge and information. We know, finally, that modern science and modern technologies (including many socio-technologies, which also find military application) are thoroughly mathematicized.

Instead, we shall turn our attention to the second problem: the implications for the sciences, especially the mathematical sciences, of their military involvement.[2] The problem may briefly be so formulated, whether mathematics is paid in good or counterfeit coin when prostituting itself to the military sector. Without metaphor: Is the production of mathematical knowledge, and the overall progress of mathematical science, furthered or impeded by the military involvement of mathematics?

At first the question is approached by means of a ladder of historical steps. Section 1 presents the phenomenology of the past, i.e., the factual development of the connections between mathematics and military concerns through the eighteenth century, concentrating on a set of paradigmatic examples. Section 2 discusses the basic structures that appear to materialize from this pre- and early Modern phenomenology. Section 3 deals with the prelude to the scientific-technological revolution in the nineteenth century: the attempt, made in the French revolutionary wars, to wage war on a scientific basis; and the creation of the preconditions for the later science-based reconstruction of broad societal practices—preconditions involving the sciences, general and specialists' education, technology, and social, economic, and political structures.

After this prelude, Section 4 considers the beginning of the real drama, the era of the World Wars, where scientific warfare was first implemented; where wars approached "totality," in the sense of a total utilization in the service of war of the societal resources made available by the incipient scientific-technological revolution; and where this "revolution" made two major, blood-soiled steps toward maturity, thanks to the determined application of all scientific assets.

No simple picture emerges from history. It does not present mathematics as a diabolic undertaking inseparably bound up with war, nor does it—after the opening of the scientific-technological revolution—suggest any gullible exculpation. In order to achieve analytical understanding of the present relations between mathematics and its military sponsors and applications, one has to investigate the actual phenomena of today within a theoretical framework considering mathematics as a science, as a subject for teaching, and as a tool for applications; the functioning of the social community of mathematicians and of military structures; and that societal totality in which both are embedded. Such intricacies are confronted tentatively in Section 5.

Of course, the question of the influence of war on the development of mathematics—"mathematics through war or through peace?"—is only the subordinate half of the issue. For mathematicians as for anybody else, the fundamental problem is that of survival, "mathematics for war or for peace?" In principle, everybody ought to agree on this point. But what can and what should the mathematician do? This is the topic of the closing Section 6, which in concrete detail attempts to derive moral imperatives and corresponding practical strategies from the preceding five analytical sections.

## 1. The Past

Often have we been told that the development of modern sciences and especially of modern mathematics is intimately connected with the interests of armament and war. We are even told that this involvement is the *sine qua non* of scientific progress.

If that were true, little hope would be left for the future of mathematics, of science, and of our modern technical civilization. Either continuous armament will lead us into the final catastrophe (anyone who risks constantly his total capital in Monte Carlo is going to lose everything some day); or disarmament will lead to an era of stagnation. It is, however, obvious, that the 5–10 billion people of the late second and early third millennium can obtain no satisfactory life either through the mere suppression of modern science or through reliance on already-available technology. Scientific stagnation will thus lead us into an untenable situation—better, will not allow us to free ourselves of present untenable conditions. If progress is conditioned by armament, the dilemma *war, or stagnation and ecological catastrophe* seems inescapable and without solution.

But is the arms race really a necessary condition? Was it always the condition, as claimed in certain quarters, and not only in the Galbraithian parody *Report from Iron Mountain*?[3] Critical examination of the historical claim may help us to discern better the contemporary situation and to assess the authenticity of the dilemma.

A number of historical episodes seem to support the thesis of intimate dependency.

Already in the early second millennium, the Babylonians possessed what Neugebauer (1933 and elsewhere) has labelled "siege computation" (*Belagerungsrechnung*). At closer analysis, however, siege computations are no different from other computations—they are but one of several comparable fields of application of the same mathematical techniques: bricks remain bricks, and volumes of earth are computed the same way if they belong in a siege ramp or a temple building or are to be dug out from an irrigation canal. Babylonia offers no example of a mathematical technique inspired specifically by military requirements, nor is any military stamp to be found on the global structure and style of Babylonian mathematics.

The Greeks, too, knew that war is better waged with mathematics as part of the train. Plato's Socrates explains that the commander needs arithmetic and geometry for displaying his troops optimally (*Republica* 525b). Certain "armchair tacticians" (*tacticiens en chambre*—thus Aujac 1975: 163) even tried to parade their hobby as a mathematical discipline, among other reasons because knowledge of the isoperimetric problem is of use when you try either to impress the enemy or to hide your true force. Only familiar mathematical lore, however, was implicated—as pointed out by Geminos already in antiquity (fragment on the mathematical sciences, ed., trans. Aujac 1975: 114).

To be taken more seriously is the systematic development of military technology in Hellenistic Alexandria (see Gille 1980). This technology employed knowledge available from existing mechanics and elementary mathematics *and* combined it into a new, consistent branch of knowledge. Even certain parts of the Heronian corpus can be interpreted as attempts to improve actual practitioners' ways through the development of a "practical mathematics" integrated in the circle of mathematical disciplines.[4]

The Islamic Middle Ages took over this complex of applied mathematics and (occasionally militarily inspired) techniques and transformed it; in this way it came to have some effect in the early Modern period.

Medieval Western Europe, on the other hand, regarded mathematics only as a subject belonging to good education and as a tool for gaining knowledge of Nature; medieval Latin mathematics was neither influenced by nor of consequence for military techniques. First the Renaissance presents us with something similar yet immensely superior to the old Alexandrian synthesis between *technologies submitted to theoretical reflection* on one hand, and *applied mathematics* on the other. The list embraces architecture; painting and the theory of the central perspective; artillery and ballistics; cartography; bookkeeping; merchants' calculation and algebra. We also find a very high appreciation of the possibilities of mathematics in every practice—at times a phantasmagoric overrating. Already Tartaglia may have overrated the practical value of his new (erroneous) mathematical ballistic theory when hiding it away for years as a "damnable exercise, destroyer of the human species, and especially of Christians in their continual wars" (*Nova scientia,* trans. Drake and Drabkin 1969: 68f.). Few today, at least, will deny that the power of mathematics was overrated when the believers in scholarly "white magic" claimed to be able to control both Nature and angels by means of arcane numbers and symbolic geometrical figures. Our modern scepticism *vis à vis* numerology and magical geometry suggests that even the Renaissance belief in the efficiency of mathematics in other, genuine technologies should be taken with a measure of sound scepticism.

As far as the development of mathematics is concerned, however, the issue of efficiency is not decisive, if only contemporaries believed in the technical significance and productivity of mathematics. The precise *way* practical concerns influenced the development of Renaissance and early Modern science has been discussed amply, as has *the extent* to which practical concerns did so. Nobody denies, however, that technical and social practice *did* influence the development of the sciences, including the mathematical sciences. But since warfare was certainly a major constituent of consciously planned societal practice, this implies concomitant military influence on the development of the sciences, not least mathematics.

This diffuse interaction between the totality of societal practice and the network of more or less mathematically founded sciences survived the Re-

## II. Tartaglia, ballistics, and the responsibility of the scientist

Through these discoveries, I was going to give rules for the art of the bombardier.[ . . . ] But then one day I fell to thinking it a blameworthy thing, to be condemned—cruel and deserving of no small punishment by God—to study and improve such a damnable exercise, destroyer of the human species, and especially of Christians in their continual wars. For which reasons, O Excellent Duke, not only did I wholly put off the study of such matters and turn to other studies, but also I destroyed and burned all my calculations and writings that bore on this subject. I much regretted and blushed over the time I had spent on this, and those details that remained in my memory (against my will) I wished never to reveal in writing to anyone, either in friendship or for profit (even though it has been requested by many), because such teaching seemed to me to mean disaster and great wrong.

But now, seeing that the wolf [i.e., the Turkish Emperor Suleiman] is anxious to ravage our flock, while all our shepherds hasten to the defense, it no longer appears permissible to me at present to keep these things hidden.

Niccolò Tartaglia, *Nova scientia*, the preface (1537), ed., trans. Drake and
Drabkin 1969: 68f.

naissance, and still survives, and we shall not discuss it further. Restricting ourselves for the moment to the period prior to the French Revolution we can, however, point to some examples of intentionally furthered scientific—in particular mathematical—development aiming at military or quasi-military advantage.

First of all, the fifteenth-century Portuguese court (the court of Henry the Navigator) took up systematic development of navigational mathematics.[5] Horst-Eckart Gross comments upon its results as follows:

Whether Portugal's policy of strict secrecy contributed to a lack of interaction with other European centres, affecting the development of mathematics negatively, or the narrow aspiration to care for practical needs led to exclusive concentration on urgent problems of the day and thus to neglect of broader development: In any case, the favourable conditions surrounding this first attempt at a close binding of mathematics to practice neither inspired nor made possible any remarkable impulses to the development of mathematics.[6]

Like the Portuguese maritime and colonial expansion, that of Elizabethan England built on planned and systematic collection, development, and (selective[7]) diffusion of scientific knowledge—not only concerning mathematical navigation and cartography, but broadly. Mathematicians like Harriot took part in the expeditions; but so did even botanists and other scholars, and Harriot's tasks in 1585 encompassed not only surveying the coastal area of Virginia but also (and mainly) ethnographic accounts of the culture and language of the inhabitants (Lohne, "Harriot," p. 124; cf. Boas 1970: 191). Francis Bacon's program for social improvement through learning was drawn up toward the end of this period and stands as a theoretical embodiment of the actual policies.

The Elizabethan encyclopedic policies and their Baconian apotheosis are reflected in the ideas behind the English Royal Society and the various continental royal scientific academies of the seventeenth and eighteenth centuries. Their task was to attend to the systematic production of scientific knowledge—thus serving the strength of state and economy. One of the means to this end was the announcement of prize subjects, which would often require mathematical answers to practical problems.[8] Since the strength of the state was militarily defined, and since the prizes might equal the yearly salary of a professor, the academies thus became efficient transmitters and interpreters of military needs for the mathematical sciences.

Research was also made for order. "Hooke's law" was in all probability a spin-off from empirical inquiries in the elasticity of wood, ordered by the Royal Society on behalf of a Navy wanting to cut down the consumption of wood in shipbuilding (see Merton 1970: 178f.).

Finally, the training of military officers should be mentioned (a similar story could be told about naval officers and officers of the mercantile marine). Most regularly, fictional as well as technical literature of the seventeenth and eighteenth centuries mentioning military matters will point to the importance of mathematics. Who learns the officer's job well learns mathematics (thus also Defoe's Robinson, cf. Box III). But which mathematics, and for what purpose?

Two branches are important. One is fortification mathematics, the technique of optimizing the complex polygonal fortification structures of the seventeenth and eighteenth centuries under the conditions posed by the object to be protected and by available artillery (cf. Schneider 1970: 223–27). The technique was based on mathematical tables, and was taught at officers' schools, especially the French *Écoles d'artillerie*.

Another branch, taught with no lesser theoretical ambition at the artillery schools, was ballistics based on the Galilean parabola (see Charbonnier 1927: 1018–40, and Schneider 1970: 222f.). Even here, extensive mathematical tables were needed, among other reasons because the velocity of the projectile as a function of the quantity of gunpowder was not known, even with the assumption of uniform projectiles. Toward the end of the eighteenth cen-

---

### III. Robinson and mathematics

In Defoe's novel, all we are told about Robinson Crusoe's education as a sea-man is that

> [ . . .] from my Friend the Captain [ . . . ] I got a competent knowledge of the Mathematicks and the Rules of Navigation, learn'd how to keep an Account of the Ship's Course, take an Observation; and in short, to understand some things that were needful to be understood by a Sailor.

Explaining how he survives and manages to build up a civilization *en miniature* on his island he observes

> that as Reason is the Substance and Original of the Mathematicks, so by stating and squaring every thing by Reason, and by making the most rational Judgment of things, every Man may be in time Master of every mechanick Art. I had never handled a Tool in my Life, and yet in time by Labour, Application, and Contrivance, I found at least that I wanted nothing but I could have made it [ . . . ].

> Defoe, *Robinson Crusoe,* quoted from Defoe 1927: 1, 18, 77

"Who learns the officer's job well learns mathematics." And who learns the fairly simple navigational mathematics learns—according to Defoe—practically everything.

---

tury, the calculus was introduced in the curriculum in order that air resistance might be considered (on the basis, it is true, of a theory that was wrong but which led to solvable differential equations; ambitions of teachers rather than the real fulfillment of officers' needs appear to have been the vehicle bringing infinitesimal mathematics into the military curriculum).

In both branches, all sorts of geodesics and their basis in trigonometry, logarithms, and instruments of analogue computing were fundamental.

## 2. The Implications of History

Hellenistic antiquity did present us with an interaction between the development of mathematics and that of military technology. But was it really important, seen from the point of view of mathematics? After all, Hero's *Metrica* (or any other work of his) ranks far below the works of Eudoxos, Euclid, Archimedes, Apollonios, Ptolemy, and Diophantos, both regarding importance for the development of ancient mathematics and, especially, if we think of later scientific *and technological* influence.

A similar point holds for the Renaissance. Of those many branches of applied mathematics which evolved during the Renaissance, only algebra proved really fundamental for the overall scientific progress of mathematics.

On the limit between the Renaissance and the Modern era (and on the limit between mathematics and mechanics) we encounter Galileo's explorations of the free fall and of the strength of materials in the *Discorsi*. Whereas *algebra* has no relation whatsoever to military questions, Galileo's work has an overall, but nothing more than *overall,* connection to problems of physical technology; when discussing firearms he points out explicitly that his ballistic theory applies to mortars only, other weapons producing too great air resistance (*Discorsi,* 4th day, theorem 1—trans. Crew and Salvio 1914: 256).

Our observations on the Renaissance are of general validity: The ultimate significance of an individual practical problem for the global long-run development of mathematics is random.[9] The solution to the problem may prove completely peripheral (as Hero's *Dioptra*), or it may call forward fundamental new developments (as did late medieval merchants' algebra). Of course, all intermediate situations are also possible.

The demands of military practice to science may be direct. More often, however, they will be indirect (cf. Galileo). In the latter case, practical challenges are no different even if they regard practices with warfare application. Moreover, even though their possession of simple names might tempt us to regard them as indivisible entities, and though any search for "influences" suggests the idea of unilinear causal chains, *science* as well as *practice* are *networks.* Looking at Galileo's *Discorsi,* one will easily see that Euclid and Archimedes and the ambition to come to grips with the medieval theory of motion were quite as important for Galileo as practical problems of fall.

The broader a certain practice in its demands to science, the greater is the probability that a broad and coherent scientific development will be called forth (e.g., the commercial calculations of the late Middle Ages and the Renaissance). The more punctual, the greater the chance that the problem (if not recurrent in new forms) will either find an isolated solution—or, if no such solution can be found, will be shelved.

Let us look at the instances of systematic sponsoring of scientific development from the fifteenth to the eighteenth century in this perspective. The Portuguese development of navigational mathematics was already characterized as punctual and hence infertile. Even in later years, cartography and navigational mathematics were on the whole barren with regard to further mathematical development—as the mathematics employed by Plato's commander it based itself on a level that had become elementary. Still, navigation was one of the driving forces behind the development of new computational techniques, including the invention of logarithms; eventually it thus became part of the network behind modern numerical mathematics. Navigation was also in need of reliable techniques for precise empirical determination of geographical latitude. This could only be done either through the construction of accurate chronometers or via the development of a precise theory of the lunar movement. These were focal interests in the development of theoretical mechanics in the seventeenth and eighteenth centuries, inspiring (together with other concerns) Huygens' theory of the cycloidal pendulum, Newton's celestial mechanics, and the further refinements of that theoretical structure (see e.g., Mason 1962: 270f.).

Of those practical preoccupations which were communicated as prize subjects, many proved fruitful *because of* the interpretation provided by the academies relating them to the network of actual scientific knowledge and theory.

Once more, however, the practice of war turns out to possess no privileged position: shipbuilding remains shipbuilding, as bricks remain bricks. In the end, the internal structure of actual science came to determine what could develop—and what not: Hooke made no mathematics, and no fruitful generalization, out of the question whether trees to be used in shipbuilding were best cut down in one or the other season of the year; in the seventeenth century, this issue did not lend itself to mathematization. The sole result of his investigation to be remembered today was his "law," the determination of the relation between load and bending, which *could* be mathematized; the fruitfulness of the law, however, only resulted when his law of proportionality was integrated into the structure of Newtonian mechanics. Precisely this circumstance illustrates the necessity of expert interpreters (the staff of the academies) translating *preoccupations* into scientifically relevant *problems*.

Officers' mathematics was, to an even higher degree than navigational mathematics (because the requirements of accuracy were higher in navigation, and because techniques were developed by cartographers and geographers before being used by the officers), "canned mathematics." Officers' mathematics was mainly influential through being an important agent for the spread of mathematical literacy—proportionally with the general social weight of officers (which was certainly heavy). No *new* mathematical developments sprang from the officers' schools until Monge.

## 3. Nineteenth-Century Prelude

Regarding the relation between science and technical practice, as in so many other ways, the nineteenth century was opened by the French Revolution. The engineering education at the École Polytechnique announced fundamental renewal. The engineer, in the contemporary sense of a scientifically trained technician, hardly existed before. His predecessors in the seventeenth and eighteenth centuries were the mathematical practitioners[10], taught rather as "higher artisans" than scientifically. At the École Polytechnique, which was founded in 1794 as a civilian institution and transferred in 1804 to the

Ministry of War, future military engineers (who in later life would often end up as civil engineering officials) were taught fundamental science, i.e., mathematics, for two years, before specializing at other schools (cf. Klein 1926: I, 66, and Box IV).

The École Polytechnique was an institutional reflection of some of the carrying ideas of the Revolution, according to which education and science constitute the fundament of social progress—ideas rooted as well in utopian-rationalistic Enlightenment thought as in unsentimental bourgeois strategies in the struggle for societal ascendancy. In the context of the revolutionary and Napoleonic wars, which for the first time in Modern history attempted the total mobilization of societal resources, the idea of scientifically founded societal life was naturally transformed into a notion of scientifically founded warfare. One manifestation of this scientific foundation of war (and of the "militarization of reason") was the just-mentioned transfer of the École Polytechnique to the Ministry of War in 1804. It was, however, only a manifestation, a symptom of a general trend: a trend whose principles were formulated by Clausewitz,[11] and which had its decisive breakthrough in nineteenth-century Prussian staff planning (which, so to speak, dealt with wars to come as complex engineering tasks—see Addington 1984: 45–49). Today, it has become the essential element in the strategic planning of all great powers (where it may, if really applied in full consequence, lead to the acknowledgment that war can *no longer* serve politics as a tool in the nuclear age, but where conventional application of its principles—as "brinkmanship" or in other versions—may on the contrary lead directly into the ultimate holocaust).

---

**IV. École Polytechnique: Mathematics, and nothing but**

During its first decennia, the curriculum of the École Polytechnique contained nothing but mathematical and semi-mathematical subjects:

|  | *Double lectures* *(1½ hours each)* |
|---|---|
| *Pure analysis* | 108 |
| *Applications of analysis to geometry* | 17 |
| *Mechanics* | 94 |
| *Descriptive geometry* | 153 |
| *Drawing* | 175 |
| *TOTAL* | 547 |

(Klein 1926:I, 66)

Not only with regard to the triumphal progress of the engineers is the
École Polytechnique thus a portend. In the early nineteenth century, however,
"time" was not ripe for successful implementation of the ideas underlying
the school. "Time"—that is, the structure of society; the level of productive
forces, and the state of the sciences. As intimated by the polytechnician Carnot in the preface to his *Réflexions sur la puissance motrice du feu* (ed. Fox
1978: 61f.), England had won the war against France because of French lack
of industrial capacity (whence not, we may add, because of French mathematical or military deficiency).

The case of Gaspard Monge is illustrative (see Wolf 1952: 59f.; Loria
in Cantor 1908: 626ff.; and Taton, "Monge"). Before the Revolution he had
been a highly appreciated teacher at the artillery school of Mézière and
developed there his descriptive geometry, which in that context must be regarded as an important though random spin-off from fortification mathematics, permitting that its complex arithmetical computations be replaced with
elegant and rigorous geometric constructions. In the institutional and mental
context of the early École Polytechnique it ripened as something much different: one of the main subjects of teaching, which came to influence deeply
future engineers, and hence also a generation of future mathematicians. And
yet, in the final instance, nothing really new came out of the "case Monge."
The intimate interaction between mathematical fundamental research and the
applications of mathematics disappeared from the École Polytechnique after
10 to 20 years' Restoration. From descriptive geometry sprang (apart from a
rather isolated though indispensable auxiliary technical discipline) an ultimately random even if epoch-making *pure-mathematical* spin-off: the modern theory of projective geometry, created by the polytechnician Poncelet
while a Russian POW. All in all, we may conclude, an interaction between
mathematical theory and societal practice of rather traditional nature.

The genuine innovations of the nineteenth century are to be found at
other fronts. First of all, the preconditions for the scientific-technological
revolution of the twentieth century were established: the full display of industrial capitalism and the institution and institutions of the modern State; an
industry able to make use of the scientifically trained engineers, i.e., of the
actual level of science of any time; *science itself* understood as the business
of systematic research, attached to universities, technological universities,
and other institutions of higher learning; and a maturation of mathematical,
physical, and chemical knowledge leading to direct industrial applicability.
Prerequisites for this maturation were, in addition to an immense quantitative
growth, a complete reorganization of knowledge and a far-ranging division of
intellectual labor. Thus, only the early nineteenth century witnessed the birth
of physics as one interconnected field of knowledge; during the following de-

> ### V. "Pure, unapplied mathematics"
>
> Already the first specialized mathematical journal, founded by Gergonne in 1810, had the title *Annales de mathématiques pures et appliquées*. The ideal that mathematical science was to embrace both theory development and advances in applications was later expressed in the titles of Crelle's *Journal für die reine und angewandte Mathematik* (1826) and Liouville's *Journal de mathématiques pures et appliquées*. Nonetheless, the nineteenth-century tendency toward specialization and crystallization of "pure mathematics" as a separate field was soon reflected in the practice of the journals, and Crelle's *Journal* was rapidly given the pet name *Journal für reine, unangewandte Mathematik*.
>
> (see Klein 1926: 1, 95)

cades came engineering sciences (see, e.g., Channell 1982) and the conceptual and institutional separation of *pure* and *applied* mathematics.[12]

That close association between mathematical science and military practice which seemed to be inaugurated by the École Polytechnique, on the other hand, came to nothing at this first attempt. Throughout the nineteenth century, military needs played only an indirect role for the development of mathematics, mediated through industry and through the general claims on mathematics raised by industry.

## 4. The Era of the World Wars

The modern war is the world war. We have known two of these. World War I was fought with old-fashioned armament—and yet it was a war of unprecedented character. It was a war involving all leading powers of the world. At stake was the worldwide distribution and redistribution of spheres of influence and the ranking order of the leading imperialist powers. It was waged at all fronts of the world. Finally, it was a total war: It was not a mere confrontation of military force in the field, but also a trial of strength comparing the total productive capacities of the belligerent powers, in which not least the social and political cohesion of these powers was tested. The war was thus directed at the entire society of the adversary, towards his productive ability, as also against the cohesion of his society, against the reproduction of his social life, and against his whole social order. (Jarvad 1981:6f.)

In the scientific-technological revolution, and even during its early years, *science* is and was an essential constituent in the productive capacity

---

**VI. The sciences in World War I**

Through the summer of 1918 the German physical scientists, like the rest of the German public, continued to look forward with confidence and satisfaction to a victorious conclusion of the war in which they had been engaged four years. They, perhaps more than any other segment of the German academic world, also felt *self*-confidence and *self*-satisfaction due to their contributions to Germany's military success and to their anticipation of a postwar political and intellectual environment highly favorable to the prosperity and progress of their disciplines.[ . . . ] The chemist, the physicist, the mathematician,[ . . . ] emphasizing the great practical importance of their subjects during the war and the desirability and inevitability of still closer collaboration with technology in the future, looked forward to yet more, larger, and better stocked institutes and to substantially increased public esteem and academic prestige.

(Forman 1971: 8f.)

---

of society. This was reflected in the marshaling of science in the war effort during World War I.

The scientists, though rarely fundamental science as such, were incorporated in the military machine as highest-level technical personnel. An attempt was made, more systematic than ever before, to conduct technical development in large scale on the basis of the actual stance of fundamental science as mastered by fundamental scientists. Most important were chemistry and metallurgy, less important the physical sciences; mathematics only came into consideration as an auxiliary discipline, as "applied mathematics."[13] (That the Engineer Corps would now as before make use of mathematics does not concern us here; as in the eighteenth century, or in ancient Greece and Babylonia, for that matter, its techniques were based on what had long since been reduced to the elementary level.)

Accordingly, the anti-chauvinist G. H. Hardy could justly defend his work in the "pure" theory of numbers in his famous comment from 1915 upon the regimentation of other sciences in the war (and upon their services for industrial capitalism): "A science is said to be useful if its development tends to accentuate the existing inequalities in the distribution of wealth, or more directly promotes the destruction of human life" (auto-quotation in Hardy 1967: 120). Only some 20 years later, in the preparations for the next inferno, would his cherished analytical theory of numbers become "useful" as a secret tool for code construction and code breaking (had Hardy known about it, he might perhaps have approved reluctantly; he was no less an antifascist than a pacifist, and morally involved in mankind rather than in mathematics[14]).

After World War I, the attempts to organize technical development on a large-scale scientific basis were given up step by step.[15] The scientific-technological revolution was still too young for society (and especially for private industry) to venture into such undertakings unless in total war.[16] Only the collaboration between the Nazi State and the German chemical industry for the production of substitute raw materials, the steady expansion of industrial research in the United States during the 1930s, and the era of five-year planning in the U.S.S.R. re-established the direct coupling of science to technological practice ("science as a direct productive force," as formulated in a slogan). Not until the Second World War were things to become really serious.

From the First World War it was well known that science could also be a *direct destructive force*. According to a widespread anecdote, the U.S. Secretary of War had declined the services of the American Chemical Society during World War I, having discovered "that it was unnecessary as he had looked into the matter and found the War Department already had a chemist" (James B. Conant, quoted from Greenberg 1969: 88). No ministry of war desired to repeat that blunder (even though the Wehrmacht almost managed to do it in Zuse's case, cf. Box IX); researchers were therefore moved to other work in the laboratories, but not sent to the front.

The most famous scientific development project of World War II is the Manhattan Project, the creation of the atomic bomb.[17] This was no pure development project in the style known from World War I: This time, even the leading fundamental scientists were not yet in possession of all the knowledge required from their side; hence, systematic and extensive fundamental research was required. Some 150,000 persons came to be engaged in the project; a plentitude of researchers from the most diverse disciplines took part, along with development engineers from public institutions and private industry; the budget ($2 billion) has been estimated to equal roughly the total expenses for research and development in world history until 1940.

The Manhattan Project was the largest but not the only development project that involved fundamental research. So did the development of radar (begun already in the thirties), of penicillin, of the jet motor, and the realization of new levels in metallurgy, sonar technology, and other fields. In or near the domain of mathematics special notice may be taken of new developments in hydrodynamics and the building of the first efficient American electronic computer—the ENIAC, planned initially for ballistic calculations—as well as the construction of other American computers utilizing relays; developments in electromagnetic field theory and in the theory of networks associated with the development of the radar; Zuse's German computers, some based on electron tubes and some on relays (and used among other things for optimization of the wings of the "flying bomb"; the development of modern coding theory and the building of the British computer

---

### VII. Mathematics in World War I

Note 13 cites H.-E. Gross for the suspicion that the participation of mathematicians in the development of aerodynamics and airplanes may have inspired the postwar development of mathematics. Indications that the suspicion may be justified will be found in W. H. Young's address to the International Mathematical Congress of 1924: Mathematicians were not only, in his opinion, necessary for the further development of hydrodynamics; both "in the theory of Ballistics and in Airplane theory," important problems for theoretical mathematics still remained unsolved (and some of them remain so to this day—BB) (Young in *Proc. . . . 1924:* 156).

During and immediately after the War a global change of mathematicians' attitudes to their subject can also be perceived. At the Fifth International Congress in 1912, even the results of applied mathematics were understood in the spirit of neohumanism, as "truths about the universe in which we live" (G. H. Darwin in *Proc. . . . 1912:* 35). The next Congress was summoned by French mathematicians in Strasbourg in newly reconquered Alsace, barring the "criminal" and "unworthy" German and Austrian mathematicians, who had worked for the enemy. For all his enthusiastic integration of mathematics with the French cause, Picard worried in his opening address that the young generation would concentrate on the applications of mathematics and neglect the development of pure theory in the years to come; evidently, the importance of their science in the victorious military machine impressed the youth (*Proc. . . . 1920:* xxviii). In the more serene climate of the following International Congress in Toronto in 1924, H. S. Béland, Minister of Health and of Soldiers' Civil Reestablishment and representative of the Canadian Government, stigmatized the application of Science as "di[s]figurement of science's divine and celestial role [misprinted as "rule"], *viz.*, "the improvement of the condition of humanity, morally, intellectually, economically, and socially." In 1924, the applications of mathematics were still understood in specific technical and not in any neohumanistic sense (cf. Young as cited above); but there was not full confidence that mathematical fundamental research and pure mathematics were legitimate partners in the enterprise (*Proc. . . . 1924:* 52, 156).

---

Colossi used for that purpose; several branches of operations research (even the very concept of mathematical research in the effective operation of military systems originated in the wake of the British radar project).

Many of these fields have proved vital in the formation of postwar mathematics (even though some of them have gained institutional and proclaimed epistemological independence). In this sense, war had a lasting influence on what was going to happen in mathematics. It can justly be asked, however, to which degree the influences really originated in the war, i.e., whether the essential breakthroughs *made productive* during the war were also *created* by the war. One the whole, the answer is no. The basic mathematical ideas making possible the determination of the "inner ballistic" of the exploding atomic bomb, for example, were taken over from the theory of the Wiener-Hopf equation, coming thus from astrophysics—*viz.*, from Norbert Wiener's prewar determination of the radiation equilibrium at the surface of a star (Wiener 1964: 142f.). The key notions on stochastic processes did not spring from the branching processes of nuclear physics; they go back to Markov's purely theoretical investigations, the results of which he had illustrated on the alternation of vowels and consonants in Pushkin's *Eugene Onegin* (A. A. Youschkevitch, "Markov," p. 129). Both Zuse and Stibitz had built their first computers before the war, and the simultaneous building of computers in Germany, Britain, and the U.S. demonstrates to what extent the ideas were in the air (cf. Box IX).[18] Precursors of linear programming and of the simplex method, finally, had been devised in 1938 for use in Soviet economic planning (see Section 5).

A second question is whether the war ripened the various mathematical techniques to that point where they would become influential in mathematics and in the modern civilization of computers, automation, and systems planning. Once more, the answer is negative. Only when the techniques were transferred to the civilian domain and were used more variedly did they reach the point where they motivated the development of theory—and only when a fruitful interaction between theory and techniques had arisen did the latter attain their societal importance. In the earlier postwar era the rapidity with which IBM conquered the market demonstrates (as later does the Japanese ascent) that only the needs of civilian business, industry, and administration were broad enough to permit unfolding of computer production and utilization; no wonder, since only the civilian market called for and permitted mass production and continuous reduction of costs (cf. Boxes IX and X).

The same observation holds for operations research, systems control, and the other mathematical technologies of their generation. Only the broad and varied civilian applications changed their governing paradigm from "Portuguese navigational mathematics" to "Renaissance algebra."

We may hence conclude:

1. Even if the military needs of World War II promoted fundamental research, only "oriented basic research" was in fact involved, which brought no essential breakthroughs. Long-term needs of science (both regarding its own development and its ability to meet long-term societal demand) were not fulfilled to any remarkable degree.

### VIII. Mathematics and mathematicians in World War II

Already during World War II one Dr. Jewett, then president of the U.S. National Academy of Sciences and vice-president of American Telephone and Telegraph, had declared that "Without insinuating anything as to guilt, the chemists declare that this is a physicists' war [whereas World War I was known as a "chemists' war"]. With about equal justice one might say it is a mathematicians' war." This was quoted by Marston Morse (1943: 51), creator of the modern calculus of variations and at the time a member of the advisory committee on war application of mathematics (Reid 1976: 237), in an article on "Mathematics and the Maximum Scientific Effort in Total War."

Morse motivates this by the "machine nature of modern warfare," which not only "places engineering skill at a premium" but also, because it is "a war of invention," calls for "a new and more mathematical use of machines."

Often, "more mathematical uses" of technology presuppose no new mathematical research. This was also the experience of most of the mathematicians who worked for the War, as told by J. Barkley Rosser (whose later career as an army mathematics manager during the Vietnam War is amply documented in *The AMRC Papers*, p. 95 and *passim*):

> I have written to practically every [U.S.] mathematician still living who did mathematics for the War effort (there are still close to two hundred) and I asked for an account of their mathematical activities during the War. Many did not answer. And many who answered said they did not really do any mathematics. I had a one-sentence answer from a man who said that he did not do a thing that was publishable. If we equate being mathematics to being publishable, then indeed very little mathematics was done for the War effort. But, without the unpublishable answers supplied by several hundred mathematicians over a period of two or three years, the War would have cost a great deal more and would have lasted appreciably longer. (Rosser 1982: 509f.)

The various examples presented by Rosser demonstrate that what was needed was the—unpublishable—common sense of trained mathematicians as combined with their routine and comprehensive perspective. In reality, this is nothing but the old fortification computation and officers' mathematics transposed into the twentieth century.

Most of what Morse discusses in his article on mathematics in total war belongs to the same *genre*—or below ("swift, accurate mathematical computation" and "solution of problems of elementary algebra, plane geometry and plane trigonometry," together with fundamental mechanical physics and "good health and hard physical condition").

*(continued)*

VIII. **Mathematics and mathematicians in World War II** *continued*

The experience of World War II is of general validity for the scientific-technological revolution: Full utilization of science-based technologies presupposes broad scientific literacy, including mathematical literacy. "The problem of navigating a plane among the islands of the Pacific is very difficult. It is possible to lose as many men by faulty navigation as through enemy fire. It is clear that we must have tens of thousands of navigators. Are our students ready for this task?" (Morse 1943: 52).

In the War, as in the scientific-technological revolution in general, this is only one aspect of the use of mathematics. The other is the integrated development of mathematical theory and application. Above, and again below, we discuss the development of sequential analysis and of computers. In her discussion of "The Mathematical Sciences and World War II," Mina Rees (1981) concentrates on this aspect, the research-oriented projects. In 1943, Morse could evidently not discuss their work publicly, even though few were in a better position to know them than he.

2.  In the long run, military technological development receives more from civilian technology and science than it has to offer.
3.  Even if the budget of the Manhattan Project should be only approximately comparable to the sum of all earlier costs of research and development, we must (leaving aside the purpose of the project) characterize the scientific productivity of this gargantuan project as outrageously low: Certainly it did not double the knowledge and technical ability of humanity!
4.  On a planet governed by human reason (and not merely through Clausewitzian strategic rationality), the direct passage "from *Eugene Onegin*" to industrial application of Markov processes would have been possible, bypassing the scandalous squandering of resources in the Manhattan Project. Only in a militarized world (if at all) is militarism a precondition for scientific progress.

### 5. Our Present Situation

"Only in a militarized world (if at all) is militarism a precondition for scientific progress." Very nice. But what is the relevance of this seemingly comforting dictum for *our* world?

Let us start with the mathematicians' point of view, as expressed in their conception of mathematical excellency. The mathematical analogue of the Nobel Prize, the Fields Medal, is awarded every four years to two to four

### IX. The first computers

The first steps toward the modern computer were taken by Stibitz in the U.S. and by Zuse in Germany.

Stibitz describes the stepwise development of his machine as follows: In his job as a mathematical engineer at the Bell Telephone Laboratories he worked in autumn 1937 with electrical relays. From "curiosity" he became interested in their logical properties, and built a simple adding device on the kitchen table. Over the next three year, this led to the construction of a device (still based on relays) for the multiplication of complex numbers, which was by then becoming important in the telephone industry in filter theory and in the calculation of transmission lines. The costs ($20,000) prevented the construction of further machines—until another one was ordered by the National Defense Research Committee for use in air defense calculations. Further models followed, but until the end of the War air defense remained their only application (Stibitz in Metropolis et al. 1980: 479–83).

Even the development of Zuse's devices shows that time was needed for the computer to unfold its potentialities, and that this could be done neither in prewar civil application nor in World War II. After private preparations Zuse built a binary calculator based on relays in 1936, and in 1937 got a manufacturer to take interest in his prototype (though only with great difficulty). At the beginning of the war he was not excepted from ordinary mobilization, in spite of a declaration of the above-mentioned manufacturer that his invention would be important to the Air Force. "The German aircraft is the best in the world. I don't see what to calculate further on," as Zuse quotes his major. Only after half a year was he transferred—not as a computer inventor, but as an engineer—to the aircraft industry, which gave him the opportunity to continue his work on computers. Even in this case, only a few machines were built, and they were applied only in a single context: aviation computation, and process control in the aircraft and missile industry. Zuse's very advanced theoretical ideas did not get beyond his own mind and his private notes (Zuse in Metropolis et al. 1980: 611–15).

young mathematicians for epoch-making research. The selection of laureates since the first award in 1936 and the eulogies delivered by leading mathematicians show what is considered epoch-making by mathematicians. Deemed worth reward were only

- The solution of old *mathematical* problems,
- The unification of several *mathematical* fields through the discovery of transverse connections and new conceptualizations, and
- The opening of ways to new *internal developments* in mathematics.

---

### X. Univac and IBM: The costs of too-intimate relations with the Department of Defense

Engineering Research Associates (ERA; later integrated in the Sperry Rand Corporation and known as Univac), was founded by former reserve officers with a wartime background in cryptographic computing. For a long time the firm worked exclusively for the military on classified tasks. According to a representative of the firm, the cost was orientation toward the solution of already-defined, isolated problems and not toward the investigation and analysis of complex situations; lack of experience with creative interaction between users and manufacturers; and thus eventually delay of "its entry into commercial activities" and of "its maturation into a total computer systems supplier." The description is confirmed by events from 1953 onward. At that time IBM offered its first general purpose computer, the 701. ERA was now also oriented toward the civilian market, and a few months later it offered a technically comparable machine, the 1103, and gained certain initial advantages thanks to its longer experience with electronic computing; very soon, however, IBM took over the dominant position (Tomash in Metropolis et al. 1980: 485–90; cf. also Goldstine 1972: 325–29).

Even the IBM 701 was originally conceived as a defense calculator, and the first customer machine went to the nuclear Los Alamos Laboratory. From the beginning, however, it was planned deliberately to possess all-round applicability—in agreement with the customs of IBM, which for decennia had produced office machines and had well-rooted traditions for employment of broadly qualified scientists and for interaction with the university environment (Hurd in Metropolis et al. 1980: 390–92; cf. Goldstine 1972: 329–32).

---

These criteria hold even in those two to four cases where the laureate has worked in fields somehow connected to applications (cf. Box XI).

*If* applications are important to the development of mathematics, then not so important, we may conclude, that their influence cannot be suppressed from the consciousness of even the brightest mathematicians. Influences from societal practice on the development of their field must hence, insofar as they exist, mainly be of an indirect nature. They *do* exist, and the outlook of the "leading mathematicians" is thus distorted and narrow and a reflection of only one aspect of reality: This becomes obvious, e.g., if we consider the example of nonlinear analysis. M. S. Berger's *Nonlinearity and Functional Analysis*, which can be regarded as the modern standard textbook of the discipline, lists not only extradisciplinary yet intra-mathematical sources in differential geometry and variational calculus, but also, in considerable detail, an abundance of sources for the central problems of the discipline in

### XI. The Fields Medal

From the *première* in 1936 until 1986, 30 Fields Medals have been awarded.

| | | |
|---|---|---|
| 1936 | L. V. Ahlfors | Quasiconformal mappings, Riemann surfaces |
| | J. Douglas | Plateau problem of variational calculus |
| 1950 | A. Selberg | Analytic number theory |
| | L. Schwartz | Theory of distributions |
| 1954 | K. Kodaira | Harmonic forms on complex manifolds |
| | J. P. Serre | Homotopy theory of spheres |
| 1958 | K. F. Roth | Algebraic and analytical theory of numbers |
| | R. Thom | Topology of differentiable manifolds |
| 1962 | L. Hörmander | Linear differential operators |
| | J. W. Milnor | Differential topology |
| 1966 | M. F. Atiyah | Topological methods in analysis |
| | P. Cohen | Logic and set theoretical foundations of mathematics |
| | A. Grothendieck | Algebraic geometry |
| | S. Smale | Differential topology |
| 1970 | A. Baker | Theory of transcendental numbers |
| | H. Hironaka | Resolution of singularities in algebraic geometry |
| | S. Novikov | Geometric and algebraic topology |
| | J. Thompson | Simple finite groups |
| 1974 | E. Bombieri | Theory of numbers, real and complex analysis |
| | D. Mumford | Algebraic geometry |
| 1978 | P. Deligne | Algebraic geometry |
| | C. Fefferman | Convergence of Fourier series and Fourier integrals |
| | G. A. Margulis | Discrete subgroups of Lie groups |
| | D. G. Quillen | Geometric and topological methods in algebra |
| 1982 | A. Connes | Topology of operator algebras |
| | W. P. Thurston | Topological analysis of three-manifolds |
| | Sh.-T. Yau | Analytical methods in geometry (and *vice versa*) |
| 1986 | S. K. Donaldson | Topological analysis of four-manifolds and mathematical physics |
| | M. H. Freedman | Topological analysis of four-manifolds |
| | G. Faltings | Arithmetic algebraic geometry |

Of these, Laurent Schwartz built his works directly on mathematical problems belonging with various applications. His theory of distributions provided a mathematical basis for apparently paradoxical methods used currently in quan-

*(continued)*

---

**XI. The Fields Medal** *continued*

tum physics and in electrical engineering and physics. Donaldson had applied techniques borrowed from mathematical physics and has also worked on the theory of magnetic monopoles. Hörmander's and Fefferman's works belong in fields that originated in classical "applied mathematics"; their actual accomplishments, however, have very little to do with practical calculation, a fact that was emphasized in the eulogies. So in Hörmander's case: "Questions of this nature have no physical background but a very solid motivation: mathematical curiosity" (*Proc. . . . 1962:* XLV). Fefferman's work belongs in the domain of classical analysis, which in the 1940s and '50s was considered closed and hence "dead." He got the medal for the "unification of methods from harmonic analysis, complex variables and differential equations"; the realization that "in many problems complications cannot be avoided" (*Proc. . . . 1978:* 53) was praised.

Atiyah's eulogy of Donaldson (*Proc. . . .* 1986: 3–6) mentions and praises the mutual fertilization of pure mathematics and theoretical physics. That, however, is quite exceptional. Even when we go to Laurent Schwartz, whose achievement is so obviously connected to non-mathematical practice, this is no theme for Harald Bohr. Instead, his ability "to shape the new ideas in their purity and generality" is underscored ("the physicists and the technicians" *are* mentioned, it is true, but only so that it may be told that their methods are illegitimate) (*Proc. . . . 1950:* 130–33, quotations pp. 133 and 130). In all other cases, extra-mathematical connections have been mediated historically through multiple conceptual and disciplinary reconstructions. Nonetheless, the results of several medal winners have stirred interest in other sciences (so Thom's in biology, Atiyah's in quantum dynamics, Smale's in economics). The inner dynamic and the dominance of internal interactions does not impede the applicability of mathematics; on the contrary, the existence of a dynamic inner structure is one (of several!) conditions that qualitatively *new* applications may turn up.

Connections between Fields Medal winners and the military cannot be traced at the level of mathematical substance, but only at the level of NATO and similar grants.

(Sources: *Proc. . . . 1936*–*Proc. . . . 1986*; *Notices of the American Mathematical Society* 29 (1982): 499–502; Albers et al. 1987).

---

classical and modern physics, in economy, and in biology (Berger 1977: 10–18, 60–63).

Precisely in the domain of nonlinear analysis, it is true, are strong influences from modern applications to be expected. Nonlinearity is (together with mathematical statistics), so to speak, *the* mathematics of the scientific-technological revolution. Classical mechanics and classical mathematical economics build largely on linear approximations to reality without bothering about determining the higher-order deviations from linearity (cf. Hooke's law, d'Alembert's wave equation, and Walras's "pure political economy"); similarly, they build on the fiction of exact and complete predictability. Application of such theories in practice, then, presupposes the *ad hoc* rules of thumb of the practician knowing *when* and *how* to apply them *and when not.* If instead, as characteristic of the scientific-technological revolution, the breach between idealizing classical theory and real material or social practice is to be bridged systematically and scientifically, mathematization of non-linear relations and of the incompleteness of information must be appealed to.

If instead we look at disciplines such as modern topology or modern algebra, such direct inspiration from problems of practice is not to be expected, nor to be found. In another sense, however, even these disciplines can be regarded as characteristic of the scientific-technological revolution. They mathematize not the complexity of the real world, but *complexity as such.*[19] In which time, if not in ours, would research deal with groups with 808 017 424 794 512 875 886 459 904 961 710 757 005 754 368 000 000 000 $(= 2^{46} \cdot 3^{20} \cdot 5^9 \cdot 7^6 \cdot 11^2 \cdot 13^3 \cdot 17 \cdot 19 \cdot 23 \cdot 29 \cdot 31 \cdot 41 \cdot 47 \cdot 59 \cdot 71)$ elements?[20] Which century but the present would come to consider the infinitely irregular Mandelbrot set and no longer the supremely regular circle the most beautiful of geometrical figures?

Already nonlinear analysis is thus subject quite as much to influence from the characteristics of societal practice as a whole as from specific problems. In the cases of topology and algebra, the same assertion can be made with greater emphasis.

Will all this—Fields Medal, nonlinearity, topology—go together in a common picture of the relations between mathematics, overall societal practice and, not to forget, military influence? It will, *viz.,* under the heading *organized segmentation* (we could also speak metaphorically of a "division of labor").

We must remember that the number of mathematicians, as well as the number of people applying mathematics, is greater today than ever before. As D. J. de Solla Price once observed, the majority of scientists and engineers ever born are still alive. So many mathematicians and so many technologists are active today that strong *inner* interactions are possible and may dominate in each of the two domains. Both domains have also developed so far beyond the terrain of common sense that most questions posed by a mathematician can be understood (let alone answered) only by other mathematicians, and most of a mathematician's work will be directed only to disciplinary colleagues (in a narrow sense). *Mutatis mutandis,* the situation of the technologist should be described in similar terms. In general, only the accumulated result of the work of many mathematicians and the airplane on which many technologists have cooperated are of any use to outsiders. Mathematics, technology, and even their single subdisciplines are "open systems": For each average element (mathematician or theorem, engineer or machine component), interactions with other similar elements dominate in number as well as importance. For each system, these internal bonds generate both the inertia of the system and its ability to "produce" efficiently. External interactions are marginal compared to the internal ones; as in a biological system they can, however, be said to provide the necessary energy; like the pilot on the ship they may determine the direction taken by the inertial mass and thus also ultimately the character of the production.

These metaphors borrowed from systems theory can be filled out. The Fields Medal reflects the primary dominance of the inner dynamics. Even in the rare cases where an impact of system-external factors can be tracked, the inward-pointing aspect of the work alone is rewarded. The role of the marginal external factors, on the other hand, is obvious when we turn to nonlinear analysis. This discipline is certainly not a mere heap of mathematical answers to discipline-external problems; most theorems relate to problems actualized or created by other theorems—which is the reason that we can at all speak of "a discipline." The fundamental issue, on the other hand, and the overall aims pursued belong outside the domain and, to a considerable extent, even outside mathematics.[21] In this way, system-external influences thus provide both energy and direction. The same could be said of mathematical statistics and probability theory.

Disciplines such as topology are better shielded from extra-mathematical influences. The external sources of topology belong almost exclusively in other branches of pure mathematics: theory of numbers, algebra, complex analysis, partial differential equations, differential geometry, and finally, of course, the very abstract character of contemporary mathematics in general and its interest in complexity as such.[22]

Already in Galileo's time, we remember, both science and technical practice had to be understood as networks; the differentiation between physics and engineering science and between pure and applied mathematics, furthermore, were among the preconditions for the nineteenth-century beginning of scientific technology. Then, however, it is scarcely a miracle that comparable features characterize the situation in the ripening scientific-technological revolution even more clearly. The epistemological arguments that "organized segmentation" is necessary for the productivity of the total scientific-technological system are thus supported by historical continuity.

It should be observed already in this place that organized segmentation not only is responsible for the productivity of the system (which is after all a very abstract concept as long as we do not discuss *what* is being produced) but also for the distorting trends, i.e., for the insensate and unscrupulous destruction of nature and fellow beings. Each participant tends to know only his own subsystem and its inner (social and cognitive) structure. Its outward interactions, its role in the greater whole, tend to be obscured, as demonstrated with astonishing clarity in the case of the Fields Medal. The professional conscience of the single ("naive") participant is only related to the subsystem, and the ignorance of its more global role will easily express itself as unscrupulousness with regard to global end effects. From the overall viewpoint, "naive" participants are *elements*, comparable to the elements of an engine, and not carriers of moral responsibility.

If only the "naive" participants were present, segmentation would not be "organized"; the effectivity of each subsystem would then be measured by arbitrary local, inertial standards, and the total system would soon not work efficiently at all. But subsystems, as we have seen, are not as isolated as they seem from the perspective of local participants. Interaction between subsystems, however, is often governed by blind or half-blind economic forces, by political decisions funding one area or type of activity and not the other, or by intentional canalization of scientific information (calculated by those in power in big business or in bureaucratic or military structures) in select directions and forms.[23] In as far as this is actually the case, the standard according to which the total system is "efficient" will be determined accordingly. As long as the conglomerate of big business and bureaucratic and political power is not humanly and ecologically responsible, neither will a scientific-technological system dominated by "naive" participants be.

Until this point, this section of the chapter has only considered the links between mathematics and overall technical and even societal practice. What is to be said about the connection to the military sector, which is, after all, our theme? What is predicted by historical continuity?

Never in history was there anything *specific* in war and armament, seen from the point of view of mathematics. *War* was, as one of the constituents of global societal practice, comparable to other constituents. When war could boast of societal priority, it would also be a major external influence on the way mathematics was made (cf. the École Polytechnique or the preface to the *Dioptra*). Rarely, if ever, was the influence of war and warfare of specifically military character. Hooke's law would have looked no different if the Royal

Society had wanted to make economies in the construction of merchant vessels; if anybody had been willing to pay for an investigation of air resistance to sugar balls, the same questions would have been posed to mathematics as in military ballistics.[24]

Is the situation different today? Let us look at the case of sequential analysis, as it was developed in World War II. Evidently, the statistical analysis of the reliability of a production line was important in the mass production of weapons and munitions. As it appears, moreover, it had never been important enough in prewar civil production for anybody to develop the technique. In real history war was thus the precondition for the formulation of the problem as a scientific question and for its solution—as once was the English Royal Navy for the study of wood. Today, however, the standard examples in the textbooks deal with quality control in civilian production and other harmless—perhaps even useful—random samplings. In epistemological principle, these might just as well have presented the occasion for Abraham Wald to open up the field of statistical decision theory. The practical elaboration of the technique[25] was taken care of by members of "what surely must be the most extraordinary group of statisticians ever organized, taking into account both number and quality."[26] At least until 1941, we have to conclude, civilian life had not matured to a point where it could organize a comparable group.[27]

The possibility of a civilian development of sequential analysis is an unhistorical speculation, though supported by the detailed course of its history.[28] In *real* history, however, another mathematical technique promoted by war was developed independently in the civilian domain. As touched upon already, the first beginning of linear programming, created by T. C. Koopmans for the optimization of transport in the Pacific war, was presented in a no less developed form by L. V. Kantorovich in his Russian *Mathematical Methods of Organizing and Planning Production* in 1939 (see Rees 1980: 618), related to the Soviet five-year planning.

During the postwar period, the *not specifically military* character of the military influence on scientific and technological development has been highlighted by the genesis of higher programming languages. COBOL (COmmon Business Oriented Language), until today the language most widely used for administrative data processing (hence most widely used *at all*), was hatched by a commission organized by the Pentagon, in which the Air Force and the Navy cooperated with representatives of the largest industrial corporations (see Sammett 1969: 330 ff., and Box XII). The language also owed its rapid spread to military initiative, *viz.*, to a Pentagon declaration to buy in future only computers equipped with a COBOL compiler or with demonstrably equivalent facilities.[29] The military customer carried such weight that all manufacturers had to offer COBOL.

FORTRAN (FORmula TRANslating system), still most widely used for scientific and technical calculations, conveys the same message in other terms. The language appears to have been developed by IBM on its own initiative from 1954 onward. The undertaking was closely bound up with the attempt of the office machine giant IBM to enter the market of scientific-technical computation and rapid electronic computers. All institutional and personal connections indicate, however, that this market was largely identical with the fraternal association of military establishment with military and nuclear technology: University of California, Radiation Laboratory; United Aircraft Corporation; Join Meteorological Committee of the Joint Chiefs of Staff; Oak Ridge National Laboratory; and U.S. Navy (see Goldstine 1972: 338–31, and Sammett 1969: 143).

In more recent times, finally, the Pentagon is known to have undertaken at huge cost the NATO-wide development of the programming language Ada, intended for real-time control and command of subsystems. Just as COBOL can be applied to many other administrative tasks than the management of spare parts for military airplanes; and as FORTRAN was only "by chance" developed for military and not for civilian weather forecast; so real-time communication between subsystems equipped with disparate software is not a requirement of the NATO integration alone but a characteristic need of any large system in the present phase of the scientific-technological revolution.

Ada may be overloaded for civilian purposes in its actual execution. That it is so is claimed by many experts. The integration of all arms systems during the one decisive hour of a third world war is, indeed, qualitatively different from the lasting integration within a great corporation or a national economy. The possible overloading of Ada is hence a parallel to the exorbitant reliability standards for military electronics, which Joachim Wernicke has characterized as a "tribute to an absurd understanding of technology" (the abundance of components, for example, in a modern tank being so great that it will only run approximately 200 km between thorough servicings, in spite of the extreme standards for single components).[30] However, what we might designate the "civilian-progressive aspect" of Ada—that civilian utility which purchasers of the language as a commercial product expect to find—is only affected negatively by the overloading. A civilian-progressive version of Ada free from much of the overloading might just as well have been called forth directly by civilian needs.

Ada demonstrates once more that the direct passage "from *Eugen Onegin* to industrial application" is only impossible in a world militarized to a point where nationwide scientific development projects are organized solely under the aegis of the armed forces. Confrontation of the apologies that insist upon the progressive role of military research with the conceivable overloading of the programming language also suggests the general question whether

---

**XII. Who needs COBOL?**

The Pentagon-initiated commission that in 1959 decided to develop COBOL (and did it) was composed of representatives of these institutions and firms:
   Air Material Command, United States Air Force
   National Bureau of Standards, U.S. Department of Commerce
   Burroughs Corporation
   David Taylor Model Basin, Bureau of Ships, U.S. Navy
   Electronic Data Processing Division, Minneapolis-Honeywell Regulator
      Company
   International Business Machines Corporation
   Radio Corporation of America
   Sylvania Electric Products, Inc.
   Univac Division of Sperry-Rand Corporation
That is, Air Force, Navy, Bureau of Standards, and six computer manufacturers. In the further work of the "Maintenance Groups," the following manufacturing and user corporations were also represented:
   Allstate Insurance Company
   Bendix Corporation, Corporation Division
   Control Data Corporation
   Dupont Corporation
   General Electric Company
   General Motors Corporation,
   Lockheed Aircraft Corporation
   National Cash Register Company
   Philco Corporation
   Standard Oil Company (N.J.)
   United States Steel Corporation

(S. Rosen 1967: 121f.)

---

the detour over military research is *a risky and murderous but all in all economically sound* or even efficient way to civilian progress, or it is, like the Manhattan Project, in addition to risk and moral monstrosity, also a waste of resources and perhaps a source of generally perverted technology?

If we compare the resources put at the disposal of military research since 1945 with the score of known gains for civilian life, the term *waste* seems appropriate and even mild. The discussion of the spin-off effects of World War II research points in the same direction. So also do theoretical arguments:

First, military research is bound to secrecy. This obstructs the "natural" growth process of science, where every open question and every important result is in principle presented to the worldwide community of

---

### XIII. ADA

[ . . . ] no great computer corporation and no group of scientists with university background sponsored Ada. The U.S. Department of Defense (DoD) set fire to the rocket, which according to the intent and the firm conviction of its creators will ascend the heaven of programming languages like a comet. In the beginning of the seventies the investigations of the DoD ascertained that the production, maintenance and use of all its different software products absorbed astronomical sums. For 1973 alone this chapter in the budget was calculated to amount to 3000 000 000 $. It was believed that the reason for this horrifying consumption was the abundance of different programming languages in use. From then on everything went on in military style. (Barth 1982: 141)

As a matter of fact, the military requirements were directed not only at rationalization and standardization of "the abundance of different programming languages in use," but also at qualitatively new performances, as described, for example, in the foreword to *Military Standard 1815*, the official description of the language (from December 1980; reprinted in Horowitz 1983):

[ . . . ] the real-time programming language Ada [is] designed in accordance with the United States Department of Defense requirements for use in embedded systems. Such applications typically involve real-time constraints, fail-safe execution, control of non-standard input-output devices and management of concurrent activities. Ada is intended as a common high order programming language and has the mechanisms for distributing large libraries of applications programs, packages, utilities and software development and maintenance tools. Machine and operating system independence is therefore emphasized throughout its design.

The Ada Language is the result of a multinational industry, academic and government effort to design a common high order language for programming embedded computer and real-time defense systems.[ . . . ]

(quoted from Horowitz 1983: 419)

Obviously an invitation to overload!

*(continued)*

---

colleagues for criticism and further work. Vital information concerning military research circulates in narrow circuits only, not even defined by national borders but by the single institution or even working group: Portuguese navigational science is still a valid paradigm. We may also think of the development of nuclear plants, where the use of the reactors for the manufacture of bombs caused "the earliest development to take place in strict secrecy,

**XIII. ADA** *continued*

The computerization of the NATO-integration dressed up as a beautiful young upper-class girl. (from *The Mathematical Intelligencer* 2 (1979–80): 16):

*The United States Department of Defense*

*Invites You to Attend the Debut*

*of the*

*Ada Programming Language*

*September Fourth and Fifth, Nineteen Hundred Eighty*

*U.S. Department of Commerce Auditorium*

*Constitution Avenue at Twelfth Street, N.W.*

*Presentation by Dr. Jean Ichbiah*
*(Principal Language Designer)*

*Starting at Nine o'Clock Each Morning*

*R S V P*

whereby the responsible engineers and physicians and the technologies which they created were shielded from the critical scrutiny of the vast majority of their colleagues''—thus according to Burhop (1980: 2), who considers this original sin against the norms of science to be an essential source of the problems that still ride nuclear technology (cf. also Box X on IBM and Univac).

---

**XIV. Project Hindsight**

In 1963, the Pentagon had the technology of 20 essential and advanced weapons analyzed: various nuclear warheads, rockets, radar equipment, a navigation satellite, and a naval mine. As far as possible the contributions of separate scientific and technological advances made since 1945 to each weapon were traced. In this way, 556 separate contributions were found. "Of these, 92 per cent came under the heading of technology; the remaining 8 percent were virtually all in the category of applied research, except for *two*," which came from basic research (Greenberg 1969: 59).

The methodology of the project was not without problems. Thus, all semiconductor technology was reckoned as *one* contribution (the transistor). Still, the overall result is not subject to doubt: Technological innovation takes most of its nourishment from the technology subsystem.

---

Next, problems posed by the military are mostly of punctual character, and to be answered within a brief delay. "Better to have a fairly satisfactory answer now than to wait two years for one that is theoretically worked out" (the Portuguese paradigm once more!). Especially after the early sixties, where a Pentagon-conducted evaluation project "Hindsight" found very little gain for the armed forces in the entire bulk of postwar fundamental research (see Box XIV), the trend has been for military development projects to build increasingly on existing fundamental knowledge alone.[31]

For both reasons, resources given to military research tend to produce "less knowledge per dollar" than resources used in open civilian or fundamental research. Science *as a body of knowledge* is thus less influenced by its military involvement than might be expected from the weight of military financing. Since, moreover, not all knowledge produced for its destructive relevance is equally valuable for constructive purposes, "less knowledge per dollar" becomes *even less civilly productive knowledge per dollar.*

Some critical observers of the military economy go further, claiming that *no*, or almost no, civilly productive knowledge derives from contemporary military research. Thus Mary Kaldor (1982), according to whom military technology is, since the late fifties, so specialized and the "autistic drive" of the military-industrial complex so precisely focused on this specialization, that spin-off from the "baroque arsenal" can lead only to total technological degeneration.[32] The American Marxist Victor Perlo (1974: 172) argues on similar lines, whereas the Soviet economist Viktor Kudro (1981), as well as West German and Japanese competitors of the American corporations, fear that the armament expenditures may after all contribute to stabilization of the technological leadership of the United States.

Competitors have good reasons not to assess the potentialities of their rivals wrongly. We may therefore trust them, in that "less knowledge" remains *knowledge*. Kaldor's arguments, on the other hand, are valid for many if not for all domains of armament technology—from which follows that "less knowledge" remains *less*.

Military attachment, it is true, is not the only source of secrecy, needless duplication, inadequate orientation, and punctualization. Even the economic competition in capitalism,[33] as well as the publish-or-perish degeneration of the scientific community itself, produce such malfunctions.

Against these effects of economic forces and social structures, however, other forces are at work: public registers of patents, the social norms of scientific life, citizens' protest movements against baroque technology, and even the interest of the great corporations themselves in appropriate commercial exchange of knowledge. In military research, these forces are largely thwarted[34]—see Box XV on Reagan's Executive Order on Classification of National Security Information, classifying in principle all research results that have not been explicitly declassified, and on the resistance from both industry and scientists.

All this holds true for every kind of research associated with military development: All sciences, and not only mathematics, are frustrated by a military attachment. But it *also* holds true for mathematics. In addition, at least punctualization may affect mathematics even more than other sciences. Questions asked by technological practice to (for instance) physics will often presuppose an inclusive understanding of the physical phenomenon concerned; this applies to the connection between the front of computer technology and solid-state physics, as well as the relation of fusion reactors to plasma physics and that of meteorology to hydrology. On the other hand, military as well as most other practical claims on mathematics will ordinarily be mediated through technical problems involving mathematics as an auxiliary discipline answering isolated questions. Here "mathematics no longer appears as the 'Queen of Sciences,' but takes over the function of an ancillary science for other disciplines or for application in daily life."[35] The function of ancillary disciplines, however, is essentially that of offering ready-made answers on punctual questions; services rendered in this capacity by mathematics (or by any other science in a similar position) are not likely to stimulate or to participate in the internal development of the field.[36]

Things *need* not be like this. In the recent history of mathematics examples *can* be found where a whole branch of mathematics has had to be developed in order to solve a practical problem—according to insiders in the field, for instance, Pontrjagin derived his branch of control theory from the problems of missile guidance.[37] Thus, things *need* not be like this. Yet as a

## XV. NO ADMITTANCE! SCIENTIFIC ZONE!

Like many others, changing U.S. governments have classified certain scientific results as important for national security. Yet from Eisenhower through Nixon and Carter, they have done so less and less. Since 1980, the trend has changed violently. Already in Carter's last year, the National Security Agency, the National Science Foundation, and the American Council on Education initiated a commission to prepare a precensorship on all publications in the domain of cryptology. In April 1982, this resulted in an invitation to the members of the American Mathematical Society (loyally communicated by the Society) to accept until further notice a voluntary preview—"We would welcome the opportunity to review and comment on papers, manuscripts or related items of any individual who is performing research in or related to the field of cryptology and believes such research may have national security implications" (*Notices of the American Mathematical Society,* 29(4) (1982): 322f).

In the same month, the Reagan administration issued its Executive Order on National Security Information, which makes the classification of all scientific, technological, and economic information possible if the government considers it to touch national security. In principle, all scientific information is classified until otherwise decided.

In the final version, it is true, the major but not clearly delimited part of basic research is exempted from compulsory preview; still, administrative praxis shows that a very wide concept of "security interests" prevails. Thus, in February, 1980 the organizers of an international conference on bubble memory were threatened by fines of $10,000, 10 years in prison, and an additional fine of five times the value of any equipment seen or demonstrated if participants from socialist countries were not disinvited (which they were).

In 1982, the deputy director of the CIA told the American Association for the Advancement of Science how global classification should be organized: By including "in the peer review process (prior to the start of research and prior to publication) the question of potential harm to the nation" (quoted in Gerjuoy 1982: 34).

The new mania for secrecy will not only damage American science, American economy, and American democracy in general; it will also hamper the development efforts of the American armaments industry. This was the conclusion at which arrived a panel appointed by the National Academy of Sciences, the National Academy of Engineering, and the Institute of Medicine, consisting of senior members of university faculties and administrators, former federal officials, and executives of high-technology companies, and supported by the Department of Defense, the American Association of the Advancement of Science, the American Chemical Society, the American Geophysical Union, and a consortium of private foundations (*see Scientific American* 247(6) (December 1982): 65f., "Secrecy v. Security"). This and similar advice forced the DoD to reach a certain accommodation with the opponents of secrecy; at the same time, however, other agencies continued the old policies (see *Scientific American* 251(6) (December 1984): 60f., "Secret Struggle").

rule they are. If we refer back to the introductory aphoristic question, at least part of the reward received by mathematics for prostituting itself to the military sector is paid out in counterfeit coin.

Up to this point we have discussed the involvement of mathematics with the military system as motivated by the applicability of mathematics. Especially in the U.S., but in other places too, another source of involvement exists that may perturb the picture: The funding of research, including mathematical research, in order to further the integration of universities into "the life of the nation"—to use the nice words by which Vice-President Hubert Humphrey described the incorporation of universities in the military-industrial complex.[38] The issue is thus purchase not of know-how of military relevance but of ideological, political, and moral loyalty. This, however, is mostly not realized by critical observers—nor, in many cases, by those who are bought. Some mathematicians enjoy telling about how they have palmed off their pure-mathematical hobbies on the "stupid colonels" as possessing military potential—caring solicitously, however, not to pique the man with the check book by politically offensive conduct. In private they are proud of having tricked the destructive apparatus out of funds by parading their Hilbert-space inquiries as space research. Others, conversely, are proud of their participation in "the life of the nation," and are delighted that their research, the futility of which they had regretted, seems after all to be of national importance. Critically minded observers, finally, see the prevalence of military financing and grants as a proof that all science is thoroughly militarized.

At times it may be difficult to know whether a grant is given in order to obtain the ideological compliance of the scientist (and *university teacher*, which may in fact be the key point), or because the integration of a seemingly esoteric detail in a particularly sophisticated military project is anticipated. In both ambiguous and unambiguous cases it is obvious, however, that those resources which are now channeled to fundamental research through military budgets in order to serve the intimidation and ideological vassalization of mathematicians are no more scientifically productive than if they had been given directly.

### 6. What Can—and What Should—Be Done?

Mathematics as it developed historically until the beginning of this century was no offspring of war, this is what follows from Sections 1–3. The increasingly total character of military preparation and warfare, in combination with the scientific-technological revolution, has led since then to escalated military influence in mathematical research and to important military exploitation of mathematical knowledge, though until today without determining the overall development of mathematical thought and theory formation—cf. Sections 4–5.

*As far as only the logic of research is concerned*, it also emerged from the discussion in Section 5 that mathematics could develop quite as well or better in a peaceful world, if supplied with the same or even with somewhat smaller resources (a trivial conclusion, had it not been for the contrary claims of too many apologists). Mathematicians are thus not dependent on the commerce of killing in their striving for the advantage and the progress of their science. That may ease the bad conscience of those who believe themselves to belong to a morally corrupt science.

Mathematics, however, does not exist in abstract logic alone but in the real world with its very real threats and military blocks. Similarly, mathematicians belong to a profession and not only to a science—and at least if we include in the class of mathematicians not only those doing genuine mathematical research, but also those trained in mathematics at university level and applying it more or less creatively, a large proportion of this community will be found on the military payroll. Let us consider the implications of these realities, concentrating upon the situation of mathematicians of our own block—those of the socialist countries are in a better position to assess their own dilemmas.

Mathematicians calculate the "decapitation" of the Soviet Union in a nuclear first strike (and not only U.S. mathematicians take part!). They calculate the techniques of the strike, especially the precision of "our own" missiles and thus their ability to destroy those of the adversary in their silos, and the acoustical underwater localization of adversary atomic submarines (two necessary measures if an adversary is to be robbed of a deadly second strike); they calculate the miniaturization of H-bombs and the optimal maximization of explosive force or radiation effect through precise calibration of the temporal progress of ignition, fission, and fusion processes; they calculate the "vertical proliferation," the proliferation of nuclear vehicles that from one year to the next becomes increasingly unverifiable,[39] and augment the range of missiles through improved thrust chamber geometry and that of long-distance cruise missiles through the development of diminutive, high-efficiency jet motors; finally, they calculate the total strategy of the first strike, accepting the death of 30 to 60 millions of Americans as "admissible losses," regarding the destruction of Western Europe as consumption of a "dispensable item," and passing over the killing of perhaps 150 millions of Soviet civilians as "collateral damages."

This is probably the most scary example of the involvement of mathematics in the preparation of war, yet only one example. As already touched on, since the 1940s many mathematicians have been employees of the armed forces, have taken part in military research projects, or have worked on military grants; many mathematicians were one way or the other brought to ideological and political compliance; and a few theoretical specialities were even created or experienced essential new developments in interaction with military needs. On the whole, mathematics may be the most important constituent in the infrastructure of the scientific-technological revolution and of contemporary scientific advance. The kind treatment of mathematics in the midst of cutbacks in all areas of civilian research (see Box XVII) implies that the Reagan administration took the military commitment of mathematics for granted, and probably not quite unfoundedly.

Observation of the privileged treatment of mathematics as compared to most sciences by an administration oriented so unequivocally toward what Hardy caustically labelled "utility" introduces another challenge to the abstract logic of research (which may worry *some* mathematicians): the question whether in a peaceful world mathematics would be "supplied with the same" or with only "somewhat" smaller resources.

"Why not?" we might ask naively. Mathematics would be useful for meeting all sorts of human needs, and not only those of the strategists and generals. Still, economists teach us to distinguish effective demand from need, and sociologists will tell that not everybody "needing" is entitled to "demand." We shall therefore have to consider the conditions of the real world, which—as far as its "Western" part is concerned—is a highly organized monopolistic capitalism. Certain dogmatics will maintain that this system is driven *inevitably* toward imperialism and aggression, and that this is the only reason it invests in research and development. A thorough discussion of these two questions would lead too far; here we shall restrict ourselves to two brief observations. First, the decisive question even for mathematicians must be that of the feasibility of a non-aggressive monopolistic capitalism (or, if that will not be possible, of the abolition of monopolistic capitalism), not that of the scale of mathematical research budgets once a peaceful world has been attained one way or the other. Second, armament is surely an important incentive for the U.S. to invest roughly 2 percent of its national product in research and development. The mechanisms of economic competition between firms and between states under the conditions of monopolistic capitalism are, however, of a nature that even without the incentive of so-called "defense" would secure automatically some two-thirds of the present R&D budget.[40]

In the real world, as it exists and as it will go on existing according to current trends, mathematics is thus bound up with the military and the arms race; inasfar as it is at all possible to make this world peaceful, mathematics could, on the other hand, be disentangled without suffering loss of theoretical

## XVI. The loyalty market

The regular use of research funding as a means to buy loyalty is demonstrated by Serge Lang (1971). He presents the connections between the Department of Defense, public research funding in general, and universities through a wealth of cases where attempts were made to scare or discipline American mathematicians during the Vietnam War—and he points out what could be done through collective opposition.

One case is that of L. Lecam and J. Neyman, two famous statisticians from Berkeley, who during the fifties and sixties had carried out unclassified research in statistics that was published in standard journals and supported by contracts with the Army and the Office of Naval Research (*inter alia*). In 1968, they signed, together with others, an advertisement in the *Notices of the American Mathematical Society,* stating that

> Job opportunities in war work are announced in the *Notices* of the AMS[ . . . ]and elsewhere. We urge you to regard yourselves as responsible for the uses to which your talents are put. We believe this responsibility forbids putting mathematics in the service of this cruel war.

Even though the military value of their research could in no way be influenced by their signature, the two "disloyal" contractors were threatened by both the Army and Navy that their contracts could not be prolonged—under the pretext that continuance would place them "in a most uncomfortable, and perhaps untenable" conflict of conscience. (Upon which Neyman commented that he "did war work during the Second World War. For 16 years, I have not done any war work. I prove theorems, they are published, and after that I don't know what happens to them").

The two statisticians made the threat public, which led to scandal, and it was then withdrawn. They made renewed applications to the ONR, Neyman for "weather modification studies promising benefits for Nation and humanity," and both got new contracts without submission. Less famous scientists, however, were not as lucky, and in order to avoid future defeats the Director of Defense Research and Engineering issued a memorandum that all contracts *where the "non-technical" situation* (read: *the political loyalty of the contractor*) was *"uncertain"* should be submitted to special control of quality and productivity; for one thing, diplomatic reasons required that rejections should not emphasize political but only technical issues; besides, even politically disloyal researchers might happen to contribute "significantly to the country" (pp. 53–58, full text of the memorandum p. 57).

substance and inspiration (the question of the "logic of research") and without major "social" reductions (budget, manpower, social prestige, and whatever else may interest the members of a profession). Disentanglement need not, however, wait until an evangelical Peace on Earth has been achieved; on the contrary, disentanglement may contribute to pacification.

This imposes the obligation to realize at our best that which *can* be realized, and raises the question *how* it can be realized.

To such questions physicists are traditionally more receptive than mathematicians. Since Hiroshima and Nagasaki the whole mythology of their field (understood as those parts of its history which are known by everybody) tells them that they stand "one foot in jail." Some years ago, one of their number formulated sarcastically that "a member of the physics enterprise can minimize his possible contributions to military needs by either not teaching or by teaching poorly, and by either not doing research or by doing research unrelated to winning advances in basic or applied knowledge" (Woollet 1980: 106).

The second half of the aphorism refers once more to the network character of scientific knowledge: Concepts, methods, and techniques may well have been developed in order to gain command of one, perhaps totally abstract, domain; yet no warranty can be given that in military contexts they will not suddenly win concrete, destructive significance. The first half, however, points to something which we only touched in passing so far. Until this point, indeed, we only considered mathematics as a body of knowledge and as a researchers' community. Many mathematicians, though, also teach and

---

**XVII. The Reagan administration and mathematical fundamental research**

The Budget of the U.S. National Science Foundation (see next page).

The table appears in an article on the budget of the National Science Foundation for 1983. Other tables in the same article show that the bulk of funding for mathematical research will continue to flow to basic research, which in real value is not increased, but anyhow is less severely cut than all fields on average (not to speak of programs in educational science). Also exempted from reductions are computer science, physics, chemistry, and materials research. The real increase in mathematics falls exclusively on "special projects," research activities that combine several mathematical domains and which seem to be decoupled from teaching.

*(continued)*

**XVII. The Reagan administration and mathematical fundamental research** *continued*

**TABLE I. NATIONAL SCIENCE FOUNDATION BUDGET**

(Millions of Dollars)

| | 1979 Actual 1/28/80 | Change (79–80) | 1980 Actual 1/15/81 | Change (80–81) | 1981 Actual 2/8/82 | Change (81–82) | 1982 Plan 2/8/82 | Change (82–83) | 1983 Request 2/8/82 |
|---|---|---|---|---|---|---|---|---|---|
| **(1) Mathematical Sciences Research Support** | **$ 22.8** | **9.6%** | **$ 25.0** | **13.2%** | **$ 28.3** | **10.3%** | **$ 31.2** | **8.7%** | **$ 33.9** |
| (2) Other Research Support (Note A) | 761.0 | 7.5% | 817.7 | 6.8% | 873.7 | -0.3% | 870.8 | 9.5% | 953.4 |
| (3) Education, Information, Foreign Currency Program (Note B) | 88.4 | 7.9% | 95.4 | -15.5% | 80.6 | -61.9% | 30.7 | -27.0% | 22.4 |
| (4) Program Development and Management ("Overhead") (Note C) | 54.7 | 6.4% | 58.2 | 1.7% | 59.2 | 7.3% | 63.5 | -0.6% | 63.1 |
| **(5) Totals** | **$926.9** | **7.5%** | **$996.3** | **4.6%** | **$1041.8** | **-4.4%** | **$996.2** | **7.7%** | **$1072.8** |
| (6) (1) as % of (1) and (2) | 2.91% | | 2.97% | | 3.14% | | 3.46% | | 3.43% |
| (7) (1) as % of (5) | 2.46% | | 2.51% | | 2.72% | | 3.13% | | 3.16% |

NOTE A. Scientific research and facilities (excluding mathematics and science information). National and special research programs, and national research centers. Support for mathematics has been excluded, cf. items (1) and (3).

NOTE B. The programs in this group are ones in which there is some support for projects in every field, including mathematics. The foreign currency program involves both cooperative scientific research and the dissemination and translation of foreign scientific publications. Foreign currencies in excess of the normal requirements of the U.S. are used.

NOTE C. This heading covers the administrative expenses of operating the Foundation; the funds involved are not considered to constitute direct support for individual projects.

*(Notices of the American Mathematical Society* 29 (1982): 238).

---

**XVIII.** *Diplom-Mathematiker*: **Mathematicians for "State, Defence and Economy"**

*On this question, Horst-Eckart Gross writes to us:*

The curriculum for students in physics and in mathematics (instituting the Diploma Examination as well as the *Diplom-Mathematiker*) was announced by an instruction from the Reichsminister für Wissenschaft, Erziehung und Volksbildung August 7, 1942, and published, e.g., in *Studium und Beruf* 12(9) (September 1942): 97–100. As reason for the innovation it is stated (p. 97) that "the increasing claims of the State, the Wehrmacht and the Economy on physicists and mathematicians necessitate that the education of future representatives of these disciplines be put on a new basis." The reform was the result of a long discussion, though beginning, it is true, only after 1933.

In the article "The New Syllabus and Examination System for Industrial Mathematicians in the Greater German Empire" by Lois Timpe, published in *Zeitschrift für die gesamte Versicherungswissenschaft* 43 (1943): 65–71, the author explains: "The moves toward realization of the *Diplom-Mathematiker*, which had already been begun by the beginning of the War but then got stuck because of war conditions, were driven forward anew from Autumn 1941 thanks to the praiseworthy initiative of the responsible commissioner and has now been brought to a successful conclusion." From this it is obvious that plans were already in store, but that the Ministry regarded the question to be important enough for giving it the necessary attention even in 1942, that is—given the date—that the reform was considered important for the War. If the lack of mathematicians at the time and the increasing use of mathematicians and mathematics in armament is taken into account, this is indeed obvious.

(Private communication)

---

thus educate students, of whom many, once they have graduated, go neither into fundamental research nor to school as teachers but instead into applications—not least with the military. In this connection it is worthwhile remembering that the German *Diplom-Mathematiker*, the mathematician trained to go into applications, was invented in the Third Reich (cf. Box XVIII).

The responsibility of the mathematician must hence be discussed from two points of view: the responsibilities of the researcher, and those of the university teacher. On the first question one preliminary remark should be made. Many of those scientists who discuss the question of responsibility insist that "the basic reason for the irrationality of the whole process" of the arms race is the influence "of different groups of scientists and technologists." Thus, for example, Sir Solly Zuckerman,[41] for many years scientific advisor first of the British Minister of Defense, later of the Prime Minister. Such self-gratifying opinions on the part of a scientist are understandable, but nonetheless exaggerations. The scientist who accepts (or even seeks) the role of an insensate and unscrupulous instrument is *co*-responsible. Still, responsibility is not diminished by being shared. Co-responsibility implies *responsibility*.

These observations do not abolish sociologico-structural explanations of the involvement of science and scientists. Questions of responsibility are, however, always personally directed: *I* am responsible, because nobody but I can answer for the way *I* act under the given conditions. Knowledge of causal explanations will not exonerate us from responsibility; it tells us how to act in order to meet our responsibility as efficiently as possible.

How, then, can the mathematician use that latitude for personal action that may be wide or narrow, but which is always there?

As long as we inhabit States where we cannot be sure they will not use their military potential for aggression, to subjugate other peoples and States, and as a tool for political blackmail, one should begin at the negative pole: What can be avoided?

One can avoid having one's mathematical problems formulated by the military. That is: Some mathematicians work directly in the military establishment or in armament corporations to translate the problems of their employers into mathematical questions (as once the scientific academies). One need not be one of them.

Others, often highly qualified university mathematicians, take up these mathematical questions, take care that many mathematicians contribute to work them out theoretically, and eventually combine the results and canalize the information back to the military sector (even this was part of the classical function of the academies). Mathematical results that can be argued to be neutral in themselves are thus, through selective communication and adequate combination, made partisan. Nor does one need to belong to this group.

Finally, a large number of university mathematicians and graduates are caught in the net of the second group.[42] Whether one belongs here may at times be difficult to ascertain. Often, however, keeping open eyes and ears may give hints whether one's work fits into a larger whole of military interest.

One may also avoid transforming international scientific cooperation into a battlefield of the cold war. This last negative imperative leads, through reversal, directly to the first positive possibility: The protection and amplification of international scientific cooperation, aimed at mutual benefit and mutual understanding, may contribute to stabilize or re-establish the *détente*.

As formulated by Oswald Veblen, then-president of the American Mathematical Society at the opening of the International Congress of Mathematicians in 1950 (*Proc . . . 1950:* 125): "To our non-mathematical friends we can say that this sort of a meeting, which cuts across all sorts of political, racial, and social differences and focuses on a universal human interest, will be an influence for conciliation and peace."

Even the next possibility arises by reversal: counter-expertise. One example of this is presented by the critical analysis of the calculation of the first strike. When the advisors of the President and of the Pentagon claim that a first strike is possible at the maximal condition of 30 million American civilians killed, then independent, highly qualified experts are required, for one thing in order to make the calculation and its conditions public—nobody can expect the Pentagon to advertise the scheme of 30 million fried Americans. But they are also needed if anybody is going to uncover the real dimensions of the catastrophe stored in the military incubator: Only those at least equalling the strategic planners in competence will be able to document that the "decapitation" of the Soviet Union will probably imply the death of 150 million people in that country, and maybe even more in Western and Eastern Europe. Finally, only those in better command of global questions than the strategic planners will substantiate the abstract insanity of the whole business: abstraction from possible effects of the first H-bomb explosion on one's own communication systems and electronics; from the consequences of possibly wrong estimates of the survival percentage of adversary missiles; from consequential effects (such as mass panic, hunger, epidemics, etc.) of the sudden death of 30 million Americans and many more severely injured by radiation; and from global climatic and ecological breakdown.[43]

All this cannot be done by mathematicians alone. No discipline can do it alone, however, and mathematicians are indispensable when the many uncertainties of the project are to be evaluated—e.g., for evaluating the uncertainty in the temporal and spatial precision of missiles when fired simultaneously in large numbers (as discussed by J. Edward Anderson [1981]).

Once more, the first strike was only one example, even though a most urgent issue during the Reagan era. But also in other fields will the competence of scientists be useful, not least that of mathematicians. Both in the example just discussed and in many others, they may perform necessary new research; they may popularize known research results and so translate abstract knowledge into directly applicable information for peace movements—results not only from their own discipline but also from such areas where their professional training allows them to penetrate faster than laymen; and they may use still other aspects of their specific competence, be it familiarity with libraries and library use, training in mediation between abstract and concrete

---

**XIX. The Kamke Appeal**

September 23–28, 1946, the Mathematical Institute of Tübingen University organized a scientific symposium, "the first after the end of the War in Germany." In his opening address, E. Kamke stated among other things that

The physician receives not only a technical-medical training but also a moral education which allows him to use the most dangerous aids— knives, narcotics, poisons—only for the benefit of the ill. Similarly, it is urgent that scientists use their immense power, which can make them the masters of the life or death of nations, and even of Mankind as a whole, only for the benefit of nations and Mankind. In earlier epochs, qualification for genuine scientific research was the most prominent characteristic of the scientist; in future, however, this must be supplemented by something different; a particularly elevated professional ethos, an utterly sensitive consciousness of the consequences of research for mankind. It should be contemplated whether these moral requirements to the researcher's personality should be supplemented by organizational measures, of which the mildest would be the establishment of an international information bureau where all research in specific domains would have to be notified, no restrictions ensuing for the freedom of research.

These problems are of such importance that they should be discussed at all occasions where scientists meet. We must engage ourselves with our total strength and our whole person, that in the future science will never more serve destruction but only the welfare of mankind.

This quotation from p. 11 of the *Bericht über die Mathematiker-Tagung in Tübingen vom 23. bis. 27. September 1946* (Mathematical Institute, Tübingen University, n.d. [1946/1947]) was communicated to us by Horst-Eckart Gross.

---

thought or in the systematic collection of information, or still other skills of the intellectual by profession. Most of it will certainly not be describable as mathematics. Nor was, however, most of that which mathematicians made in World War II ("not one thing that was publishable," cf. Box VIII), which did not prevent it from being most useful for the purpose.

Related to the question of counter-expertise is that of peaceful perspectives for mathematical research. As discussed already, large areas of mathematics are completely unspecific as regards practical affiliation. Other areas, however, are of specific relevance for military application, and—more important—certain institutional habits favor information exchange between the warlords and specific mathematical research environments. Not only the military, however, knows of problems inviting to mathematical solution. Procurement of raw materials and resource conservation; the all-encompassing

**XX. International scientific cooperation: A way to restore the *détente?***

Several international congresses these last years have witnessed how international scientific cooperation was used as a pretext for a continuation of the political conflicts of the "international class struggle."

That happened before in the history of our science. The exclusion of German mathematicians from the Congress in Strasbourg (formerly Straßburg) in 1920 was mentioned already. The solemn opening and closing addresses were held by Émile Picard: "How should we forget in this place the admirable conduct of so many of our teachers in the War which has just ended; their patriotic faith has contributed to the common victory which today allows us to meet in the city of Strasbourg." Keeping open the question whether "sincere repentence," would allow later generations of German mathematicians to re-enter "the concert of civilized nations," he added that in his opinion "to forgive certain crimes [*viz.*, those of the enemy] would make one an accomplice" (*Proc.-... 1920*, xxvif., xxxii]. At 70 years' distance we see how this double standard contributed to preparing the ground for the still worse derangement of "German Mathematics" and "German Physics"—as the hunger blockade and the exorbitant reparations of the same years created the "Versailles Complex" and thus made it easier for German fascism to gain a mass basis.

Not every mathematician needed 70 years to understand that the direction taken was insane, and that scientific communication should not be used to continue the World War. Already after the French-German war of 1970–71 had the Swedish mathematician Gösta Mittag-Leffler founded the journal *Acta Mathematica*, among other things as an appeal to and a means for conciliation between French and German mathematicians. As described in detail by Joseph Dauben (1980), he used the journal once more during and immediately after the Great War in order to "restore as rapidly as at all possible the cooperation between scientists, independently of political and national points of view," as he wrote to Max Planck in October 1919 (Dauben: 281; further quotations from Mittag-Leffler's letters will be found in Box XXII).

Similarly was Hardy, whose personal abhorrence at the "useful" service of science for mass murder and exploitation was noted already, active in making science useful for conciliation—against "the many imbecilities printed during the last year by eminent men of science in England and France" (letter to Mittag-Leffler of January 7, 1919; further information on Hardy's activities and on the conflict between chauvinism and internationalism in science is given in Cock 1983).

As we see from the present quotations and from those in Box XXII, Mittag-Leffler and Hardy aimed at general, broad cooperation. Attempts like those of recent years to use the interest in specific "cases" (be it quite justified) as pretext to interrupt broad international cooperation and thus to undermine the climate of *détente* are quite in the spirit of Picard and against that of Mittag-Leffler and Hardy.

## XXI. When science goes advertising: Embarrassment and frankness

Since the Vietnam War, the public-relations agencies of the scientific establishment have become discreet on the question of military application. Close reading of their publications will, however, disclose the rattling of the hidden skeleton. Thus in a report from the U.S. National Science Foundation:

[ . . . ]. The problem of visual perception by computer is a minefield of sub-problems. At the simple end of the scale, systems already exist for comparing what the robot's eye would see with stored pictures; these are used today to guide a *missile* toward a target. The problem of interpreting objects in photographs collected routinely by weather and *surveillance* satellites is somewhat more difficult, but still feasible. Reliable techniques have been developed to automate parts of these photo-interpretation tasks. Similar systems can also be used to spot *roads, bridges* or *railroads* in aerial photographs.

(*Science and Technology* . . . 1979: 248; emphasis added)

On p. 249, the same report speaks of the computers—once publicized as prototypes of the beneficial results of war research and development—as "initially the esoteric tools of a small scientific community"! Who could find a more innocent name for the Oak Ridge, Los Alamos, and Argonne atomic-bomb laboratories.

climatological and ecological problems, both local and global; animal and plant production; medicine; and many fields of fundamental research, from physics to linguistics, can be presumed to procure for mathematics an abundance of problems for centuries (cf. Booß & Rasmussen 1979). Just as mathematicians may orient their research, in content or institutionally, toward the armament pole, they may look in these directions.

Finally, mathematical researchers are also participants in the social process of research. That gives everybody the responsibility and the possibility to make clear to her- or himself, to colleagues, and to the public, what goes on in his or her own division:

- Which are the possible applications of the research that is pursued?
- In which ways can it be applied: directly/specifically or indirectly/unspecifically? Is the subject, e.g., "harmonic analysis"; "fast Fourier transforms"; broadly applicable procedures for "pattern recognition"—or is it the specific problem of automatic terminal control of missiles in search of adversary missile silos?

- In which way is the applicability communicated, i.e., which are the sources for possible external inspirations, and to which addressee and in which form are results channeled? Does one send offprints of published papers to colleagues, or do results go to the Air Force (as offprints or as classified reports) with careful explanation of the relevance of a new algorithm for the mining of seaways?
- Which are the personal and institutional affinities, loyalties, and dependencies created, for example by marginal funding?

Usually, such questions are hidden under a thick cover of academic discretion. If we are to rid our world of war and our science of corruption it is, however, crucial to end this silence—even at the cost of consideration and tact among colleagues.[44]

We shall not discuss the options of mathematicians as researchers at greater length. There is, however, another side of their professional life, that of teachers at the university level. Should we say with Woollett that the only ways not to contribute to military needs are to abstain from teaching or to teach poorly?

If somebody abstains from teaching, the occupational situation of the day will guarantee that somebody else will soon be found to do it. General-purpose teaching cannot be boycotted as can participation in SDI research.[45] Abstention may ease bad conscience, but has no further effect. Poor teaching is no better. Who teaches poorly will only undermine respect for himself as a person, and thus also for his political position, among students as well as colleagues.

According to these considerations, in order to be able to act efficiently for peace one has to teach well. Good teaching, however, is only a necessary background, and has of course no effect in itself. What should one then *do* as a teacher?

First, one may once more *avoid*. Truly, one cannot avoid teaching mathematics of military relevance. Mathematics teaching (as long as it remains *mathematics* teaching—the teaching of specific mathematically founded techniques, for example, to bomber crews is irrelevant here as not being the task of mathematicians) always transmits multivalent and flexible concepts, methods, and techniques, which are neither coupled to specific applications nor shielded from them. One can, however, avoid presenting the mathematical enterprise as something associated organically or inevitably with military inspiration, application, or funding, and the career of an army mathematician as a natural and laudable option for a young colleague. One need not, as is done so often, illustrate the problems of Bayesian statistics without comment by the targeting of gunnery, nor use in the same way the

## XXII. Mittag-Leffler

*Mittag-Leffler to Ludwig Bieberbach, April 23, 1919*

. . . Soeben erhielt ich das Heft 3/4 des 3. bandes von "Mathematischer Zeitschrift," wo ich Ihre schöne Abhandlung "Ueber eine Vertiefung des Picardschen Satzes bei ganzen Funktionen endlicher Ordnung" finde. Bitte schicken Sie mir gütigst einem Separatabdruck. Ich möchte Ihnen auch einen Vorschlag machen, welcher ich glaube sowohl in Ihrem Interesse als in Interesse Deutschlands liegt. Schreiben Sie mir einen Brief, welchen Sie zum Beispiel ungefähr so anfangen: "Ich habe neulich eine Untersuchung ausgeführt, die ich unter dem Titel 'Uber eine Vertiefung des Picardschen Satzes. . .' in Mathematischer Zeitschrift publiziert habe, und Sie vielleicht interessieren wird." Hiernach teilen Sie mir *ausführlich* die schöne Untersuchung mit, die Sie über meine $E_a(x)$ Funktion vorgenommen haben. Es wäre auch gut, wenn Sie mir gleichzeitig die Untersuchung mitteilten, die Sie über $\alpha > 2$ angestellt haben (cf. pag. 185 in Ihre Abhandlung). Es wäre sehr Zweckmässig, wenn Ihr Brief in französisch abgefasst wäre oder, noch besser, in englisch.

Ich habe mir die Aufgabe gestellt die jetzt abgebrochenen internationalen Wissenschaftlichen Beziehungen, so viel an mir ist, allmählich herzustellen. Mein Vorschlag an Sie bildet ein Gelenk in diesen Bemühungen. Ich bin der Ueberzeugung, die Mathematiker müssen die Leitung für ein solches Streben übernehmen. Meine Zeitschrift ist für diese Aufgabe in einer günstigen Lage. Die Bände, die ich in den Kriegsjahren publiziert habe, enthalten Artikeln von den beiden Kriegführenden Gruppen.

> *As we know, this touching letter from the 73-year-old Mittag-Leffler to the young Bieberbach did not prevent the latter from becoming some years later the leading figure in the racist "German Mathematics."*

*Mittag-Leffler to Max Planck, October 7, 1919*

Die Hauptasche ist zuerst die wirkliche Stimmung in den höchsten wissenschaftlichen Kreisen kennen zu lernen um dann später so zu handeln, dass man in so kurzer Frist wie nur möglich das Zusammenarbeiten der Männer der Wissenschaft, unabhängig von politischen oder nationalen Gesichtspunkten, wieder herstellen mag.

Ich gehöre einer Wissenschaft, die sich besser als jede andere für die Aufgabe eignet an die Spitze von solchen Bestrebungen zu treten. Ich bin auch in der glücklichen, wenn auch nicht sehr verdienten Lage überall in den Sitzungen der ersten gelehrten Gesellschaften in jedem Lande teilnehmen zu können, und dadurch auch in den übrigens sehr seltenen Fällen, wo ich nicht

*(continued)*

---

**XXII.** **Mittag-Leffler** *continued*

persönliche freundschaftliche Verbindungen seit Jahren habe, doch überall mit den leitenden Persönlichkeiten in Verbindung treten zu können. Offenbar sehe ich es deshalb als eine Pflicht an diese Umstände zu benutzen um den Ziel näher kommen zu können, welches jedem Manne, für welchen die Förderung der Wissenschaft die höchste Aufgabe seines Lebens ist, ganz besonders am Herzen liegt.

*Mittag-Leffler to G. H. Hardy, January 25, 1919*

. . . I agree with you that we as mathematicians need to be at the head in "the task of the reestablishment of friendly relations" between the men of science of all countries. I also hope to be able to aid such reestablishment of scientific relations through my journal *Acta Mathematica*, which in the area of mathematics has been able to maintain such relations during the last four terrible years.

*Mittag-Leffler to Max Planck, March 30, 1919*

. . . Die Männer der Wissenschaft müssen sich vor aller Politik abhalten und nur auf die rein wissenschaftlichen Gesichtspunkte denken. Wenn die Wissenschaft nicht hoch über das jetzige politische Elend aufrecht erhalten werden kann, geht alles zu Grunde. Mein ganzes Streben geht dahin so viel wie nur möglich für dieses Ziel zu wirken.

> *In the context of 1919 the expression "political misery of the day" shows that Mittag-Leffler's efforts to keep mathematics aloof from politics was not meant as an ivory-tower policy but the necessary consequence of sincere moral responsibility.*

(All quotations are from Dauben 1980)

---

perturbations of a missile trajectory to exemplify the application of infinitesimal methods. One need not flaunt operations research and computer technology as paradigms of the blessings of war research (though covering up their ties to institutionalized mass murder is certainly no better).

So much about the ideological impregnation to be avoided. Similarly, one can and should avoid the corresponding real-life behavior: one should avoid the part of an impresario introducing students via their dissertation work to the armament industry. On this point, the dependence of students imposes very strict circumspection upon the teacher.

It is also possible to act positively. One cannot, it is true, solve those moral dilemmas for the students to which they will later be exposed; but one can assist them in developing the sensitivity to discover the dilemmas and the ability to solve them.

It is well known that many of the physicists taking part in the Manhattan Project opposed the use of the atomic bomb against Japanese cities, and that even more turned against the military use of nuclear energy after Hiroshima. The *Bulletin of the Atomic Scientist* and the World Federation of Scientific Workers both owe their origin to this opposition movement. Less commented upon is the absence of such documented feelings and initiatives among the engineers from Dupont, Union Carbide, and General Electric carrying the project to technical completion. To be sure, the physicists' protests did not save Hiroshima and Nagasaki. But they have contributed to making large strata of the world population sensible to the dangers of nuclear war, and so perhaps averted new Hiroshimas in Korea, Vietnam, or elsewhere.

We have no possibility to know the reason why the engineers remained mute at least as a group: Was it fear of dismissal or of repression on the job? Lack of a comprehensive outlook? Absence of the aspiration to work up such outlook? Isolation on the job not permitting organized communication and manifestation? Nor do we know the circumstances under which our students will eventually have to work, whether they will be similar to those of the "physicists" or those of the "engineers". We cannot protect them against repression on the job or against that deliberate blindness to the consequences of one's work which one may develop to protect oneself in this situation; we can, however, take care that blindness will not be the only option open to them. We cannot prevent that scientists are employed in industry as narrowly specialized functionaries; but we can make them see the wider connections of science and thus also the real inner connections between scientific subjects and those between science, applications, and consequences, and so keep open for them the psychological possibility not to be absorbed by narrow specialization. We must make them discover the double character of scientific-technological work: Not only knowledge and application of knowledge in abstract technical contexts, but knowledge that is known by specific people under specific circumstances, and technology which is operated under specific societal conditions and applied for specific purposes.

Mathematics teaching will hence contribute to the conservation of peace *by being humanistic,* by caring that the future applier of mathematics will not be a passive transmission link between commissioner and work product but an active human person who understands his or her own activity in broader context and consequence. If it is not to remain an empty phrase this "broader context" must, however, be presented *concretely* to the students during their studies. The broad perspective must be present in and integrated with the complete course of studies—a separate course on "humanistic values" or "mathematics as a liberal art" will have little value if counteracted by the implicit message of the main body of teaching.

In the wake of these considerations another look at the concept of "organized segmentation" will prove illuminating. It was argued previously that the segmented character of the scientific-technological system, in combination with organizational principles set not by conscious participants but by the conglomerate of economic, bureaucratic, and political power would cause the standard according to which it is so indubitably efficient to be set by those instances. Now, both bureaucratic and commercial structures tend to protect their own interests without considering global concerns. Public authorities, even when not themselves dependent upon bureaucracies or big business, are rarely competent to go into the subtle mechanisms of the system. Only participants have a chance of knowing it sufficiently well for doing so— but only if they are able to transcend their "naive" specialist's role. The condition that participants may interact with the public and with democratic public authorities to avert global catastrophes hence coincides with the condition that was sketched for mathematics teaching to operate in favor of peace.

In the end we should remember that mathematicians are also citizens. That imposes the same responsibility upon the mathematician as upon everybody else. In the situation of the day, however, one only honors one's moral obligations by doing one's best. The mathematician is thus responsible for using also his or her *specific* possibilities—that is, the *civic* duties of the mathematician imply that one discharges also one's moral obligations as a researcher and as a teacher *and* integrates the citizen, the researcher, and the teacher with one another.

That may sound easy, and quite a few mathematicians will heartily agree that they possess special qualifications for being good citizens. They need not be wrong: Analytical and synthetical thought, distinction between and mediation between the abstract and the concrete, fantasy and stubbornness—all these abilities belong to the job, and all are useful for participation in democratic political life. Many mathematicians will also be acquainted with a number of diverse applications of their subject and so have accumulated useful insights outside their specialty.

Mathematicians, however, also tend to have a handicap. The particularly important position of the logical argument in mathematics easily leads to the opinion that everything not belonging to mathematics, particularly political and moral thought and convictions, is illogical and beyond argument.

Furthermore, it is not uncommon that mathematicians mistake this episte-mological dichotomy between demonstration and subjectivity for a social dichotomy and, confusing that which belongs to the *mathematician* with that which belongs to *mathematics,* take their own inveterate persuasions and prejudices for objective truth.[46] In order to make their special qualifications fertile mathematicians must surmount this professional arrogance. They should not dismiss participation in broad political movements as unworthy of their intellectual merits. They should realize that the rationality of mathemat-ics is but one embodiment of a more general category of rationality, allowed and conditioned by the specific object of mathematics; and they should ac-cept that rational discernment is also possible in the moral and political do-main, but that it presupposes, here no less than in their own field, systematic engagement and sobriety in the argument.

Mathematicians, if they want to engage themselves for peace, must un-derstand themselves as participants in a larger, common enterprise: Neither better than others nor inferior, but their equals in rights, in merit, and in responsibility.

# NOTES

## Preface

1. On this I commented in a footnote that "I do not agree that 'the belief that mathematics is unique has exactly the same status as the belief that there is a unique moral truth' (Bloor 1976: 94). Furthermore, I would add that to the extent that this major premise is true it may cast doubt on the implicit minor premise: That moral truth is *nothing but* historical arbitrariness."

2. This formulation makes the "West" include the Islamic core area, as it should if the term is meant to describe, however vaguely, the long-term cultural divisions of our world, and not to be a mere projection of early twentieth-century capitalism into perpetuity. Not least the history of mathematics demonstrates that it is possible to speak meaningfully about a cultural sphere encompassing the Islamic and Christian core regions and—backwards in time—the ancient Near East and Greco-Hellenistico-Roman antiquity, but not about a spatiotemporal entity excluding the Islamic world. To see this, it is sufficient to regard the extraordinary effect of the reception of Euclid's *Elements* precisely within this region, compared with the relative indifference of Indian and Chinese mathematicians to Euclid. Obviously, *they* were in possession of an independent mathematical tradition and style that was too strong to let itself be displaced by a Greek classic even when incorporating specific results (though not strong enough to make an adoption of Modern European mathematics seem superfluous). Much the same could be said concerning the reception of Aristotle and the history of philosophy or the adoption of Galenic medicine.

## 1. Varieties of Mathematical Discourse in Pre-Modern Sociocultural Contexts: Mesopotamia, Greece, and the Latin Middle Ages

1. See, e.g., Chandler Davis (1974); Booß & Niss (eds.) (1979); and Struik (1936).

2. Since the term *discourse* is used in slightly different ways by different authors, I shall betray the explanation which my translator in Thessaloniki forced out of me:

> You will recognize that the modes of communication in School, in the Church, and in the Army are completely different; in the selection of themes or subjects about which one communicates, but in many other respects too. The didactical, ecclesiastical/religious, and military *discourses* are different. Philo-

sophically speaking, the discourse can be considered the *form* (in an Aristotelian sense) of the totality of communications of a certain institution or type of situation.

3. The ascription of a "molding" role for the proto-Sumerian Temple does not imply acceptance of the traditional (Wittfogel-related) *Tempel-Stadt* thesis, according to which the Temple had complete societal dominance, and retained such dominance until c. 2400 B.C. Cf. B. Foster (1981) and Nissen (1982).

4. The detailed information on which I build this and most of my interpretation of Sumero-Babylonian mathematics is presented in Høyrup (1980: 14–29, 80–94). References contained in that paper are repeated only to a restricted extent.

Addendum 1992: In 1980, almost all inferences about the character and development of Mesopotamian third-millennium mathematics had to be made indirectly, at best from recent investigations of the metrology (Powell 1971, 1972) but often on more shaky foundations. Only one publication (*viz.*, Powell 1976) and two prepublications (Friberg 1978, 1979) dealt directly with mathematical texts. During the twelve years that have passed since then much more information has become available (partly published, partly discussed at the series of workshops "Concept Development in Babylonian Mathematics" organized at the Seminar für Vorderasiatische Altertumskunde and the Max-Planck-Institut für Bildungsforschung, both in Berlin). This new evidence has allowed the addition of shades and details—cf. Chapter 3. The global picture as presented in 1980 and reflected in the present article, however, is unaffected or confirmed.

5. The current neo-Sumerian "balanced account" did not possess the automatic built-in controls of later double-entry bookkeeping; but automatic control was known and used in certain cases—cf. Struve (1969: 147f.). In surveying, too, built-in control was in use, perhaps as a check and in any case as a means to restrict the effect of measuring errors. This is seen in the field plan discussed in Thureau-Dangin (1897), where the value of the area results as the average between the outcomes of two different computations.

6. Only scattered administrative texts have been found from the reign of the first king and from the earlier years of the second; then, all at once, thousands and thousands turn up, in a way that suggests a thorough administrative reform, organized via the scribal school (see Nissen 1983: 207–13).

7. Cf., for instance, Diakonoff (1971). Further details in Høyrup (1980: 87 n. 51).

8. For various aspects of this rise of the individual as a private person, cf. Klengel (1977; 1980: 83–88), and Kraus (1973, *passim*). On religious change, see Jacobsen (1976: 147 ff.).

9. One text among many that identifies the King as the upholder of affluence and justice is Hammurapi's famous "law code", more specifically its prologue and epilogue (translation, e.g., in Pritchard 1950: 163–80). Recently, Marvin Powell

(1978: 143) has argued that the very rise of the royal State in the mid-third millennium should be sought in its mediating role in the face of violent class conflicts inside the city-states (in a way reminiscent of Engels' analysis in *Der Ursprung der Familie, des Privateigentums und des Staates*). If Powell's assumption holds good, the "royal ideology" known to the scribes can be considered an embellishment of facts, but still a reflection of the real *raison d'être* of the state.

10. For this description of the Old Babylonian scribal culture, cf. Landsberger and others, in Kraeling and Adams (1960: 94–123); Kraus (1973: 17–32); Kramer (1949); Falkenstein (1953); Sjöberg (1972, 1973, 1975, 1976).

11. See Landsberger in Kraeling and Adams (1960: 100f.) and Sjöberg (1975).

12. The Mesopotamian written record is full of random lacunae; still, the beginnings of the new style of Mesopotamian mathematics in the early Old Babylonian period appears to be established beyond reasonable doubt. For this, statistical arguments can be given (see Høyrup 1980: 87). Further, the formulation in the "new" Akkadian literary language is important evidence: the Old literary traditions were transmitted for almost two millennia in Sumerian, cf. Hallo (1976). Finally, the basic technique (*viz.*, the quadratic completion) for the treatment of second-degree equations, which form the backbone of the new style, seems to have carried the name "the Akkadian method," according to a hitherto badly understood late Old Babylonian text (cf. Høyrup 1990: 326).

13. The character of the techniques is investigated in detail and depth in Høyrup (1990); according to this analysis of terminology, formulations, and concepts, the Old Babylonian "algebraic" texts rest on a basis of mostly geometric justification. They are not just recipe-descriptions of algorithms, as is often inferred from misleading translations.

An overview of the results, focusing in part on the comparison of Old Babylonian "algebra" with post-Renaissance algebra, is Høyrup (1989). In the following, terms in quotes ("equations," "coefficients") refer to entities that fulfil roles corresponding to the equations and coefficients of modern algebra.

14. The problem occurs in two different texts, BM 85 194, Rev. II, 22–33, and BM 85 210, Obv. II, 15–27—incidentally solved by two different methods. A third, still more artificial, problem concerning the same ramp is BM 85 194, Rev. II, 7–21. All three will be found with a discussion in MKT I. Cf. also Høyrup (1985: 53–61).

15. In the middle and late Old Babylonian examination texts, "the whole character of the dialogue changes. The master becomes more boastful, and the examinee assumes a merely secondary role of giving the master an opportunity to display his knowledge" (Landsberger, in Kraeling and Adams 1960: 112). In the mathematical domain, especially the higher-degree problems can be taken as examples of the same phenomenon: No general methods for the solution of such equations were known. As pointed out by Thureau-Dangin (1938: xxxviii), the Babylonians "demonstrate their inability to solve the third-degree equation" by the very methods they employ to do so. The higher-degree "equations" that occur are thus all constructed so as to be solv-

able by a special trick—a trick that no student would be able to transfer to analogous problems lacking the specially prepared "coefficients." Such problems do not train the abilities of the student; they display the fake abilities of the teacher.

16. Two Old Babylonian texts appear to exhibit attempts to describe a method in general terms: IM 52 301, Edge, and AO 6770, Obv. 1–8—see, e.g., Gundlach and von Soden (1963), and S. Brentjes and Müller (1982). A few other texts indicate by their formulation that a certain procedure *is* regarded as an exemplification of a general method; thus *TMS* IX, 17, see Høyrup (1990: 326, 114f.); and BM 85 200 + VAT 6599, rev. I, 12, see Thureau-Dangin (1938: 17).

17. This relation between the aims of teaching and the ensuing mode of mathematical thought, as well as the relation between non-scholasticized and scholasticized practitioners' mathematics, is dealt with in more detail in Chapter 2.

18. The term *algebraic* for once taken in the sense given to it in twentieth century mathematical theory.

19. Strictly speaking, two types of problems should be distinguished; cf. note 15: The huge lot of problems that train the solution of first- and second-degree "equations" and problems; and the by far fewer problems of a higher degree, which are constructed so as to be solvable by special tricks. The former train methods, which may be of no practical value, but may then serve the display of virtuoso ability; the latter train nothing at all—they *are* a means of display. In both cases, the methods (and tricks) at hand determine which problems shall and can occur.

20. At least in Denmark, such complaints are mainly directed against medical research. This drives the parallel even further, since it is precisely the medical establishment that is characterized by emphasizing publications when choosing candidates for promotion without being really able to evaluate the scientific content or importance of specialized research. Publications therefore serve precisely as a means to display professional virtuosity.

21. Both translated in Czwalina (1931). Interestingly, the Islamic astronomers were dissatisfied with this formal purity of Greek spherics. In most Islamic paraphrases of the Greek works, commentaries are inserted explaining what the theorems are *really* about in astronomy—see Matvievskaya (1981). In the Latin Middle Ages, the same thing came to happen: The formally pure thirteenth-century mathematician Jordanus de Nemore wrote a treatise on the stereographic projection in Greek style; it achieved a rather wide circulation, but only in versions with extra commentaries explaining the astronomical substance of the work; see R. B. Thomson (1978), and Chapter 4, Section 10. Evidently, the formal purity of substantially applied mathematics must be considered a special characteristic of Greek mathematics; in other cultures it was intentionally avoided.

22. A brief exposition of the theme will be found in Høyrup (1980: 29–36, 94–97). Cf. also Chapter 3, Sections 11–12.

23. As evident, e.g., from the Old Babylonian text YBC 8633 (in *MCT*, 53–55, cf. discussion in Høyrup 1985: 105.6–105.11).

24. *TMS*, texts IX and XVI, according to the new interpretation of these texts in Høyrup (1990: 299–305, 320–28).

25. Without being so formulated, this is the conception behind the prominent place of mathematics in the classically oriented neohumanist educational ideal of nineteenth- and earlier twentieth-century Germany.

26. *In primum Euclidis Elementorum librum commentarii*, 65[16f]. Of course, Proclos should not be considered as serious a source as is sometimes done. It itself, a commentary written in the fifth century A.D. is late. Furthermore, Proclos the neoplatonist is biased in his sympathy for Pythagoreans and for neopythagorean sources; the statement in question is in fact not taken from Eudemos' mathematical history (fourth century B.C.) but from the late and not very reliable neopythagorean Iamblichos—see E. Sachs (1917: 30f.).

27. Once again, a more detailed discussion of this evidence will be found in Høyrup (1980: 37–52, 97–108).

28. "Was created" at least in the sense that it acquired its final form and place; much in Euclid's *Elements* was taken over in only partially reworked form from earlier works and traditions; W. Knorr (1983) even argues that Book X may have been taken over in virtually unchanged form.

Euclid should not be taken quite literally as the watershed. His *Elements* played a decisive role; but he also happened to live (we assume) at a time when things were anyhow taking place: As shown by Germaine Aujac (1984a), wide-range codification down to single formulations had occurred by the late fourth century B.C.

29. The *quadratrix*, named so in later times when it was also used for the squaring (or, better, the "rectification") of the circle. See Heath (1921: I, 226–30), and Bulmer-Thomas, "Hippias of Elis."

30. Aristotle, *Analytica posteriora* 75[b]41–76[a]3, and *De sophisticis elenchis* 171[b]16–22, 172[a]3–8. Cf. Thomas (1939: 1, 310–17); Kerferd, "Antiphon"; and Merlan, "Bryson of Heraclea."

31. As a background to his personal conclusions, with which one may agree or disagree, van der Waerden (1979) offers a balanced presentation of the central sources. Cf., however, Schuhl's cautious attitude (1949: 257–60) concerning the range of early Pythagorean mathematical knowledge and Burkert's very careful argumentation (1962, *passim*) to the same effect.

32. See Archytas, Fragment B.1, ed. Diels/Kranz (1951).

33. More subject to doubt is the traditional ascription of the geometrical "application of areas" and the whole "geometric algebra." See Knorr (1975: 199f.). In his review of the book, van der Waerden (1976: 498) argues in favor of the opposite

point of view, referring to Neuenschwander's analysis (1973); Neuenschwander's argument for the ascription of the contents of *Elements* II to the Pythagoreans is, however, only a reference to van der Waerden (1956), where everything is spun out from a quotation from Proclos's commentary to *Elements* I stating that Eudemos ascribes the application of areas to "the Pythagorean muse" (according to the English translation of van der Waerden's *Erwachende Wissenschaft*—1962: 123). The quotation, however, is a mistranslation, the Greek text speaking *not* of Eudemos but of *hoi peri tòn Eudemon*, "the Eudemean circle" (*In primum Euclidis*, $419^{15}$), which indicates that Proclos did not use Eudemos's history on this point but some vaguer source or tradition; furthermore, there are good reasons to believe that the mathematical substance if not the formulation as an "application with excess" or "with defect" was not only known but also taken over in naive-geometric form from Near Eastern sources—see Høyrup (1990a).

34. The (partly incoherent) evidence in ancient sources is displayed in van der Waerden (1979: 64ff.), and discussed more in depth in Burkert (1962: 187–202).

35. Except, perhaps, that the cumulation seems not to have gone very far in the later fifth century, if we are to judge from the fragments that can be ascribed to the Pythagorean Philolaos; see von Fritz, "Philolaos of Crotona," or Burkert (1962: 222–56), ending in the conclusion that "only in a very restricted sense is it possible to speak of a specifically Pythagorean science before Philolaos."

This agrees well with the most plausible conjecture concerning the origin of the *mathematikoí, viz.,* that this branch of the movement is a secondary formation, and that serious cumulation had only begun by the mid-fifth century.

36. First, a (probably authentic) fragment from a comedy written by the poet Epicharmos at latest in the earlier part of the fifth century shows that some sort of abstract discussion of odd and even numbers was supposed to be well known to the general public (Diels/Kranz 1951: I, 196). This agrees only badly with the idea that the doctrine of odd and even was created from scratch by *mathematikoí*, select members of a Pythagorean "Inner Party" (or for that matter by the Master himself). The claim that Epicharmos himself should be a Pythagorean appears to be ill-founded—cf. Freeman (1953: 132–35).

Second, the Epicharmos fragment shows beyond doubt that the early doctrine of odd and even built on arrangements of *psêphoi*, stone calculi (cf. also Lefèvre 1981; Knorr 1975: 135ff.; and Becker 1936). The particular contribution of the Pythagoreans to arithmetic appears, on the other hand, to start from the connection between arithmetic and harmonics, as revealed in the theory of ratio and proportion—cf. A. Szabó (1969: 136–56) and von Fritz (1971: 47–52). Such a start agrees well with the ethical and religious attitude of the Pythagoreans, but it presupposes the existence of an arithmetic that permits and invites one to discover the arithmetical properties of musical harmony.

Third, Szabó (1969: 352–61) argues that the basic characteristics of Pythagorean arithmetic—its distinction between "one" and "numbers" and its replacement of fractions by the theory of proportions—must be understood as results of a reformulation of the theory of numbers on Parmenidean and Eleatic foundations.

This, however, would place its development well after the testimony of Epicharmos, and agree with a gradual mathematization of the doctrine influenced by the developments of philosophy and mathematics outside the order. Even the indirect proof, a favorite technique of the supposedly Pythagorean arithmetic included in the *Elements*, would be a borrowing from Eleatic dialectics, as argued by Szabó (1969: 328–42) and Lloyd (1979: 110). So, it cannot have been used in any postulated earlier Pythagorean investigations of the odd and the even.

Fourth and finally, even the arithmetization of harmonics may have been borrowed from outside. A recurrent anecdote in late sources makes Pythagoras discover the simple arithmetical ratios corresponding to the musical octave, fifth, and fourth by weighing the hammers of a blacksmith, the ring of which differed by these intervals. Taken literally, the story is physically impossible; another version, however, makes the Pythagorean Hippasos (often connected to the formation of the branch of *mathematikoí*) make experiments on metal disks with the same diameter and different thickness; a third version (involving Hippasos and the Pythagorean Lasos) may speak of resonating jars as used in the Greek theater (for all three versions, see Burkert 1962: 354–57). Now, the catalogue (*Fihrist*) written by the tenth century Baghdad court librarian al-Nadīm mentions a work *Kitāb al-juljul al-ṣiyyāḥ* (possibly to be read *al-ṣayyaḥ*), *Book on the octave chime* (possibly *clamorous chime*) written by one Sāʿāṭus (see Farmer 1931: 61, and Dodge 1970: 643), identified by Dodge as Sacadas, a composer contemporary with Anaximander (a composition of his was written in 586 B.C.; cf. Albert, "Sakadas aus Argos.") If this ascription is true, musical theory (seemingly related to precisely such "experiments" as those ascribed to Pythagoras, Hippasos, and Lasos) was already committed to writing when Pythagoras was a young man.

37. *In primum Euclidis*, 78$^9$. This passage belongs to one of the parts of the commentary that build on the first-hand knowledge of Aristotle's disciple and co-worker Eudemos. It is thus contemporary testimony.

38. Conversely, Plato's polemics against the sophists are built on the fact that they live not *for* philosophy, as philosophers who would incidentally teach, but *off* philosophy, as paid teachers; cf. *Protagoras* 313c–314b.

39. See the fragment in Thomas (1939: I, 234–52). Cf. also Heath (1921: I, 183–200).

40. See Bulmer-Thomas, "Oenopides of Chios," 180f.; and Szabó 1969: 369–73), commenting upon Proclos referring to Eudemos (*In primum Euclidis*, 283$^{7ff}$, 333$^{5f}$).

41. Cf. Bulmer-Thomas, "Oenopides of Chios," 179. Bulmer-Thomas claims that the ancient tradition must refer to a determination of the obliquity of the ecliptic, since "the Babylonians no less than the Pythagoreans and the Egyptians must have realized from early days that the apparent path of the sun was inclined to the celestial equator." However, neither Babylonian nor Egyptian astronomy would describe the ecliptical movement as a movement in a *plane* cutting the celestial *sphere* in an inclined *great circle;* but the earliest reference to Oenopides' discovery (*Amatores* 132a–

b, a pseudo-Platonic dialogue from c. 300 B.C.) suggests that it consisted precisely in a coupling of astronomical phenomena to abstract geometrical description.

42. The counterposition of the two was made already in antiquity. The Socrates of Plato's early writings discusses both the sophist master-student relation, where ready-made doctrines are poured into the mind of the defenseless disciple (*Protagoras* 314b), and the non-hierarchical relationship practiced by Socrates himself, who was never the teaching master of anybody (*Apology* 33a-b).

43. The reader may observe a certain convergence with the Kuhnian distinction between "normal science" and "scientific revolution," and with the Piagetian distinction between the quiet periods of (infant) cognitive development dominated by assimilation, and the transitions from one cognitive stage to another. The convergence is not accidental; nor is it, however, complete.

44. In *Works and Days*, verses 11ff., a discussion of two different sorts of "struggle" (*eris*).

45. *Odyssey* IV, 762–766 is one clear instance. In *Odyssey* I, 60–62, a related argument is used by Athena in the Assembly of the gods. In the *Iliad*, too, several instances can be found, of which the coupling between I, 394f. and I, 503f. (Achilles asking Thetis to implore Zeus for revenge, and Thetis actually doing it) suggests that this was to be recognized by the listening public as a standard format for prayers.

46. I shall leave aside the connections between the particular coloring of Greek philosophical rationality (*viz.*, the quest for "pure" and the neglect of "productive" knowledge); the prevailing contempt for manual work; and the spread of slavery. On this account, Farrington's studies (e.g., 1938, 1965) are still valuable, and much less mechanistic than his critics try to make us believe.

47. This counterposition is not meant as a full support to the ancient Greek counterposition of the "democratic spirit" of the Greek city-state and the "despotism" of the Orient; see, in addition to Herodotos' *Histories, passim*, and Aristotle's *Politica* 1285[a]10ff. and 1327[b]23ff., the contrasting passages in Aeschylos' *The Persians*, 211–14; *The Suppliant Maidens*, 370–401; and *Seven Against Thebes*, 6–9. At least in the earlier third millennium, a "primitive democracy" with popular assemblies existed in the Sumerian city-states (cf. Jacobsen 1957), and throughout the history of ancient Mesopotamia, expressions of popular discontent and even rebellion are to be found (cf. B. Brentjes 1966). One Old Babylonian epic myth contains even a precise description of a violent strike among field workers who have to be bought off, transposed into the world of the gods (*Atra-ḥasīs*, ed., trans. Lambert and Millard 1969; see also Jacobsen 1976: 116ff.). But whereas the philosophical movement in Greece was ultimately connected with the political climate of the *agora*, the Mesopotamian scribal culture had no such connections. Heraclitos, himself a descendant of kings and aristocrats, could make eternal movement and change the very basis of his philosophy; in contrast, the scribe telling about a popular rising in Babylon around 1600 B.C. could only consider the ensuing social changes on a par with the catastrophes of a plague, which the gods had to stop (see Brentjes 1966: 36); moreover, even

though the rebellious minor gods of *Atra-ḥasīs* are admitted to have been subjected to too hard suppression while digging the Euphrates and the Tigris, their strike is not given approval by the progress of the story; instead, the instigator of the rebellion is slaughtered and his flesh and blood used for the creation of mankind— and a world order is created where mankind has the duty to take over obediently the toil of the gods. Thus, political expression on the part of the people was known to the scribes, but it was not tolerated.

On the other hand, the dissimilar social affiliation of literacy seems to make up only part of the difference between Mesopotamia and Greece. As early as the mid-eighth century, the Greek assembly and *agora* appear to have permitted a sort of free critical speech hardly known in Mesopotamia. Not only are deliberating assemblies described time and again in the Homeric epics (*Iliad* II, 211–93 and *passim; Odyssey* I, 26–95 and II, 9–259); after all, such assemblies where the aristocrats deliberate in front of the people are well known both to historians of ancient Germania and to anthropologists. But numerous passages in the *Iliad* refer to War and the Assembly as *the* two places where an aristocrat gains honor, presenting them always as equally important (thus III, 150ff.; IX, 440ff.; XII, 211ff.; XV, 282–85; XVI, 630f.; XVIII, 105f. and 245ff.; XVIII, 490ff. opens up a wider perspective, presenting the two types of cities, one celebrating weddings and deliberating in the assembly, the other in war). Moreover, the Thersites episode (*Iliad* II, 211–68) shows that even if the assembly was definitely no constitutional democratic organ, commoners speaking up against the rulers must have been a well-known phenomenon to the bard's aristocratic public (as also observed by Aristotle, *Politica* 1285ᵃ11f.); further, the way Thersites is brought to silence by a blow of irregular kingly violence makes it clear that no regular procedures for doing so were at hand; finally, the presentation of Ulysses' irregular use of his staff as an appeal to the feelings of the public demonstrates that the poet expects it to appreciate the methods by which the ugly commoner was put in his place in the Good Old Days. So we are told that commoners speaking up against their betters were not only well known but also felt to be a nuisance by the latter; and probably also that those days had already passed away when a commoner could be beaten up in the assembly for speaking critically.

In *Seven Against Thebes*, 5–8, Eteocles complains that in case of disaster the king's name will be "on many a citizen's tongue, bruited up and down in mutterings and laments." It has been argued that this critically minded political attitude was produced by the "hoplite reform" of c. 650 B.C. (first perhaps by Aristotle in *Politica* 1297ᵇ22–25; references to modern authors in Lloyd 1979: 258 n. 142). The *Iliad* tells us that although the military reform may have enhanced the hoplite infantrymen's criticism of the prevailing aristocratic power, it did not introduce it—and so, that the political discourse that put its stamp on Greek philosophy has its roots far back in Greek history.

48. εἰς ἐπιστημονικωτέραν σύστασιν— *In primum Euclidis*, 66[17f]. The expression should be understood in agreement with Aristotle's idea of *epistēmē* (rendered in scholastic Latin as *scientia*, whence the somewhat misleading translation *science*) as set forth in *Analytica posteriora*, i.e., a branch of knowledge dealing with a specific subject and organized axiomatically from subject-specific principles.

49. According to Proclos (*In primum Euclidis,* 379[2ff]), such a proof is attributed by Eudemos to ''the Pythagoreans.''

50. Cited from Aristotle, *Metaphysica* 998[a]3–4, who quotes Protagoras's polemics with the geometers to make clear that even the geometry of the astronomers deals with something different from perceptible points, lines, and circles. The autonomy of mathematics appears to be the paradigmatic case for Aristotle's general idea of autonomous single sciences (cf. note 48).

51. Another illustration of the separation of mathematical discourse from general discourse is Aristotle's distinction between the *pseudographos,* producing fallacies by means of wrong but yet *geometrical* figures, and the sophist producing his by arguments not belonging with the subject matter (*De sophisticis elenchis* 171[b]14ff. and 36ff.).

52. Curiously enough, this attempt to close discourse in the ''Platonic reform'' of mathematics (a term coined by Zeuthen) did not prevent Plato from advocating ''discovery learning'' in mathematics for children, as can be seen in his description of Egyptian arithmetic teaching in *The Laws* VII, 819. If *Truth* itself is the authority, the teacher can do without authoritarian methods. Only the imposition of socially necessary *lies* requires the cruel measures proposed toward the end of the same work (X, 907ff.); cf. Farrington (1938: 437f.).

53. Some hints concerning India can be found in Høyrup (1980: 52–55, 108f.). The Islamic situation is dealt with in Chapter 4.

54. Once more, factual background information can be found in Høyrup (1980: 62–77, 113–15). Cf. also Chapter 5 and 6.

55. ''With the Greeks geometry was regarded with the utmost respect, and consequently none were held in greater honor than mathematicians, but we Romans have restricted this art to the practical purposes of measuring and reckoning''—*Tusculanae disputationes* I, ii, 5.

56. A striking example is Gerbert's joyful announcement (c. 983) to his friend Adalbero that he has found ''8 volumes: Boethius' *On Astronomy,* some utterly splendid [volumes] with geometrical figures, and some others not less to be admired'' (letter in Bubnov 1899: 99–101, cf. note 6).

57. In 1090, we are told by Hermann of Tournai, the squares of his city were filled by curious crowds when Master Odo discussed philosophical questions with his students, and ''the citizens left their various employments so that you might think them fully attached to philosophy''—see Werner (1976: 57 and 93 note 358). Werner's excellent monograph is fundamental to the understanding of the rationalism of the eleventh and twelfth centuries, its background and its implications. Briefer statements in the same avowedly historical materialist spirit have been formulated by the Dominican Father M.-D. Chenu (1974 and 1977).

58. On the persistent problem presented by uneducated priests, se Mandonnet (1914).

59. Related interpretations are discussed by Thorndike (1955); Lemay (1962: xxiiff. and *passim*); Birkenmajer (1930); and Gregory (1975, especially 203ff.). The theme is dealt with in some depth in Chapter 5.

60. See, e.g., Augustine, *De civitate Dei* V, i–viii; Isidore of Seville, *Etymologiae* III, xxviii; and Hrabanus Maurus, *De clericorum institutione* III, xxv.

61. The basic information on the translations will be found in Steinschneider (1904) and Haskins (1924). Cf. also Steinschneider (1893).

62. It is characteristic that the Middle Ages conflated the two Latin words *auctor,* "source of a written text," and *auctoritas,* "source of power."

63. This becomes particularly clear in the case of the translations of Greek authors made directly from the Greek, which, in Haskins' words (1924: 150–52) were "closely, even painfully literal, in a way to suggest the stumbling and conscientious schoolboy. Every Greek word had to be presented by a Latin equivalent. Sarrazin laments that he cannot render phrases introduced by the article, and even attempts to imitate Greek compounds by running Latin words together. [ . . . ] This method, *de verbo ad verbum,* was, however, followed not out of ignorance but out of set purpose, as Burgundio, for example, is at pains to explain in one of his prefaces. The texts which these scholars rendered were authorities in a sense that the modern word has lost, and their words were not to be trifled with. Who was Aristippus that he should omit any of the sacred words of Plato? Better carry over a word like *didascalia* than run any chance of altering the meaning of Aristotle. Burgundio might even be in danger of heresy if he put anything of his own instead of the very words of Chrysostom." In the preface cited, Burgundio speaks literally of his translation of Chrysostomos as *scriptura sancta* (Haskins, 151 note 36).

64. In a way suggesting a (freely invented) student couple of 1968 giving the name *Che* to a son, Abelard and Héloïse called their little son *Astralabius,* the name of the main astronomical observational and calculational device (Abelard, *Historia calamitatum,* ed. Muckle 1950: 184f.).

65. Stephen of Tournai, quoted from Grabmann (1941: 61). On p. 60, Grabmann quotes Absalon of Springersbach (like Stephen, late twelfth century) to the effect that "the spirit of Christ reigns not where Aristotle's spirit is Lord."

66. Cf. Grabmann (1941) and van Steenberghen (1966).

67. The works and attitudes of Jordanus de Nemore, the greatest (or only really great) traditional mathematician of the Latin thirteenth century, are analyzed in Chapter 6. (Leonardo Fibonacci was no "traditional" mathematician, in a sense referring to the medieval Latin tradition). Other "mathematicians by inclination" are discussed in Chapter 5.

68. Though otherwise directed, this outward loyalty holds even for those branches of mathematics which were bound up with astronomy and hence with "medico-astrological naturalism," a phenomenon that is discussed in Chapter 5.

69. Of Witelo's *Perspectiva*, Book I was introduced into many university curricula as an optional alternative to the *Elements*. This is the part of the work that deals *not* with optics but with general geometry. Cf. also note 21, on the conflict between pure and explicitly astronomically oriented spherical geometry.

70. See, e.g., the presentation of the current in Chapter 5, and the much fuller discussion in J. E. Murdoch (1969).

71. And, in contrast to what happened in the Islamic world (where astronomical teaching was organized in other patterns—see Chapter 4), the interest in astrology gave rise to little mathematical activity beyond the writing of compendia before Regiomontanus.

72. On the abacus school, see Fanfani (1951), C. T. Davis (1965) and Goldthwaite (1972). Leonardo Fibonacci is one link to Islam, but not the only one.

73. On the relations between the humanist movement and mathematics, cf. Rose (1975: 26–56 and *passim*). M. D. Davis (1977: 1–20) deals with the teaching of artists' and architects' mathematics in the abacus school. Cf. also Chapters 5 and 7.

74. Described in some detail by Ore in Witmer (1968: xviii–xxii).

75. The parallel can be carried into mathematical details. As mentioned above, higher-degree equations occur in the Old Babylonian texts in a way that suggests their exclusive use for the display of fake professorial virtuosity. Similarly, the abacus masters showed their ability by solving such equations. But ''the rules given [by Piero della Francesca, abacus master and painter—JH] for solving equations of the third, fourth and fifth degrees are valid only for special cases of these equations. The rule for solving the equation of the sixth degree is altogether false,'' as formulated by Jayawardene (1976: 243). The algebraic virtuosity displayed by Piero was at least as fake as that of the scribe school teachers, and Tartaglia and del Ferro (who found the solution first) could claim to be the first real virtuosos in a field where imposture had reigned for 3500 years.

76. A somewhat more detailed discussion of the relations between contemporary mathematical development and the scientific-technological revolution will be found in Chapter 8.

77. In this respect, the story can be regarded as an elaboration of an important theme from the antifascist historical sociology of science, as it was dealt with, e.g., by Farrington (1938), Sigerist (1938), and Merton (1938a, 1942).

## 2. Subscientific Mathematics: Observations on a Pre-Modern Phenomenon

1. That those who were inspired by Hessen's study read more shades into it is evident, e.g., from Joseph Needham's review (1938) of Robert Merton's *Science,*

*Technology and Society in Seventeenth Century England.* That Hessen's own thinking was different from the vulgar Marxism imputed to him is clear from his professional biography.

2. Very briefly stated (too briefly!), "scientific" knowledge thus aims at *truth*, whereas subscientific knowledge is directed toward utility. In the mathematical domain this might make us conclude that "scientific" knowledge is mathematics built on proofs, whereas subscientific mathematics builds on receipts and empirical rules. This conclusion is false for several reasons. First, as we know, e.g., from cosmogonies, statements may be considered supremely true because they are old and revered or because they are supposed to stem from sacred revelation. At times, the latter source is encountered even in mathematics: in one of his works, al-Bīrūnī tells that the Indians hold the ratio of circular circumference to diameter to be as "the ratio of 3,927 to 1,250, because it was communicated to them, by divine revelation and angelic disclosure" that such were the proportions of the world—trans. Pingree (1970: 120). Second, Nicomachos' "science" is argued solely on the basis of empirical rules. Third, complex mathematics, not least complex mathematics aimed at many-sided *application*, can only be transmitted successfully if supported by some level of understanding, and hence only by some sort of argument or proof. Basically, the need for proofs in mathematics comes from *teaching;* the elevation of the need into a transcendental philosophical principle is historically secondary (cf. Høyrup 1980).

3. The Old Babylonian mathematical texts were produced between c. 1800 B.C. and 1600 B.C. When speaking here of "Babylonian mathematics" I shall refer exclusively to this corpus.

4. This is Morris Kline's position (1972: 11).

5. When, for instance; will you know the area of a trapezoidal field and the portion which broke off from your measuring reed but not the reed itself (cf. below)? The evidence could be multiplied *ad libitum.*

6. So, *grosso modo,* Carl B. Boyer (1968: 45).

7. Ed., trans. Dodge (1970: II, 634ff.). In other parts of the catalogue, al-Nadīm mentions many non-Greek authors; their absence in the chapter on mathematics can hence not be explained as a result of personal prejudice on his part.

8. This follows, e.g., if one compares Indian and early Islamic algebra; cf. Chapter 4, Section 2.

9. On the whole, this holds even for the mathematical training of astrologers. Part of the Indian mathematical techniques that they took over was incorporated in books—but as technical chapters in astronomical books, only indirectly derived from the great mathematicians, where, e.g., Āryabhaṭa's value for $\pi$ would occur as something communicated "by divine revelation and angelic discourse" (cf. note 2; a general impression of the mathematics of early Islamic astronomy can be gained from Kennedy 1968, and Pingree 1968, 1970).

10. This word, like any other modern term, is not completely adequate. We might also speak of "crafts," if only we keep in mind that no guild institution need be involved, and that the groups in question were composed of "higher artisans."

11. Notable exceptions to this rule are made up by Sumerian and Babylonian accounting tablets and the Mycenaean Linear B tablets. Others could be mentioned.

12. Bhascara II, *Lílávatí*, 53—ed., trans. Colebrooke (1817: 24).

13. Rudolph 1540, the introduction to the chapter "Schimpfrechnung" (my emphasis).

14. *Liber abaci*, ed. Boncompagni (1857: 228).

15. Problem 52, version II, ed. Folkerts (1978: 74).

16. I shall not go into details concerning the distinction between *geometrical* and everyday (practical) problems, only refer to Aristotle's polemics against various sophists' approaches that simply miss that distinction and thus permit trivial solution (*Analytica posteriora* 75$^b$40–76$^a$3; *De sophisticis elenchis* 171$^b$16–22, 172$^a$3–7; *Metaphysica* 998$^a$1–4). Cf. also Chapter 1, Section 2.

17. See, eg., Heath (1921: I, 218–70), "Special Problems," which lists these attempts.

18. There are of course lots of individual exceptions. When mathematics *per se* has become a profession, people in need of a dissertation subject or another item in the list of publications will easily end up looking out for problems that are likely to be solved with the methods already at their disposal.

19. VAT 7532—ed. *MKT* I, 294f. The translation is mine, and builds on my reinterpretation of the Old Babylonian mathematical terminology. Without going into irrelevant details, the text should be comprehensible with the following explanations:

1.  Numbers are written in a sexagesimal place-value system (Neugebauer's notation).
2.  1 cubit = $\frac{1}{12}$ nindan; the nindan is the basic length unit and equals approximately 6 m.
3.  To "detach the igi of $n$" means finding its reciprocal ($1/n$).
4.  To "raise" means calculation of a concrete entity through multiplication.
5.  To "repeat until twice" means (concrete) doubling.
6.  To "make $a$ confront itself" means constructing a square with side $a$; if we do not care about the real (geometric) method of the Babylonians we may translate it "to square."
7.  $a$ makes $b$ equilateral" means "$b$ is the side of a square with area $a$"; in numerical interpretation, $b = \sqrt{a}$.

20. 6,14,24 · $z^2$ − 12,0 · z = 1,0,0, where z is $\frac{5}{6}$ of the original length of the reed.

21. According to the mathematical Susa text IX (ed. TMS, 63f.), See Høyrup (1990: 326f.).

22. *Book on the Chapters of Hindu Reckoning*, ed., trans. Saidan (1978: 337).

23. This is not the place to go into detail about the substance of Babylonian "pure" mathematics. I shall only refer to Høyrup (1990), which gives the reasons why its "algebra" must have been built on geometrical (though "naive," not critical) argumentation, and in Chapter 9 of which the overall cognitive orientation of Babylonian mathematics is also discussed.

24. The detailed arguments for the following discussion of composite fractions are given in Høyrup (1990b). Here I also discuss why similar usage in different cultures cannot be explained away as a random phenomenon.

25. *Arithmetica* I, xxiv–xxv (ed. Tannery 1893: I, 56–61) are unmistakably stripped versions of the "purchase of a horse."

26. Ed. Soubeyran (1984: 30). The text is discussed and compared to later versions in Høyrup (1986: 477–79). Not listed there is a significant Greco-Egyptian version from the Roman epoch: Papyrus Ifao 88 (ed. Boyaval 1971), to which Jöran Friberg has drawn my attention.

27. Connected to a tale about a peasant and his servant, whose wages are determined as successively doubled harvests from one grain of rice (reported in Stith Thompson 1975: V, 542, Z 21.1.1).

28. We observe that this oral character of genuine recreational mathematics sets it apart from Old Babylonian sub-scientific mathematics, which was carried by written texts even though the details of didactical explanation were normally given orally. Genuine recreational mathematics belongs with "lay" traditions; the methodical orchestration of scholasticized mathematics negates its recreational value, even though single problems such as the "broken reed" may betray a recreational origin.

29. *Book of Rare Things in Calculation;* German translation Suter (1910: 100).

30. Another group of encyclopedias does reflect the subscientific traditions, those concerned with the practice of science and not directly with books. A good example is al-Fārābī's *Catalogue of the sciences* (ed., trans. Palencia 1953: 39–53; cf. Arabic, 73). Here, seven branches of mathematics are distinguished, the last of which is ʿilm al-ḥiyal, "science of devices/ingenuities," which according to the description appears to refer to practical applications of "scientific" mathematics—i.e., to "applied mathematics" as defined in Section 1. At the same time, several of the other branches (so arithmetic and geometry) are subdivided into "theoretical" and "practical." No doubt this division is a reflection of Aristotle's conceptions—but Aristotle's categories are apparently understood exactly as we have done, as "scientific" and subscientific mathematics, respectively.

31. In Babylonia this is made fully clear, e.g., in line 27 of the "Examination Text A" (ed., trans. Sjöberg 1975), Kennst du die Multiplikation, die Bildung von reziproken Werten und Koeffizienten, die Buchführung, die Verwaltungsabrechnung,

die verschiedensten Geldtransaktionen, (kannst du) Anteile suweisen, Feldanteile ab-grenzen?'' In the case of Egypt, it follows from the range of subjects dealt with in the Rhind Mathematical Papyrus, as also from the Papyrus Anastasi I (ed., trans. Gardiner 1911), a "satirical letter" much used in the scribal school to vilify the poor dunce who neither knows how to calculate a ramp, nor to provide rations for troops, nor to find the number of men required to transport an obelisk (etc.).

32. A factor $^{24}/_{25}$ found in a table of constants has been assumed to have served as a correction factor, corresponding to $\pi = 3\frac{1}{8}$ ($^{24}/_{25} = 3 \div 3\frac{1}{8}$)—see *TMS*, texte III, lines 2, 3, and 30, and commentary pp. 31, 33. Most likely, however, the role of the technical factor was different or more specific.

33. See Rhind Mathematical Papyrus, Problems 41 and following (ed., trans. Chace et al. 1929).

34. *Kings* 7: 23; 2 *Chron.* 4:2

35. See, e.g., Parker (ed., trans.) 1972, Problem N[os] 32 and 42.

36. See Cantor (1907: 551) and Gandz (1932: 49).

37. Or even the ratio 3! The text is doubtful. See *De Architectura* X, ix, 1 and 5 (ed. Granger 1970: II, 318 and 322, with appurtenant notes).

38. Earlier parallels *could* be pointed out. So, when Middle Kingdom Egyptian scribes began making their accounts in unit fractions instead of metrological sub-units (as had been done in the Old Kingdom), the only reason is that they had learned this system in school—and the reason to introduce it in school will have been that it was easier to argue for (and thus teach unambiguously) the precise calculations in unit fraction notation than to defend the necessarily approximate (and thus ambiguous) solutions of practical problems in metrological units (see Høyrup 1980: 34).

39. This formulation fits "Western" civilization (the medieval center of which was the Islamic world). Indian high-level algebra is older, but it did not influence the Western development (cf. note 8).

40. *Kitāb al-jabr wa'l-muqābalah*, ed., trans. Rosen (1831). Another book on the subject and probably also the same title (ed., trans. Sayili 1962) was written by ibn Turk at almost the same time. The question of priority cannot be settled with certainty, but terminological considerations suggest that ibn Turk was at least *independent* of al-Khwārizmī (see Høyrup 1986: 474, note 28).

Addendum 1992: The precise wording of the Arabic text used for all modern editions and translations (including Rosen 1831) turns out at closer investigation to be the outcome of stylistic normalization in at least two phases (see Høyrup 1991). (Strange as this may sound, Gerard of Cremona's twelfth century Latin translation (ed. Hughes 1986) renders the original formulations more precisely as far as it goes). Rosen's translation, furthermore, is somewhat free at times. Both the Arabic revisions and Rosen's grammatical inaccuracies and slight adjustments to familiar terminology tend to mask the similarity between al-Khwārizmī's formulations and those characteristic of the sub-scientific tradition.

41. The detailed arguments for this (which also involve Abū Kāmil's *Algebra*) are given in Høyrup (1986). One correction should be given to p. 472 of that paper: The distinction between two groups of calculators each with its own specific methods, referred to Abū Kāmil is not found in the facsimile edition (ed. Hogendijk 1986). It seems to have been inserted into the Hebrew Renaissance translation, and thus to be irrelevant to the early history of the subject.

42. See Datta and Singh (1962: I, 169). In India, the term is used in connection with the extraction of a square root, and the interpretation is geometrical and thus meaningful. In the *al-jabr* tradition, the term is *not* understood geometrically, which deprives it of metaphorical meaning.

43. See problem N$^{OS}$ 13 and 17 of the Papyrus Akhmîm (ed., trans. Baillet 1892: 70, 72). It should be observed that there is nothing strange in the use of the same term for the unknown in problems of type $a \cdot x = b$ and for the second-degree term in problems of type $y^2 + a \cdot y = b$, if only we put $y^2 = x$, thus transforming the latter problem into $x + a \cdot \sqrt{x} = b$. Islamic algebras will normally give the value for both $x$ and $y$, thus regarding the *māl* as an unknown in its own right. The transformation is hence justified by the sources.

44. For other operations al-Khwārizmī *did* invent geometrical justifications for the rhetorical reductions himself, but then the wording is different ("This is what we intended to elucidate"; "We had, indeed, contrived to construct a figure also for this case, but it was not sufficiently clear"—trans. Rosen 1831: 32, 34).

45. Detailed analysis in Høyrup (1986: 456–68). It should be noted that the geometrical character of the argument only follows from indirect arguments—the figures belonging with the solutions are lost in the Latin translation.

46. Chapter X, xiii—ed. trans. Krasnova (1966: 115).

47. This *might* also be what al-Fārābī tells us when stating in his *Catalogue* (ed., trans. Palencia 1953: 52) that the science of algebra is "common to arithmetic and geometry" (other interpretations are possible).

48. BM 13901—ed. *MKT* III, 1–5; cf. translation and discussions in Høyrup (1990: 266–80).

49. Analysis in Høyrup (1990: 271–75).

50. Ed., trans. Rosen (1831: 13–15). The procedure fits the algorithm badly; when it is nonetheless used (and used *first*), the reason must be that it was familiar.

51. This "Islamic miracle" (a term coined in similitude of the well-known "Greek miracle," the creation of the autonomous "scientific" approach) is dealt with extensively in Chapter 4.

3. Mathematics and Early State Formation, or, The Janus Face
of Early Mesopotamian Mathematics: Bureaucratic Tool and
Expression of Scribal Professional Autonomy

1. The date is B.C., of course, like all dates in the following, and approximate, like all dates below!

2. See, e.g., R. Adams (1982) and C. C. Lamberg-Karlovsky (1976: 62f.).

3. A large number of case studies and further references will be found in Claessen and Skalník (1978) and in Gledhill, Bender, and Larsen (1988).

4. Childe (1950: 3). A recent comprehensive discussion of the connection between state formation, writing, and alternatives to writing is Mogens Trolle Larsen (1988).

5. Cf. for this description Nissen (1983: 36–40, 55); Mellaart (1978); and below.

6. See Wright and Johnson (1975: 269f.), and Nissen (1983: 57f.).

7. Disputed by Weiss (1977). The difference of opinion depends on different estimates for the relative lengths of archaeological periods, again dependent on different absolute datings. The most recent radiocarbon datings appear to favour Wright and Johnson (B. D. Hermansen, personal communication).

8. See Wright and Johnson (1975: 272, 282f.) and Johnson (1975: 295–306). The presumed evidence for ration distribution (the particular "bevelled-rim bowl") has been challenged by Beale (1978). In protoliterate Uruk (see below), however, the connection between the bowl in question and the delivery of rations is corroborated by its seeming appearance in the pictogram for rations ($KU_2$). Cf. also Damerow and Englund (1989: 26).

9. Writing turns up in Susa (and in fact the Iranian area at large) somewhat after its emergence in Uruk. The idea of writing seems to be borrowed, but the pictographic script itself is independent—while, on the other hand, there is clear kinship but not identity between the "proto-Sumerian" and the "proto-Elamite" counting and metrological systems. For detailed information I shall only refer to Damerow and Englund 1989 (including Lamberg-Karlovsky's introduction to that work).

10. Adams and Nissen (1972: 17–19); Johnson (1975: 310–24); Nissen (1983: 73–116, 132–34; 1986a: 330).

11. The tablets are never found in the places where they were originally made or used, but mostly in rubbish heaps. The relative dating thus relies on paleographic criteria, which, however, seem reliable. See Nissen (1986a: 319–22) for details. Because of the greater complexity and regularity of Uruk III tablets, some of the administrative features ascribed here to the whole protoliterate period may indeed only

be fully developed in the later phase.

The organization of text formats and the use of formats as carriers of information is explained and discussed in Green (1981: 348–56).

12. Details of the settlement structure, it is true, suggest that an inner core of settlements (until some 12 km from the city) was bound more strongly to the center than those farther removed (Nissen 1983: 144f.). The outer zone can be surmised not to belong to the Temple estate proper; but we have no means to ensure that all land of the inner zone was submitted uniformly to the theocratic system.

13. For a discussion of the general arguments for the presence of such communities, see Diakonoff (1975). Diakonoff (1969a) is an English summary of his epoch-making investigation of twenty-fourth century Lagaš.

14. Or ours? Our own bureaucratic conditioning, in combination with the internal rationality of the bookkeeping records may easily lead us into more Weberian readings of the text than intended by its original authors.

15. Whereas protoliterate Uruk was a full-fledged state according to Wright and Johnson (quotation [G]), it is thus far from certain that it would be so according to Carneiro (quotation [E]) and Runciman (quotation [F]). From their point of view, the control system will probably have directed not *a state* but only *an estate* immersed in and influencing a pre-state society. Especially for Runciman, who sees early seventh-century Athens as a "protostate" only, the protoliterate Uruk system can have been no more.

16. Traditionally, it is true, the opposite view has been accepted on preliminary evidence from a single, somewhat ambiguous sign combination in a single text. However, the ongoing progress on a large project on the archaic texts directed by Hans Nissen (see Nissen 1986b; the results of the project are reflected in many references in the present paper) has uncovered no supplementary testimony; for this and other reasons discussed by Robert Englund (1988: 131–33) in a two-page footnote, we must now opt for a vigorous *nescimus*.

My present favorite hypothesis (for which I argue in Høyrup 1993) is that Sumerian shares so many grammatical features with "creole languages" (on which see Romaine 1988) that it may have originated as a creole at the influx of new population segments in the later fourth millennium.

17. This aspect has been investigated by Thorkild Jacobsen in several publications (1943; 1957).

18. Basing himself on other evidence, Nissen (1982) argues for duality of the Sumerian society along several other dimensions.

19. In this connection one may also recall the oft-made observation that nobody would have guessed from the written record that Sumerian rulers might be buried with a large retinue of killed servants (as was actually the case in Ur, during the first phase of ED III).

20. This is particularly clear in a series of "reform texts" by Uru'inimgina, either elected king of Lagaš by the assembly or usurper in the late twenty-fourth century B.C., describing the abuses that had developed and his restoration of good old time, which includes giving back the temple land appropriated by the ruler to the gods (a recent though not fully convincing discussion of the obscure texts and an exhaustive bibliography is Foster 1981: 230–37; cf. also Hruska 1973). But since Uru'inimgina and his consort are to act as stewards of the gods on their reacquired estates, realities did not change at least on this point (Tyumenev 1969a: 93f.). Whether his protection of "widows and orphans" fared any better is unclear. In any case, Uru'inimgina was soon brushed aside by Lugalzagesi's conquest and unification of the whole Sumerian region.

21. In fact, the analysis reminds strikingly of Engels' (and Aristotle's) analysis of the Solon reforms in *Die Ursprung der Familie*. Even this formation of a mature state in Athens followed upon a phase considered as "military democracy"—and followed shortly after the establishment of a state in Runciman's sense.

That conflicts between the city-states became intense in late ED III is obvious from the surviving royal inscriptions. After centuries of mounting city walls combined with amazing royal taciturnity on warlike matters, proclamations of military triumph and menaces against potential aggressors suddenly abound.

22. Halfway only—many of Deimel's didactical tablets carry names of what seem to be authors, editors, or teachers, and many of the persons mentioned carry a priestly title (Deimel 1923: 2*f.).

23. Foster (1982: 7–11) distinguishes three Sargonic archive types: family or private; "household" (with a horizon restricted to a single city) and "large household."

24. When systematic writing of the Semitic Akkadian began, using the phonetic values of the Sumerian signs, orderly succession of the signs became compulsory.

25. Brief expositions are given by Nissen (1983: 207–13) and Liverani (1988: 267–83). A recent critical survey of the state of the art concerning Ur III administration is given by Robert Englund (1990: 1–6).

26. An overview of the centralized economy, as well as the exceptions, is given by Hans Neumann (1988). Cf. also Neumann 1987: 151–54 on non-statal artisanate.

27. After toiling 40 years night and day in the great marsh, the minor gods decide to confront their chamberlain (the god Enlil); they do so, armed with spades and hods to which they have set fire, and insist that the chamberlain call in the collective leadership (consisting of Enlil himself, together with the gods An and Enki). When asked for the instigator, the strikers deny the existence of such a person and declare their solidarity—thus begins the plot of the Old Babylonian *Story of Atraḫasīs* (ed., tr. W. G. Lambert and A. R. Millard 1969; this passage pp. 45ff.). The whole description is too close to the social psychology of real wildcat strikes to have been

freely invented, and the setting suggests that the author builds on experience from Ur III estates, rather than contemporary events.

In the end, the problem is solved by a "social reform": *man* is created in order to take over the toil of the gods.

28. Most likely, ecological reasons were also involved in the breakdown, accentuating the incompatibility between the costs of the state apparatus and the productivity of the work force. In any case, the political center of Iraq from now on moved northward.

29. Lipit-Eštar Hymn B, lines 21–23, trans. Vanstiphout (1978: 37).

30. A very readable narrative not only of Hammurapi's history and policies but also of the sociopolitical and cultural conditions since early Old Babylonian times is given by Horst Klengel (1980). Other works to be consulted include Dandamajev (1971), Diakonoff (1971), Gelb (1965), Jakobson (1971), Klengel (1974, 1977), Komoróczy (1978, 1979), Kraus (1973), Leemans (1950), Oppenheim (1967), Renger (1979), and Stone (1982).

31. This is in fact part of the complaint of the minor rebellious gods in the *Story of Atraḫasīs* (note 27). Whereas they were originally the "sons," i.e., the lower-ranking members of the clan community, and the "chamberlain" thus nothing but the "elder" member governing common affairs, he has now become the master and they the dumb subjects.

32. Two fairly recent presentations are Sjöberg (1976) and Lucas (1979). Older important general discussions are Falkenstein (1953), van Dijk (1953: 21–27), Gadd (1956), Landsberger (1960), and Kraus (1973: 18–45). Didactical texts illustrating various aspects of the school enterprise have been published and translated by Kramer (1949) and Sjöberg (1972, 1973, 1975).

33. Mostly in public administration. "Scribes were limited to positions connected with administration or with substantial accumulations of private capital. Perhaps, also, they filled out contracts and legal documents at the gate of the city. If I were to make an intuitive sweeping estimate, I would say that perhaps seventy percent of the scribes had administrative positions, twenty percent were privately employed, and the remainder became specialists in the diagnosis of illness, charms, magic, and other activities calling for some knowledge of writing," as formulated by Landsberger (1960: 119) in answer to a question whether the important role played by secret idioms of various crafts in the "Examination Text A" (see below) could correspond to future employment.

Employment outside the notarial, accounting, and engineering sphere was clearly secondary: "A disgraced scribe becomes a man of spells," we are told by a proverb (Lucas 1979: 325).

34. Ed., trans. Sjöberg 1975; cf. Landsberger 1960: 99–101. Admittedly, the earliest extant copies of the text are quite late (they are Neo-Assyrian); as observed by Sjöberg, however, the contents of the text seem to require an Old Babylonian origin.

35. The abstract marriage algebra of Malekula is described by Ascher and Ascher (1986: 137–39), the graph-theoretically refined closed patterns by M. Ascher (1988: 207–25). The disconnectedness between the two does not imply, of course, that the intellectual training gained through graphs cannot have made it easier for the informant to formulate the principles of marriage rules explicitly for the benefit of the ethnographer.

Ascher and Ascher (1986: 132) make the point that the ''category mathematics is our own'' but stop short of drawing the same conclusion about ethnomathematics, for fear perhaps of devaluing the non-literate cultures they discuss. This caution should be superfluous: the elements of ethnomathematical thought are no more random or isolated than our elements of mathematical thought—their connections are different.

36. Denise Schmandt-Besserat, who discovered the widespread appearance and high age of a system that until then had only been recognized in the later fourth millennium, has published a long array of papers on the subject, of which I shall only refer to the original publication (1977), an early popularization (1978), and a recent paper (1986) discussing *inter alia* social and cognitive interpretations. Another recent publication on the matter to be mentioned is Jasim and Oates (1986).

37. It may be objected that we would not expect so highly developed stratification in the beginning of the Neolithic. Some indications exist, however, that the ecology of the Near East was rich enough to support stratified settlements and to call for organized redistribution, and that ranking and even hereditability of high status had developed in the late Mesolithic Natufian (see G. A. Wright 1978: 218–21).

38. Evidently, this cannot be read out from the tokens themselves. It follows from an agreement between general ethnomathematical experience and the reflection of the token system in protoliterate metrologies.

One question that cannot be solved in this way is whether ''bundling'' was included into the system. *If*, e.g., a small disk corresponded to an animal, would a large disk then correspond, say, to a ''hand'' (5) or ''hands and feet'' (20) of animals? Would a ''sphere-container'' be supposed to contain a fixed number of ''cone-containers''? At some point in the development, such bundling was introduced, but we have no means to ensure that it had already happened in the Neolithic.

39. At this point we begin to approach hard facts. This last-mentioned use of the tokens follows from the geographical distribution between Susa and lower-ranking settlements of seals, broken sealings, bullae prepared for use but not yet closed, and dispatched bullae (see Wright and Johnson 1975: 271).

40. See Le Brun and Vallat (1978: 47, 57) for Susa and Jasim and Oates (1986: 349) for Habuba Kabira.

41. Readable expositions of the various facets of the development are given by Nissen (1985) and by Damerow, Englund, and Nissen (1988, 1988a). A global overview is presented in Nissen, Damerow, and Englund 1990.

It should be observed that the sequence bulla—numerical tablet—pictographic tablet is in the main derived from the inner logic of the process combined with indirect arguments, rather than from direct stratigraphic criteria: because only numerical and no pictographic tablets are found in Habuba Kabira, this setlement must be earlier than Uruk IV, where pictographic writing *is* attested. But then, since bullae and numerical tablets are found in Habuba Kabira, they must be earlier than pictographic writing; and so on.

42. In principle, the appearance of the signs could be an accidental result of the fact that these are the impressions that can be made by vertical and inclined impression of a thin and a thick circular stylus; the existence of bullae where the tokens actually contained are impressed (Schmandt-Besserat 1986: 256) suggests, however, that the similarity between tokens and signs is not accidental, and that the circular stylus was chosen precisely because it could so easily produce the desired impressions.

43. A sequence "for counting" is characterized by a separation of quantity from quality, as, e.g., in our "3 sheep" or "6 m." A metrological sequence, on the other hand, has quality inherent in quantity (as in "m m m m" instead of 4 m).

Throughout the history of Mesopotamian mathematics this distinction remains less clear than the historian of mathematics might prefer. Instead of our "4 m," e.g., an Old Babylonian scribe would usually have written "4," expecting everybody to know that lengths are measured in this unit.

44. The sign itself, it is true, differs from the turned picture of the cone used in the counting sequence: it might look like a picture of the half- or quarter-sphere tokens, and could thus have been present already in the token system. But like the fractional counting number, it is turned 90° clockwise, indicating that both are conceptualized as belonging to the same ("fractional") category.

45. The following description of Sumerian and protoliterate timekeeping is built on Robert Englund's pioneering work on administrative timekeeping (1988).

46. A similar, albeit weaker, observation could be made from the existence of "dependent metrological sequences" produced from those just described through addition of strokes and used to count or measure, e.g., specific varieties of the goods counted or measured by the corresponding fundamental system—for instance, to measure emmer instead of barley. In this case the innovation may go back to the late preliterate creation of supplementary token types (and token sequences?) by means of incisions.

47. See Powell (1972), the principal reference for Sumerian area measures.

48. I.e., average length times average width. This method was used in the computation of the area of not-too irregular quadrangles at least from ED III to Old Babylonian times, and even far into the Middle Ages.

49. Like the idea of writing (but not the script itself), this technique also seems to have been borrowed by the proto-Elamite culture (which had a center in Susa but

had outposts far into the Iranian East, and which was more or less contemporary with Uruk III). This follows from Beale and Carter's careful analysis (1983) of the geometry of the architectural complex of Tepe Yahya IVC (proto-Elamite, and thus contemporary with Uruk III), in which base-lines separated by integer multiples of a standard measure (equal to 1.5 times the standard brick length) define the exterior edge of outer walls and the midlines of inner walls. Apart from the ratio between the standard measure and the standard brick, moreover, the same code appears, e.g., in buildings from Habuba Kabira (the Uruk V outpost mentioned in Sections 3 and 7).

50. One field that was not yet integrated (and which never was until the modern era) was "ethnomathematical graph theory." That is was nonetheless present we may infer from somewhat later evidence: in the Fara tablets such "graphs," complex symmetric patterns drawn by a continuous line, turn up time and again—see the specimens in Deimel (1923: 21 [broken]); Jestin 1937: CLXXX, #973); and Edzard (1980: 547).

51. A beautiful example seems to be presented by the linear B tablets of the Mycenaean palace bureaucracy. Even though Mycenaean art bears witness of a strong and inquisitive interest in geometrical regularity (Høyrup 1992), there is to my knowledge no evidence whatsoever of a transformation of scribal accounting arithmetic into *mathematics*.

52. Illustrated, e.g., by this dialogue (Luria: 55):

*Luria:* (explaining a psychological test) Look, here you have three adults and one child. Now clearly the child doesn't' belong in this group."

*Rakmat:* (an illiterate peasant from Central Asia): "Oh, but the boy must stay with the others! All three of them are working, you see, and if they have to keep running out to fetch things, they'll never get the job done, but the boy can do the running for them [ . . . ]."

Situational thinking was found in Luria's investigation of prevailing modes of cognition in Soviet Central Asia to be "*the* controlling factor among uneducated, illiterate subjects," whereas both modes were applied (with situational thinking dominating) among "subjects whose activities were still confined primarily to practical work but who had taken some courses or attended school for a short time." "Young kolkhoz activists with only a year or two of schooling," on the other hand, employed the principle of categorical classification "as their chief method of grouping objects."

53. This problem of the interplay between tool and mode of thought I shall not pursue any further in the present connection, only refer to its position as the central theme in Damerow and Lefevre (eds.) (1981).

54. Import of metals will course have been a matter of bureaucratic interest. But as far as I know nothing suggests that archaeologists have come upon tablets from the archive of trade.

55. Apparently for rhetorical reasons, Friberg discards the protoliterate school exercises, which he himself had been the first to identify.

56. This conclusion is not changed by the claims and the partially new text material presented by Whiting (1984), who conflates place value notation with what I have here called "sexagesimalization". But Whiting's evidence underscores how much was in the air in the actual computation techniques in use at least since the Sargonic era, and his explanation of two apparent writing errors in a pre-Sargonic tablet of squares (OIP 14,70, transliterated and translated in Edzard 1969) suggests that an idea similar to the gin-tur was used already in the twenty-fifth century B.C.

The errors so abundantly present in the computations on which Whiting bases his argument, on the other hand, make it obvious that the system after which calculators were groping was *not yet at hand* as more than an inherent possibility—similar, perhaps, to the way the decimal place value system may have been potentially present in the Chinese use of counting rods for perhaps 2000 years before giving rise to the genesis of a genuine place value notation (see Martzloff 1988: 170f., 181–84)), and to the way it was demonstrably mimicked by the Greek idea of *pythmens* (see Pappos, *Collectio* II.1, in Hultsch 1876: I, 2).

57. In the integer range between 1 and 599, place value and "normal" administrative notation cannot be distinguished. Therefore, the scribe did not need to decide whether he used one or the other in such cases, nor can we settle the question.

A few undated tables of reciprocals probably belong to Ur III, but the paleographic distinction between Ur III and Old Babylonian tablets is not very safe for tablets containing exclusively or predominantly numbers.

58. So much was in the air, indeed, that the most difficult step was not to get the idea in itself but to find the courage to do so. For an isolated inventor (be he practical calculator or teacher) the system would be worthless. Only when backed by tables of constants, reciprocals, etc., and thus only when large-scale use made it economically feasible to produce these, were place value numbers any good.

59. See Høyrup (1980: 19f., and 85f. notes 39, 42, and 44), which contain cross-cultural comparison, whose references for Ur III bookkeeping itself, however, are partly outdated. The most recent treatment of the subject is given by Englund (1990: 13–55).

60. The details of the argument build on my investigation of Old Babylonian "algebra" (Høyrup 1990).

61. It seems likely that some specialization was present. According to Landsberger (1960: 97; cf. 1956: 125f.), indeed, the Old Babylonian "lexical lists distinguish, according to degree of erudition and specialization, fifteen varieties of *dubsar* or scribe" which, however, all disappear in the subsequent period, together with the scribe school. The evidence is insufficient, however, to decide to which extent the job specialization was reflected in specialized school curricula.

It should be observed that dub-sar NIG.ŠID, translated "mathematician" by Landsberger (1956: 125), should rather be understood as "accountant."

62. Often, of course, by means of what we would call "approximate formulae," forgetting in this distinction that even the most exact area formula becomes ap-

proximate when the terrain surveyed is hilly and no Euclidean plane.

Karen Rhea Nemet-Nejat (1993, Chapter 4) presents a survey of practical prob-lem types occurring in the Old Babylonian mathematical texts.

63. Høyrup (1990) presents the arguments for this interpretation in philological and mathematical detail, whereas Høyrup (1989) presents an overview. Høyrup (1985) is a fairly complete but preliminary and rather unreadable exposition ("It is difficult to follow the red thread—provided there is any," as Asger Aaboe put the matter).

64. BM 85194, rev. II.7–21—ed. *MKT* I, 149. The translation is mine, and builds on my reinterpretation of the Old Babylonian mathematical terminology (I have left out indications of restituted damaged passages and corrected a few copyist's er-rors tacitly). Without going into irrelevant details, the following explanations should in principle make the text comprehensible for those who want to wrestle with a real piece of fairly complex Babylonian mathematics:

1) Numbers are written in a sexagesimal place value system (Neugebauer's notation).

2) Horizontal extensions (length, breadth) are measured in the unit nindan ($\approx 6$ m).

3) Vertical extensions are measured in kùš (cubits), where 1 kùš = $\frac{1}{12}$ nindan $\approx 50$ cm.

4) Volumes are measured correspondingly, in the unit sar = nindan$^2 \cdot$ kùš, here left implicit (gán is not the unit but an indicator of category and loose order of magnitude).

5) To "append" designates a concrete addition, and to "tear out" the cor-responding concrete subtraction.

6) To "detach the igi of $n$" means finding its reciprocal ($\frac{1}{n}$)—actually looking it up in the table of reciprocals.

7) To "raise" means calculation of a concrete entity through multiplica-tion, as done, e.g., in operations involving proportionality.

8) To "double" designates a concrete process—in the actual case, the dou-bling by which a rectangle is produced from a right triangle.

9) To "break" denotes a bisection into "natural" or "customary" halves—as, in the actual case, one side of a triangle is customarily bi-sected when its area is calculated.

10) To "make ($a$) surround" means constructing a square with side $a$; if we do not care about the real (geometric) method of the Babylonians we may translate it "to square."

11) The "equilateral" of an area is the side it produces when laid out as a square; in numerical interpretation, its square root.

Some hints can be found in *MKT* I, 186. Those who want to apply the geometrical interpretation (not given in *MKT*) should be aware that a rectangle $n$ [cubit high] by $n$ [nindan long] is dealt with as a square; i.e., the units that anyhow are left implicit are disregarded. Cf. Høyrup (1985: 56).

65. More precisely: such problems were popular according to their place in the corpus of texts and thus in the curriculum. There is no particular reason to believe that

average students liked them. To the contrary, the generally suppressive character of the examination texts might suggest that mathematics was, within scribal humanism no less than in nineteenth century (A.D.) German neohumanism, also cherished because of its disciplining effects.

66. Of course, the situation was going to be different in the Middle Ages, even for professional groups resembling the Old Babylonian scribal profession. By then Greek mathematics was *already at hand,* and ''scribal'' computation could (and would, in the Islamic and Christian worlds) be seen as a special instance of that lofty enterprise. What is at stake here is the option of *inventing* something like Greek mathematics, which was a task quite different from that of assimilating the *Elements*—cf. the analysis of the former process in Chapter 1, Section 2.

67. Truly, quite a few historians of mathematics have supported the view that it was based on a toolkit of recipes found empirically and assimilated by the scribes through rote learning—a view mostly based on familiarity with one or two problems quoted in translation in some semi-popular exposition. Scholars really familiar with the sources have always known that Babylonian mathematics could only have been produced by people who understood what they were doing, and they have supposed that oral explanations will have accompanied the terse expositions written in the tablets. During my own investigation of the sources I have located a couple of texts that in fact contain this fuller explanation (see Høyrup 1989: 22–25, and Høyrup 1990: 299–305, 320–28).

68. The relation between practitioners' mathematics and recreational problems is discussed in Chapter 2, which also takes up from this perspective the ''scholasticized'' character of Old Babylonian ''pure'' mathematics.

69. Evidently, this difference in kind between recreational and scribe school mathematics does not preclude that a scribe school in need of non-trivial problems and corresponding methods borrowed them from a non-literate, recreational tradition. Evidence exists that this is precisely what happened.

First, it is characteristic that the key terminology of the early ''algebra'' texts is Akkadian (as is in principle the whole Old Babylonian mathematical corpus, even in texts where Sumerographic shorthand and Sumerian technical terms abound). In one text the quadratic completion, the essential trick in the solution of second-degree equations, even seems to be designated ''the Akkadian'' (*viz.,* Akkadian method) (Høyrup 1990: 326). No doubt, then, that ''algebra'' was no heritage from the Sumerian school tradition.

Second, at least a cognate of second-degree ''algebra'' predates Ur III. Another favorite problem, indeed, shares part of the characteristic terminology (and, presumably, the naive-geometric technique) with the ''algebra'': the bisection of a trapezium by a parallel transversal. The oldest known specimen of this problem, however, is the tablet mentioned in the end of Section 8, which was found on the floor of a Sargonic temple (see Friberg 1990: 541).

The problem is so specific that independent reinvention is unlikely. But if the school has not transmitted the problem and its solution, who has? My best guess is an

Akkadian surveyors' environment, which can quite well have existed in central Mesopotamia in the early Old Babylonian epoch, to the north of that Sumerian core area where graduates from the scribe school may possibly have had a monopoly of surveying.

Interestingly, one Sargonic school exercise (A 5446, see Whiting 1984: 65f.) seems to presuppose knowledge of a basic "algebraic" identity. It asks for the areas of two squares of side $R - r$, where $R$ is a very large, round measure, and $r$ a very small unit. Without knowledge of the identity $(R - r)^2 = R^2 - 2Rr$ (which of course follows easily from geometrical considerations), the calculation will be extremely cumbersome.

Even in the Sumerian South, it should be added, scribal monopoly on surveying and geometrical practice is not too firmly established. Krecher (1973: 173–76) points out that Fara contracts for purchases of houses involve a "master who has applied the measuring cord to the house" (um-mi(-a) lú-é-éš-gar), whereas a "scribe of fields" (dub-sar-gána) is involved when land is bought; a Sargonic document groups together the surveyor" (LÚ.EŠ.GID), the "scribe" (dub.sar) and the "chief of the land register" ($SA_{12}.DU_s$). Krecher supposes even the chief and the surveyor to be scribes, but in particular concerning the latter we cannot know for sure.

70. For one thing, the set of Sumerian equivalents for Akkadian technical terms changed—tab, once used as a Sumerogram for eṣēpum ("to double" or "repeat concretely"—arithmetically, to multiply by an integer $n$ below c. 10) came to designate addition. It thus appears that the scribes translated the language of "algebra" into their favorite Sumerian tongue for a second time without knowing that (or without knowing too precisely *how*) it had been done before.

71. Apart from the general literature, the following builds in particular on Baines (1988); Brunner (1957); and Høyrup (1990b).

72. Or, at least, only on redistribution in an utterly distorted form (cf. Endesfelder 1988): Pharaoh took hold of the societal surplus and redistributed part of it to his officials while returning perhaps promises of cosmic stability to the general peasant population.

## 4. The Formation of "Islamic Mathematics": Sources and Conditions

1. Cf. Chapter 3 as far as the Mesopotamian situation is concerned. A short but striking illustration for the case of Egypt is supplied by the opening phrase of the Rhind Mathematical Papyrus, the mainly utilitarian contents of which are presented as "accurate reckoning of entering into things, knowledge of existing things all, mysteries [ . . . ] secrets all" (trans. Chace et al. 1929: Plate 1; similarly Peet 1923: 33).

2. I shall not venture into a discussion of this conception, which is probably no better founded than its Aristotelian counterpart.

3. This status is barred to me already for the reason that my knowledge of Arabic is restricted to some elements of basic grammar and the ability to use a dic-

tionary. Indeed, the only Semitic language I know is the simple Babylonian of mathematical texts.

4. I use the term *culture* as it is done in cultural anthropology. Consequently, the Sabian, Jewish, and Christian minorities that were integrated into Islamic society were all participants in the "Islamic culture," in *dār al-Islām*.

Similarly, "Islamic mathematics" is to be read as an abbreviation for "mathematics of the Islamic culture," encompassing contributions made by many non-Muslim mathematicians. I have avoided the term *Arabic mathematics* not only because it would exclude Persian and other non-Arabic mathematicians but also (and especially) because Islam and not the Arabic language must be considered the basic unifying force of the "Islamic culture"—cf. Section 10.

5. Even the protophilosophical cosmogonies that precede the rise of Ionian natural philosophy are now known to make use of Near Eastern material—see Kirk, Raven, and Schofield (1983: 7–74, *passim*).

6. A stimulating discussion of the formative conditions for the rise of philosophy is Vernant 1982. An attempt to approach specifically the rise of scientific mathematics is found in Chapter 1, Section 2.

7. For the same reason, I shall treat this part of the subject with great brevity, mentioning only what is absolutely necessary for the following. A detailed account of the transmission of individual Greek authors will be found in *GAS* V, pp. 70–190.

8. A recent translation based on all available manuscripts is Dodge (1970). Chapter 7, section 2, dealing with mathematics, and the mathematical passages from section 1, dealing with philosophy, were translated from Flügel's critical edition (from 1872, based on a more restricted number of manuscripts) by Suter (1892; Supplement 1893).

9. Comparison with other chapters in the *Catalogue* demonstrates that the lopsided selection is not caused by any personal bias of the author.

10. A full discussion is given by Steinschneider (1865), a brief summary by Sarton (1931: 1001f.).

11. This can be compared with the list of works that al-Khayyāmī presupposed as basic knowledge in his *Algebra:* The *Elements* and the *Data*, Apollonios' *Conics* I–II, and (implicit in the argument) the established algebraic tradition (trans. Woepcke 1851: 7). The three Greek works in question constitute an absolute minimum, we are told.

12. See Woepcke's introduction to and selections from al-Karajī's *Fakhrī* (1851: 18–22 and *passim*); Sesiano (1982: 10–13); and Anbouba (1979: 135).

13. On Thābit's investigation of "amicable numbers," see Hogendijk (1985), or Woepcke's translation of the treatise (1852). Two later treatises on theoretical arith-

metic were also translated by Woepcke (1861), one anonymous and one by Abū Ja'far al-Khāzin (a Sabian like Thābit). Among recent publications on the subject, works by Anbouba (1979) and Rashed (1982; 1983) can be mentioned.

14. So, al-Fārābī's *Iḥṣā' al-'ulūm* (*De scientiis*, trans. Palencia 1953: 40); al-Khuwārizmī's *Mafātiḥ al-'ulūm* (translation of the section on arithmetic in Wiedemann 1970: I,411–28; theoretical arithmetic is treated amply pp. 411–18); and the encyclopedic part of the *Muqaddimah*, trans. Rosenthal (1958: III, 118–21).

The treatment in the encyclopedias is remarkably technical. In itself it seems highly probable that Late Hellenistic Hermeticism and Sabian, Jabirian, and Ismā'īlī numerology would mix up with "speculative" arithmetic. To judge from the encyclopedias, however, any inspiration from that quarter has remained without consequence for the contents of the subject *when understood as mathematics*. Cf. also Section 16.

15. See, e.g., *GAS* V, 191ff.; Pingree (1973); and Pingree, "Al-Fazārī." The *Zīj al-sindhind*, the Sanskrit astronomical treatise translated with the assistance of al-Fazārī around A.D. 773, was mainly built upon the methods of Brahmagupta's *Brāhmasphuṭasiddhānta*—but even influence from the *Āryabhaṭīya* is present. The original authors had become invisible during the process.

16. The discrepancy between the advanced syncopated algebra of the Indians and the rhetorical algebra of al-Khwārizmī was already noticed by Léon Rodet (1878). This observation remains valid even if his supplementary claim (*viz.*, that al-Khwārizmī's method and procedures are purely Greek and identical with those of Diophantos—p. 95) is unacceptable.

Al-Khwārizmī can be considered a key witness: He is one of the early Islamic workers on astronomy, and mainly oriented toward the *Zīj al-sindhind*, with some connection to the Pahlavi *Zīj al-Šāh* and (presumably) to Hellenistic astronomy (cf. Toomer, "Al-Khwārizmī"). As we shall see when discussing his treatise on mensuration, he would recognize and acknowledge Indian material when using it.

17. The translation conserves the traditional Islamic invocation of God (see Vogel 1963: 9), which would in all probability have been cut out before credit-giving references were touched on (as it was cut out in both Gherardo of Cremona's and Robert of Chester's translations of al-Khwārizmī's *Algebra*—see the editions in Hughes (1986: 233) and Karpinski (1915: 67), or the quotations in Høyrup (1988: 351 n. 92).

18. Cf. Saidan (1978: 14). On the later (and probably independent) origin of Islamic combinatorial analysis, see Djebbar (1981: 67ff.).

19. See Pingree (1963: 241ff.) and Pingree (1973: 34). Through the same channels, especially through the Sabians, some Late Babylonian astronomical lore may have been transmitted (cf. the Babylonian sages mentioned in the *Fihrist*). Still, the integration of Babylonian results and methods into Greek as well as Indian astronomy makes it impossible to distinguish any possible more direct Babylonian contributions.

In principle, non-astronomical Greek mathematics *may* also have been conveyed through Syriac learning. There is, however, no evidence in favor of this hypothesis—cf. Section 11.

20. According to a remark in the third part of Abū Kāmil's *Algebra* (Jan Hogendijk, personal communication).

21. Relevant passages from al-Yaᶜqūbī and al-Khazinī are translated in Wiedemann (1970: I, 442–53).

22. Published in Soubeyran (1984: 30); discussion and comparison with the Carolingian problem and the chessboard problem in Høyrup (1986: 477f).

23. Cf. Chapter 1, Section 1 on the distinction between theoretical aim and display of virtuosity in a sociological discussion of the different cognitive and discursive styles of Greek and Babylonian mathematics. Even the difference between the arithmetical Books VII–IX of the *Elements* and Diophantos' *Arithmetica* is elucidated by the same dichotomy; truly, Diophantos has theoretical insight into the methods he uses, but his presentation is still shaped by an origin of his basic material in recreational mathematical riddles (cf. Chapter 2, Section 2 for a more thorough discussion of this question).

We observe that the complex of practical and recreational mathematics can (structurally and functionally) be regarded as a continuation of the Bronze Age organization of knowledge (cf. Section 1). The two were, however, separated by a decisive gap in social prestige—comparable to the gap between the Homeric bard and a medieval peasant telling stories in the tavern.

24. The epigrams were edited around A.D. 500 by Metrodoros.

25. Trans. Kokian (1919). It should be observed that Anania had studied in the Byzantine Empire, and that part of the collection appears to come from the Greek orbit.

26. A detailed discussion would lead us too far. A wealth of references will be found in Tropfke/Vogel (1980, *passim*).

27. This is illustrated beautifully by the chessboard problem and its appurtenant tale. The *motif* turns up in the Chinese as well as in the "Eurasian" domain; the Chinese *tale*, however, is wholly different, dealing with a peasant and the wages of his servant, determined as the successively doubled harvests from one grain of rice. (Stith Thompson 1975: V, 542, Z 21.1.1).

28. It is worth noting that two arithmetical epigrams from the *Anthologia graeca* deal with the Mediterranean extensions of the route: XIV 121 with the land route from Rome to Cadiz, and XIV 129 with the sea route from Crete to Sicily.

29. The Carolingian *Propositiones* appear to form an exception. The editor of the collection (Alcuin?) was obviously not more competent as a mathematician than the practitioners who supplied the material.

30. Its distribution (from Ancient China and India to Aachen) is described in Tropfke/Vogel (1980: 614–16).

31. Saidan (1974: 358), among others, proposes Egyptian influence. Youschkevitch (''Abū'l-Wafāʾ,'' p. 40) quotes M. I. Medevoy for the suggestion of independence.

32. ''*Ša-lu-uš-ti* 20 *ú ra-ba-at ša-lu-uš-ti ú-ṭe₄-tim*''—MLC 1731, rev. 34–35, in A. Sachs (1946: 205) (the whole article deals with such phenomena). Expressions in the same vein are encountered in the tablet YBC 4652, N$^{os}$ 19–22 (in *MCT*, 101).

33. The problem in question is of typical ''riddle'' or ''recreational'' character: ''Go down I times 3 into the *hekat*-measure, ⅓ of me is added to me, ⅓ of ⅓ is added to me, ⅑ of me is added to me; return I, filled I am'' (the ''literal translation,'' Chace et al. 1929: Plate 59). It can thus be seen as a witness of a current, more or less popular usage. A close analysis of the (utterly few) instances of rudimentary unit fraction notation from the Old Kingdom (*viz.*, from the twenty-fourth century B.C.) suggests that they stand midway between this original usage and the fully developed system (space and relevance does not permit further unfolding of the argument).

34. See the cubit rod reproduced in Menninger (1957: II,23).

35. Once again, the evidence for a shared tradition is found in Islamic sources—e.g., Abū'l-Wafāʾ's *Book on What Is Necessary from Geometric Construction for the Artisan* (trans. Krasnova 1966).

36. E.g., different ways to find the area of a circle; the Babylonian treatment of irregular quadrangles and of the bisection of trapezia, and the absence of both problem types in Egypt.

37. Thus, the Demotic Papyrus Cairo JE 89127–30, 137–43 (third century B.C.) has replaced the excellent Egyptian approximation to the circular area (equivalent to $\pi = {}^{256}\!/_{81} \approx 3.16$) with the much less satisfactory Babylonian and Biblical value $\pi = 3$ (see Parker 1972: 40f., problems 32–33). The same value is also taken over by pre-Heronian Greco-Egyptian practical geometry, cf. Pap. Gr. Vind. 19996 as published by Vogel and Gerstinger (1932: 34). A formula for the area of a circular segment that is neither correct nor near at hand for naive intuition is used in the Demotic papyrus mentioned above (N° 36); in the Chinese *Nine Chapters on Arithmetic*, it is used in N$^{os}$ I,35–36, and made explicit afterwards (trans. Vogel 1968: 15). Hero, finally, ascribes it to ''the ancients'' (*hoi archaîoi—Metrica* I,xxx, ed. Schöne 1903: 72) while criticizing it (cf. the discussion in van der Waerden 1983: 39f., 174).

The Babylonian calculation of the circular area, which is a deterioration when compared to the Middle Kingdom Egyptian method, was probably an improvement over early Greek and Roman practitioners' methods—Polybios and Quintilian both tell us that most people incorrectly measured the area of a figure by its periphery. How it was done can be seen in the Carolingian *Propositiones*, N$^{os}$ 25 and 29, which find the area of one circle as that of the isoperimetric square, and that of another as that of an isoperimetric non-square rectangle.

To complete the confusion, the *Propositiones* find all areas of non-square quadrangles (rectangles and trapezoids alike) by means of the "surveyors' formula" for the irregular quadrangle (average length times average width). This formula is employed in Old Babylonian tablets. It was used by surveyors in Ptolemaic Egypt (see Cantor 1875: 34f.), and it turns up in the pseudo-Heronian *Liber geeponicus* (Cantor 1875: 43). It was not used by Hero, nor by the Roman agrimensors (nor, it appears, in Seleucid Babylonia); but it turns up again in the Latin eleventh century compilation *Boethii geometria altera* II,xxxii (ed. Folkerts 1970: 166). In the eleventh century A.D., Abū Manṣūr ibn Ṭāhir al-Baghdādī ascribes the formula to "the Persians" (Anbouba 1978: 74)—but al-Khwārizmī (who does not use it) has probably seen it in the Hebrew *Mišnat ha-Middot* II,1 (ed. Gandz 1932: 23), or possible in some lost prototype for that work.

38. So the value π = ²²/₇, represented by Hero as a simple approximation (*Metrika* I, 26—ed. Schöne 1903: 66), is taken over by Roman surveying (Columella and Frontinus, see Cantor 1875: 90,93f.) and stands as plain truth in Latin descendants of the agrimensor tradition (e.g., *Boethii geometria altera* II,xxxii—ed. Folkerts 1970: 166). The *Mišnat ha-Middot* (II,3 ed. Gandz 1932: 24) presents the matter in the same way. So does al-Khwārizmī in the parallel passage of his *Algebra*, but in the introductory remark he represents the factor 3¹/₇ as "a convention among people without mathematical proof" (Gandz 1932: 69 and 81f.)—telling thereby that he considered at least that section of the *Mišnat* (or its prototype) representative of a general subscientific environment.

Other Heronian improvements are the formula for the area of a triangle and his better calculation of the circular segment, which turn up in various places (see, e.g., Cantor 1875: 90, reporting Columella, and *Mišnat ha-Middot* V, ed. Gandz 1932: 47ff.).

39. Høyrup (1986). The essential sources involved in the argument are al-Khwārizmī's *Algebra* (trans. Rosen 1831); the extant fragment of ibn Turk's *Algebra* (ed., trans. Sayılı 1962); Thābit's Euclidean *Verification of the Problems of Algebra Through Geometrical Demonstrations* (ed. trans. Luckey 1941); Abū Kāmil's *Algebra* (ed., trans. Levey 1966); the *Liber mensurationum* written by some unidentified Abū Bakr and known in a Latin translation made by Gherardo of Cremona (ed. Busard 1968); and Abraham bar Ḥiyya's (Savasorda's) *Collection on Mensuration and Partition;* Latin translation *Liber embadorum*, ed. Curtze 1902).

On one point, my [1986] should be corrected. On p. 472 I quote Abū Kāmil for a distinction between "arithmeticians" (*baʿalei ha-mispar* in the Hebrew translation, Levey 1966: 95, i.e., "masters of number") and "calculators" (*yinhagu ha-ḥasbanim*, Levey 1966: 97, "those who pursue calculation"). The distinction turns out to be absent from the Arabic facsimile edition of the work (ed. Hogendijk 1986). (Jan Hogendijk, personal communication.)

40. The arguments for this are, as in any structural analysis, complex, and impossible to repeat in the present context—cf. Høyrup (1990).

41. True, the *Propositiones* are not pre-Islamic according to chronology. Still, they show no trace of Islamic influence, and they were collected in an environment

where mathematical development was to all evidence extremely slow. We can safely assume that most of the mathematics of the *Propositiones* was already present (if not necessarily collected) in the same region by the sixth century A.D.

42. See the discussion and the two texts in Gandz (1932).

43. In his English summary, Sarfatti (1968: ix) claims that Arabic linguistic influence, "although not evident prima facie, underlies [the] mathematical terminology" of the *Mišnat ha-Middot*. If this is true, the work must be dated in the early Islamic period. The main argument of the book is in Hebrew, and I am thus unable to evaluate its force—but since Arabic and Syriac (and other Aramaic) technical terminologies are formed in analogous ways, and since no specific traces of Arabic terms are claimed to be present, it does not seem to stand on firm ground.

44. The former is an ancient Jaina value, the second is given by Āryabhaṭa—see Sarasvati Amma (1979: 154).

45. Arabic text and translation in Gandz (1932: 69f.); *ahl al-handasa* ("surveyors" and later "geometers," translated "mathematicians" by Gandz) corrected to *ahl al-hind* ("people of India") in agreement with Anbouba (1978: 67).

46. Trans. Rosen (1831: 6–20). The whole technique of the proofs has normally been taken to be of purely Greek inspiration, partly because of the letter formalism, partly because neither the Old Babylonian naive-geometric technique nor its early medieval descendant was known.

47. Trans. Krasnova 1966. Interesting passages include Chapter 1, on the instruments of construction; and 10.i and 10.xiii, which discuss the failures of the artisans as well as the shortcomings of the (too theoretical) geometers. Consideration of practitioners' needs and requirements is also reflected in the omission of all proofs.

Though more integrative than al-Khwārizmī's *Algebra*, Abū'l-Wafā''s work is not completely free from traces of eclecticism. This is most obvious in the choice of grammatical forms, which switch between a Greek "we" and the practitioners' "If somebody asks you . . . , then you [do so and so]."

48. According to Bulliet's counting of names (1979), the majority of the Iranian population was converted around 200 A.H. (A.D. 816), whereas the same point was reached in Iraq, Syria, and Egypt some 50 years later. The socially (and scientifically!) important urban strata (artisans, merchants, religious and state functionaries) were predominantly Muslim some 80 years earlier (cf. also Waltz 1981). A correlation of Bulliet's geographical distinctions with the emergence of local Islamic scholarly life would probably be rewarding.

49. See ibn Khaldūn, *Muqaddimah* VI, *passim*, especially VI.9 (trans. Rosenthal 1958: II–III, especially II, 436–39); and Nasr (1968: 63f).

50. This particular culturocentrism is discussed in Chapter 6, Section 4. One of the rare fields where it is not clearly felt appears to be mathematics, where the

requirements of mathematical astronomy, "Hindu reckoning," and commercial arithmetic and algebra in general may have opened a breach of relative tolerance.

51. This is of course not to say that it remained totally free. The conquering Arabs, e.g., felt ethnically superior to others, as conquerors have always done. The lack of ethnocentrism is only relative.

52. The beginnings of astronomical interests in the later eighth century is different, bound up as it is with the practical interest in astrology. The same applies to the very early preoccupation with medicine—cf. *GAS* III, 5.

53. Cf. Rosenthal and Grigorian, "Thābit ibn Qurra"; and Anawati and Iskandar, "Ḥunayn ibn Isḥaq." See also Section 11.

54. The Mozarabic work is a paraphrase of Nicomachos written by the mid-tenth century Andalusian bishop Rabīʿ ibn Zaid. In his preface, Rabīʿ refers to commentaries that al-Kindī should have made to a translation from Syriac. The evidence can hardly be considered compelling; on the other hand, *some* Arabic translation antedating Thābit's must have existed. See Steinschneider (1896: 352) and *GAS* V, 164f.

55. Cf. *GAS* V, 129. Positive evidence that Syriac learning was close to mathematical illiteracy is found in a letter written by Severus Sebokht around 662 (ed., trans. Nau 1910: 210–14). In this letter, the Syrian astronomer *par excellence* of the day quotes the third-century astrologer Bardesanes extensively and is full of contempt for those who do not understand the clever argument—which is in fact nothing but a mathematical blunder, as enormous as it is elementary.

56. "The people will become subject to the people of the East and the government will be in their hands," as it was expressed by the contemporary Māšāʾallāh in his *Astrological History* (trans. Kennedy and Pingree 1971: 55). Or, in Peter Brown's modern, expressive prose (1971: 201): "Khusro I had taught the *dekkans*, the courtier-gentlemen of Persia, to look to a strong ruler in Mesopotamia. Under the Arabs, the *dekkans* promptly made themselves indispensable. They set about quietly storming the governing class of the Arabic empire. By the middle of the eighth century they had emerged as the backbone of the new Islamic state. It was their empire again: And, now in perfect Arabic, they poured scorn on the refractory bedouin who had dared to elevate the ways of the desert over the ordered majesty of the throne of the Khusros."

57. See his kindly polemical defense of a "middle range theory" whose abstractions are "close enough to observed data to be incorporated in propositions that allow empirical testing" against such precocious total systems whose profundity of aims entails triviality in the handling of all empirical details (Merton 1968: 39–72, especially the formulations on pp. 39 and 49).

58. As "institutions" of learning in the widest sociological sense (i.e., socially fixed patterns of rules, expectations, and habits) one can mention the short-lived "House of Wisdom" at al-Maʾmūn's court, together with kindred libraries; the

institutions of courtly astronomy and astrology and, more generally, the ways astronomy and astrology were generally practiced; that traditional medical training which made medicine almost a monopoly of certain families (cf. Anawati and Iskandar relating ibn Abī Uṣaybiʿa, *DSB* XV, 230); the fixed habits and traditions of other more or less learned practical professions; the gatherings of scholars; the fixed form in which science could already be found in Byzantium; the Mosque as a teaching institution (the *madrasah* was developed only much later); and practical and theoretical management of Islamic jurisprudence, including the transmission of *ḥadīth* (jurisprudentially informative traditions on the doings and sayings of the Prophet). Of only one of these institutions—Byzantine science—can it be claimed that it was really fixed by the early ninth century. Cf. Nasr (1968: 64–88); Makdisi (1971); and Watt and Welch (1980: 235–50).

59. There were at least two (tightly coupled) good reasons for this jealousy. First, traditionists and jurisprudents might easily develop into a secondary center of power; second, they might inspire, participate in, or strengthen popular risings, which were already a serious problem for the ʿAbbasid caliphs.

The destruction of the Baghdad "House of Learning" (*Dār al-ʿilm*— "Residence of Knowledge" would perhaps be a more precise translation) in a Sunnite riot in A.D. 1059 shows what could be the fate even of scholarly institutions when religious fervor and social anger combined. See Makdisi (1961: 7f).

60. The expression is quoted from the ninth-century *Mutakallim* al-Jāḥiẓ via Anton Heinen (1978: 64), who sums up (p. 57) his point of view in the formula *Knowledge (ʿilm) = kalām al-dīn + kalam al-falsafah*, the latter term meaning "the discourse of philosophy" i.e., secular theoretical knowledge.

61. In a similar way, practical charity, the management of ritual and sacraments, and religious teaching are understood as belonging *by necessity* together in Christian environments where the Church (and frequently the same priest) takes care of them all.

62. One work containing such copious references to God is Abū Bakr's *Liber mensurationum*, which was discussed previously. Normally, the invocations were abridged or left out in the Latin translations (not least in Gherardo's translations); in this case, however, they have survived because of their position inside the text (whereas the compulsory initial invocation is deleted).

Of course, routine invocation is no indicator of deep religious feeling. What imports is that the invocation *could* develop into a routine, and that it was thus considered *a matter of course* or *decorum* even in mathematical texts. You may perhaps persevere in an activity that you fear is unpleasant to God—but then you rarely invite him explicitly (routinely or otherwise) to pay heed or assist you in your sins.

63. Among the numerous examples I shall only mention Abū'l-Wafāʾ's *Book on What Is Necessary from Geometric Construction for the Artisan*, which was discussed already; al-Uqlīdisī's *Arithmetic*, the mathematical level of which suggests that the author must have been beyond the rank-and-file; and ibn al-Haytham's works on

the determination of the *qiblah* (cf. note 95) and on commercial arithmetic (N$^{os}$ 7 and 10 in ibn Abī Uṣaybiʿa's list, trans. Nebbia 1967: 187f., cf. Rozenfeld 1976: 75).

64. I use the edition in *PL* 176, col. 739–952. A recent English translation is Taylor (1961) (it should be observed that the chapters are numbered somewhat differently in the two versions).

65. *Omnia disce, videbis postea nihil esse superfluum (Didascalicon* VI.iii). Strictly speaking, "everything" is "everything is sacred history," since this is the subject of the chapter; but in the argument Hugh's own play as a schoolboy with arithmetic and geometry, his acoustical experiments and his all-devouring curiosity are used as parallel illustrative examples.

66. Quoted from Nasr (1968: 65). In this case, as in that of Hugh, the intended meaning of "knowledge" is probably not quite as wide as a modernizing reading might assume. This, however, is less important than the open formulation and the optimism about the religious value of knowledge that was read into it, and against which the opponents of *awāʾil* knowledge had to fight (cf. Goldziher 1915: 6)—for centuries with only limited success.

A major vehicle for the high evaluation of knowledge in medieval Islam (if only knowledge in the narrow sense, *viz.*, knowledge of God's "Uncreated Word," i.e., the Koran, and of the Arabic language), and at the same time a virtual medium for the spread of a high evaluation of knowledge in a more general sense was the establishment of education on a large scale in Koran schools and related institutions. (I am grateful to Jan Hogendijk for reminding me of this point, which I had not mentioned in the first version of the paper.)

A wealth of anecdotes illustrating an almost proverbial appreciation of knowledge (from the level of elementary education to that of genuine scholarship) will be found in Tritton (1957: 27f. and *passim*).

67. I.ix. "Practitioners' knowledge" translates *scientia mechanica*, whereas "moral truth" renders *intelligentia practica/activa*.

68. This picture is of course unduly schematized. Cf. the somewhat more detailed treatment of the subject in Chapter 1, Section 3, and the particular perspectives of Chapters 5 and 6.

69. The situation is expressed pointedly by Boethius de Dacia in his beautiful *De eternitate mundi* (ed. Sajo 1964: 46 and *passim*), when he distinguishes the truth of natural philosophy (*veritas naturalis*) from "Christian, that is genuine, truth" (*veritas christianae fidei et etiam veritas simpliciter*).

70. And hence, when integration was needed by a group of practitioners, as was the case in thirteenth- and fourteenth-century astronomy, the need was satisfied by means of simplifying compendia, in striking contrast to the development in Islam—cf. Section 16. Latin science, when applied, was subordinated, and hence not fertilized by the interaction with the questions and perspectives of practice. For this

reason, on the other hand, applications were bound to remain on the level of common sense. Cf. Beaujouan (1957), especially the conclusion.

71. So, according to Albert Nader, "les mu'tazila touchent à la sphère physique avec des mains conduites par des regards dirigés vers une sphère métaphysique et morale: la raison cherchant ā concilier les deux sphères" (1956: 218, quoted from Heinen 1978: 59). As pronounced enemies of the *Mu'tazila*, the tenth century Ismā'īlī *Ikhwān al-ṣafā'* are still more emphatic, claiming that the *Mu'tazila* "declare medical science to be useless, geometry to have nothing to do with the true essence of things, logic and the sciences of nature to be unbelief and hereticism, and their proponents to be people without religion" (IV.95, German trans. Goldziher 1915: 25).

72. *Muqaddimah* VI,19. It will also be remembered that "inheritance calculation" occupies just over one half of al-Khwārizmī's *Algebra* (pp. 86–174 in Rosen's 1831 translation).

73. The similarity with the *Ismā'īlī* orientation is clear; according to ibn Khaldūn, good reasons for such similarity exist through the close relations between the early Sufis and "Neo-Ismâ'îlîyah Shî'ah extremists" (*Muqaddimah* VI.16, trans. Rosenthal 1958: III, 92).

The paradoxical (or at least vacillating) attitude of the mature al-Ghazzālī toward mathematics is illustrated through a number of quotations in Goldziher (1915, *passim*).

74. The claim is even given emphasis by a somewhat clumsy repetition in the introduction to his "Discussion of Difficulties of Euclid" (trans. Amir-Móez 1959: 276).

75. In a case such as the Sunni Nizamiyah *madrasah* in Baghdad, the institutional goals were of course already quite restricted. They would permit you to teach *al-jabr* but not Apollonios.

Formally, the situation was thus not very different from that of Syriac monastic learning. The Syriac learned monk, too, merely had to stick to the institutional goal of Church and monastery. Materially, however, the difference was all-important, because the institutional goal of the Church included the defense of an already established theological opinion—we may think of the difference between the obligation to *teach biology instead of sociology* and the *prohibition to teach biology unless creationist*.

76. Cf. the anecdotes on Hulagu and Naṣīr al-Dīn al-Ṭūsī reported by Sayılı (1960: 207) and the story of the closing of the Istanbul observatory when its astrological predictions had proved catastrophically wrong (Sayılı 1960: 291–93). At most, such rulers were able to make cuts in a program that was too ambitious for their taste, or to close an institution altogether.

The relatively few rulers who were scholarly competent were active protectors of science (in particular astronomy) as it already existed and did not try to change its orientation.

77. Openings of Book IV, V, VI, and VII, and the closing formula of the work; trans. Sesiano (1982: 87, 126, 139, 156, 171), or Rashed (1984: III, 1; IV, 1, 35, 81, 120). In the last of these places, the praise that ends the work is followed by the date of copying, which is again followed by another praise of God and a blessing of the Prophet, in a way that, through comparison with other treatises with Muslim author and Muslim copyist, suggests but does not prove that the first praise goes back to Qusṭā himself.

78. ''In principle'': of course, much theory went unapplied in practice, and theory was developed regardless of possible application.

79. Several of the treatises were edited by Oschinsky (1971). Grosseteste's involvement is discussed pp. 192ff., cf. the texts pp. 388–409.

80. A discussion and a partial translation of the treatise is given by Ritter (1916). The treatise is a contrast not only to thirteenth-century European handbooks on prudent management but also to Greek common-sense deliberations such as Hesiod's *Works and Days* or Xenophon's *Oeconomica*. A similar contrast is obvious if we compare Ovid's *Ars amoris* or the pseudo-scholarly treatises to which it gave rise in the Latin Middle Ages with the development of regular sexology in Islam. Closer to mathematics, we may compare Villard de Honnecourt's very unscholarly reference to *figures de lart de iometrie* (*Sketchbook*, ed. Hahnloser 1935, Taf. 38) with the serious study of Euclidean geometry by Islamic architects (see Wiedemann 1970: I,114).

81. Cf. Hero's introductions to the *Metrika* and *Dioptra* (ed. Schöne 1903: 2ff. and 188ff.).

82. We may remember Benjamin Farrington's observation that ''it was not [ . . . ] only with Ptolemy and Galen that the ancients stood on the threshold of the modern world. By that late date they had already been loitering on the threshold for four hundred years. They had indeed demonstrated conclusively their inability to cross it'' (1969: 302).

83. A parallel case is ibn al-Haytham's investigation of the ''purchase of a horse'' (partially translated in Wiedemann 1970: II, 617–19). This treatise, too, opens with a polemic against practitioners who do not justify their procedures.

84. Latin (and German) translation in Curtze (1902: 38, 40). It should be observed that Abraham's text is meant most practically. It is in the same tradition as Abū Bakr's *Liber mensurationum*, cf. Section 6.

85. Saidan (1978: 19–31). In the following, I use Saidan's typology.

86. Listed in al-Nadīm's *Fihrist*, trans. Dodge (1970: 617). It is not clear whether the ''Introduction to Arithmetic, five sections'' also mentioned there is a finger-reckoning treatise, a commentary on Nicomachos, or the two combined.

87. To be precise, the Arabic treatise that has come down to us is written for the Būyid vizier Šaraf al-Mulūk; but we must assume that the author stays close to the

earlier treatise written in Persian of which he speaks himself in the preface. See the translation of the preface in Woepcke (1863: 492–95), and Saidan, "Al-Nasawī." The discussion of the treatise in Suter (1906) covers only the brief section dealing with the extraction of roots.

88. Quoted from Saidan (1978: 24) (pp. 24–29 give an extensive abstract of the whole work).

89. See Saidan (1978a), and Saidan, "Al-Umawī." Al-Umawī taught in Damascus; but he came from the West, where he had been taught, and he brought its methods to the East.

90. *Muqaddimah* VI.19. Ibn Khaldūn had himself been taught by a disciple of ibn al-Bannā᾽ (see Vernet, "Ibn al-Bannā᾽"). A systematic investigation of certain aspects of Maghrebi mathematics has been undertaken by Djebbar (1981).

91. A general account is given in Djebbar (1981: 41–54). The explanation given by ibn al-Bannā᾽'s commentator ibn Qunfudh is translated in Renaud (1944: 44–46). Woepcke (1854) deals mainly with al-Qalaṣādī's symbolism.

92. Compare Leonardo's various complicated fractions (ed. Boncompagni 1857: 24) with the similar forms in Djebbar (1981: 46f).

93. An alternative possibility is that Leonardo was drawing on Jordanus for the revised edition written to Michael Scot—see Høyrup (1988: 310f.).

94. An early example is a Provençal arithmetic written c. 1430. A certain affinity to Islamic traditions is suggested by an initial invocation of God, of Mary his Mother, and of the patron saint of the city (see Sesiano 1984—the invocation is on pp. 29–31).

Later examples are Chuquet's *Triparty* and Luca Pacioli's *Summa de arithmetica*. The *Triparty* tells (in its first line that it is divided into three parts "a lonneur de la glorieuse et sacree trinite" (ed. Marre 1880: 593)—perhaps a jocular reference to familiar invocations in related treatises? In any case, the same author's *Pratique de geometrie* (ed. H. l'Huillier 1979) has no religious introduction. Apparently quite serious are Luca's recurrent religious references (Pacioli 1523).

That some kind of cultural offset is involved and not a mere parallel development is suggested by the absence of God and Jesus from the mathematical part of Stifel's *Arithmetica integra* (1544), a work of similar scope, as well as from his less ambitious *Deutsche arithmetica* (1545). Stifel, indeed, was a no less sincere believer than Luca, and a closing remark in the former work promising the reader the imminent publication of the latter does contain a parenthesis "(deo dante)," "if God allows."

95. Two other practical aims for astronomy can also be mentioned: Finding the *qiblah* (the direction toward Mecca, which determined the orientation of mosques and the direction of prayer), and fixing the prayer times. None of them, however, called for astronomy of such sophistication as developed around the princely observatories—

and religious authorities, as shown by David King, rarely made use of the solutions offered by scientific astronomy.

96. Among those referred to previously, only Abū Kāmil, al-Karajī, al-Samaw'al and al-Qalaṣādī stand out as exceptions. Al-Samaw'al, however, is at least known to have written a refutation of astrology, involving both mathematical arguments and knowledge of observations (Anbouba, "Al-Samaw'al").

Strictly speaking, even ibn Turk, al-Uqlīdisī, al-Ḥaṣṣar, and al-Umawī might also be counted as exceptions, since no works from their hand are known. However, our knowledge of these scholars is so restricted that they fall outside all attempts at statistical analysis.

A last important mathematician who appears definitely to have been a non-astronomer is Kamāl al-Dīn, whose important work concentrates on optics (cf. Suter 1900: 159, N° 389, and Rashed, "Kamāl al-Dīn").

97. See Steinschneider (1865: 467 and *passim*); and Nasr, "Al-Ṭūsī."

98. See Steinschneider (1865: 464 and 457, respectively). The Greek "Little Astronomy" was to form the backbone of the *mutawassiṭāt* even in the Naṣirean canon. Naṣīr al-Dīn al-Ṭūsī, however, includes Euclid's and Thābit's *Data* and Archimedes' *Measurement of the Circle, On the Sphere and Cylinder,* and *Lemmata,* together with some other works.

99. Such general attitudes, too, remained effective—they are expressed in the praise of Archimedes' *Lemmata* formulated by al-Nasawī, who speaks of the "beautiful figures, few in number, great in *utility,* on the fundaments of geometry, in the highest degree of *excellency* and *subtility*" (quoted form Steinschneider 1865: 480; emphasis added). Cf. also al-Bīrūnī as quoted in Section 13.

100. This includes all those mentioned already, as well as others mentioned in Sabra 1969 (Qayṣar ibn Alī'l-Qāsim, Yūḥannā al-Qass, al-Jawharī) and Folkerts 1980 (which, besides some of the same, mentions al-Māhānī)—with the ill-documented al-Qass as a possible exception (Suter 1900: N° 131).

101. A good and fairly recent overview of magic squares in Islam is Cammann (1969); but see also Ahrens (1916); Bergsträsser (1923); Hermelink (1958); Sesiano (1980, 1987); and Sarton (1927, 1931, index references to "magic squares").

102. This is the most widespread assumption, judging from the *Fihrist* (translated by Dodge 1970: 733).

103. Ed. Dupuis (1892: 166). The passage shows the square
$$
\begin{array}{ccc}
1 & 4 & 7 \\
2 & 5 & 8 \\
3 & 6 & 9
\end{array}
$$

104. This is clearly the point of view of ibn Khaldūn in the *Muqaddimah,* wherever he approaches the subjects of talismans, letter magic, and magic squares (which mostly go together). In one place he also claims that a work based on such

things "most likely [ . . . ] is incorrect, because it has no scientific basis, astrological and otherwise" (III.52, trans. Rosenthal 1958: II,224).

105. So al-Karajī's summation of square numbers in the *Fakhrī* (see the paraphrase in Woepcke 1853: 60).

106. Ibn Khaldūn does not mention the subject at all during his discussion of arithmetic (*Muqaddimah* VI.19, trans. Rosenthal 1958: III, 118–29). Like amicable numbers (once investigated mathematically by Thābit but now only mentioned as a talisman producing love) it is relegated to the chapters on magic and sorcery (VI.27–28, trans. Rosenthal 1958: III, 156–227). (The silence on amicable numbers is all the more striking, since the circle of Maghrebi mathematicians was in fact interested in that subject—cf. Rashed 1983: 116f.).

107. A few exceptions, e.g., in commercial law, can be found. But the difference between the Maghrebi arithmetical textbook tradition and the Italian abacus school shows that even the institutions of commercial education could not be transferred.

108. A supplementary approach might compare the institutions of "courtly science" and the patterns of princely protection in the two settings.


## 5. Philosophy: Accident, Epiphenomenon, or Contributory Cause of the Changing Trends of Mathematics—A Sketch of the Development from the Twelfth Through the Sixteenth Century

1. I deal in somewhat more detail with this aspect of early and central medieval history of mathematics and with the concepts of a "Latin" and a "Christian" quadrivium in Chapter 6.

2. "Not until forty years after Charle[magne]s's death, when diocesan schools began to expand and the manual of Martianus Capella began to influence the curriculum of some of them, was there a study other than *computus* which dealt with the mathematical sciences," as stated empathetically by C. W. Jones (1963: 21)—maybe somewhat more emphatically than justified.

3. The full fourteenth-century biography will be found in Boncompagni (1851a: 387ff.).

4. The treatise was edited by Boncompagni (1881). See the references to Boethius' translations p. 111[19.22]; the reference to Gerbert as "having given the technique back to us Gauls" p. 91[23]; and the Boethian reference to Pythagoras p. 91[7].

5. The "first" is hypothetical: According to the dedication the treatise is written during the seven-year voyage to which Adelard refers in the beginning of the *Quaestiones naturales*, but the contents seem to belong to the intellectual luggage that

made him set out, not to anything he had learned in Sicily or in the Near East. This combination fits the beginning of his stay in Syracuse best. See the edition and discussion in Willner (1903).

6. Even though Boethius' *De musica* remains an important source—cf. Müller (1934: 25[13] and 27[23f]).

7. See Clagett, "Adelard of Bath," 63.

8. An illustrative example is Gherardo di Cremona himself. In one of his translations from the Arabic, a *Liber mensurationum* edited by Busard (1968), Roman numerals, Hindu numerals, and number words written in full are mixed up completely; even though the Arabic treatise is lost it is fairly certain that all its numbers were written as full words.

9. "Non enim in figmentis poeticis, non in opinionibus phylosophicis, in regulis Prisciani, in legibus Iustiniani, in doctrina Galieni, in oribus rhetoricis, in perplexionibus Aristotelis, in teorematibus Euclidis, in conjecturis Tolomei, summam studiorum suorum ponere et tempus suum conterere debet christianus, multominus monachus et canonicus. Et quidem artes, quas liberales vocant ad acuendum ingenium et intelligentiam Scripturarum multum valent, sed, iuxta philosophum, salutande sunt a limine." Quoted from Grabmann (1941: 61).

10. Ed. Boncompagni (1857) (from a manuscript of Leonardo's revision of the work in 1228).

11. According to the chronicler Giovanni Villani, in 1339 about 1000–1200 Florentine boys (out of a total city population of 90,000) went to one of the six schools where practical arithmetic was taught (C. T. Davis 1965: 415). See also Fanfani (1951) and Goldthwaite (1972) on Italian commercial teaching from the fourteenth through sixteenth centuries.

According to a document reproduced by Goldthwaite (pp. 421ff.), the basic curriculum in a Florentine school "consisted in seven consecutive courses": 1) arithmetical operations except division; 2)–4) division with one, two, and more digits; 5) fractions; 6) the rule of three; 7) the Florentine monetary system. Higher subjects would be reserved for the few.

A more precise idea of the teaching can be acquired from the various "abacus treatises" that have been published. A fine specimen is found in Arrighi (1973). Obviously, *part* (but only part) of the inspiration from Leonardo was alive.

12. See also Jolivet (1974), on the conspicuous absence of biblical explanations from the *Quaestiones naturales*. Discussions of Adelard in the wider context of twelfth-century naturalism will be found in Chenu (1966) and Stiefel (1977).

13. "Omina in mensura, et numero, et pondere fecisti"—Wisd. 11: 21; quoted in Augustine, *The City of God* XI.xxx (ed. McCracken et al. 1966: III, 552), and Isidore, *Etymologies* III.iv, 1 (*PL* 82, 155).

14. "Per numerum siquidem, ne confundamur, instruimur. Tolle numerum rebus omnibus, et omnia pereunt. Adime saeculo computum, et cuncta ignorantia com-

plectitur, nec differri potest a caeteris animalibus qui calculi nescit rationem"—
*Etymologies* III.iv, 3.

15. "Omnia disce, videbis postea nihil esse superfluum."

16. This is the way the material is organized in Steinschneider (1904).

17. Mathematics was not the only candidate at hand, unless we restrict the concept of "existing world" to that of "physical world." For the worlds of metaphysics, moral philosophy, canon law, and the scriptures the new *dialectical method* was the obvious choice, and it was certainly no less chosen than mathematics. Both, indeed, fulfilled the need for intellectual coherence growing out of the flourishing environment of schools and educated clerks.

18. See the "combined sermon" compiled from a variety of real sermons by Haskins (1929: 46f.).

19. See Thorndike (1954) on the continuation of computistic creativity at least until the end of the thirteenth century. In fact, Cardano, Stifel, and Clavius would still write on the subject in the sixteenth century.

20. *Opus tertium* VI, ed. Brewer (1859: 21). Given Bacon's polemical aim and irascible temper and his lack of deeper mathematical understanding, there is clearly no reason to take his testimony to the letter.

21. Discussed in Grabmann (1934). The collection may date from the 1240s, the time when Bacon was in Paris.

22. Datable to the 1250s (see Benjamin and Toomer 1971: 4f.). I used the second Basel edition from 1546 (*Euclidis Megarensis mathematici clarissimi Elementorum geometricorum libri xv*), which contains the Campanus edition in parallel with Bartolomeo Zamberti's translation from the Greek.

23. Even though we are dealing with a translation the choice of subjects is not fully fixed in advance. In fact, Campanus adds a number of extra propositions to Book V—cf. Busard (1972: 131ff.).

24. Cf. the many instances of combination of Boethian arithmetic with themes from the trivium in Evans (1978).

25. Ideals that had originally been inspired not least by the rise of theoretical geometry as an autonomous field of knowledge; cf. Chapter 1, Section 2.

26. Compare Sacrobosco's text in F. S. Pedersen (1983: $174^{1-4, 16-18}$), with Boethius, *Arithmetica* I.ii and I.iii (ed. Friedlein 1867, 12–13).

27. F. S. Pedersen (1983: 81–85). Concerning the "final cause" Petrus states that "according to the author the purpose of this art is the knowledge of everything; but *I* believe that its more immediate purpose is nothing but astronomy" (p. $82^{35ff}$).

28. In the following discussion of Jordanus I draw heavily on Høyrup (1988) (cf. also the abbreviated version in Chapter 6). It is to be observed that Jordanus's philosophical attitude must largely be read between the lines.

29. Critical editions of both treatises in Clagett (1984). The characterization of the *Liber de triangulis* as a *reportatio* is my own conclusion from a close analysis of stylistic features of the text—see Høyrup (1988: 347–51).

30. In the original version I also stated that Jordanus's *Arithmetica* presents its initial axioms as *dignitates*, in agreement with the Aristotelian tradition, and not as *communes animi conceptiones*. As revealed by Busard's critical edition (1991), however, this feature of the Renaissance edition is not original.

31. In the catalogue of his library (the *Biblionomia*, ed. Delisle 1868: II, 520–35), Richard de Fournival (b. 1201, d. before 1260) opposes the Jordanian genre *apodixis* (the Aristotelian term translated *demonstratio*) to precisely such *experimenta*. As discussed in Høyrup (1988), Richard appears to have been personally acquainted with Jordanus and collected apparently all of his works. The characterization of the genres is hence probably faithful to Jordanus's own ideas and ideals.

32. There is even some positive evidence that he didn't care. Aristotle had once distinguished the *sophist*, who when discussing geometrical questions would use arguments that by nature were alien to geometry, from the *pseudographos*, who would use legitimately geometric though misleading arguments (see *De sophisticis elenchis* 171$^b$36ff. and 171$^b$14ff., and *Topica* 132$^a$33). In thirteenth-century mathematics both figures were identified with the *opponent* in a university disputation, and so they are in Jordanus' *Liber philotegni*, prop. 18: Jordanus apparently felt no need to support himself on Aristotle's strict distinction, though obeying it himself.

The universitarian tradition continued the *quiproquo* for centuries: In a disputation from Leipzig in 1512, *falsigrafus* is used to designate precisely that argument which Aristotle uses to exemplify sophist ways (see Suter 1889: 19).

33. Both the original treatise and the different adaptations were edited critically in Thomson (1978).

34. Molland (1980: 472). The article contains many quotations from the paraphrases.

35. See the translation of the relevant Questions V and VI in Maurer (1963), or the brief discussion in Weisheipl (1975: 134–36).

36. The index in Weisheipl (1975) illustrates this beautifully by its "mathematics, *See* science and scientific method."

37. Detailed list in Minio-Paluello, "Moerbeke, William of," 434–40.

38. Ed. Baur (1912). Cf. the discussion in Crombie, "Grosseteste, Robert," 548–54.

39. See Lindberg (1971) on the connections from Bacon to Witelo and Peckham.

40. *De numeris misticis,* ed. Hughes (1985). Description also in Lindberg, "Pecham," 474.

41. Ed. Lindberg (1970). The quotations from Moses Maimonides (which are *not* found as quotations but only hinted at in Peckham's work) are from the *Guide for the Perplexed* II.x (on "the influence of the Spheres upon the Earth")—trans. Friedländer (1904: 164).

42. Risner (1572: II, 440). From the dedicatory letter, it seems that Witelo was already engaged in neoplatonic reflections before meeting Moerbeke (and, more specifically, engaged in a work *De ordine entium,* which he postponed). Moerbeke will then have explained to him the importance of *light* for understanding *that Divine light* which connects the different orders of entities—an idea that will have caught Witelo's interest because of his own physical observations. Witelo appears *not* to have brought any Baconian or Grossetestian inspiration with him to Viterbo: the terms in which the neoplatonic ontology is set forth differ from theirs, and are more orthodox.

43. In reality, it had often been so already in the twelfth century. In this connection the importance of the *Almagest* for twelfth-century translators will be remembered. As Lemay (1962) points out, astrological translations were also the first source for Aristotle's natural philosophy.

Twelfth-century scholars were also aware that astronomy was a main mobile for mathematical activity in the Islamic world. This appears from John of Salisbury's *Metalogicon* IV.vi (from 1159), as quoted in Chapter 4, Section 15.

44. See Siraisi (1973: 67f. and *passim*). The situation was similar in Bologna (see Rashdall 1936: I, 242f., 248f.). On Paris, see Lemay (1976).

45. Albert had in fact written a survey of such subjects, in his *Speculum astronomiae, in quo de libris licitis et illicitis pertractatur* (ed. Zambelli 1977).

46. In sharp contrast to what happened in the Islamic world—cf. Chapter 4.

47. I leave the discussion of merchants' mathematics and of the gradually growing mathematical abilities in non-scholarly environments to my treatment of the fourteenth and fifteenth centuries.

48. In the same connection (though with a change of emphasis from alchemy to cabala) Ramon Lull (c. 1232–1616) could be mentioned, to the extent that his works are to be counted as mathematical. His squaring of the circle (critically edited with commentary in Hofmann 1942), at least, is a typical *experimentum*—and is, by the way, written as part of a skirmish with Paris scholaticism.

49. Cf. also Beaujouan (1957) on the tendency of scholasticism to require immediate utility of its science.

50. In quotes because a modern *profession* is chosen more or less for life, whereas few medieval scholars would remain Masters of Arts for decades. Still, many insights gained in the modern sociology of professions are useful for understanding the medieval Master of Arts. Cf. Høyrup (1980: 72ff.).

51. Excellent surveys are Murdoch (1962, 1969).

52. See Sylla (1971: 13–15). It should, however, be noticed that Bradwardine refers to Grosseteste with great veneration in his theological *De causa Dei* (see Baur 1912: 108*).

53. To state the difference pointedly: No fourteenth-century user of the Aristotelian apparatus, however radically innovative his use, would claim with Galileo that "philosophy is written in this grand book, the universe, which stands continually open to our gaze," though written in the language of mathematics. On the contrary, in Mertonian hypothetical physics, as in "the *Iliad* or *Orlando furioso*, [ . . . ] the least important thing is whether what is written there is true," the attitude with which Galileo charges his opponent (*Il saggiatore*, trans. Drake 1957: 237f.).

54. See the summary of this question in Sylla (1971: 20ff.).

55. In fact, Oresme distinguishes such qualities from indivisible qualities and rejects the loose usage common among contemporary theologians applying concepts applicable only to quantifiable qualities, e.g., to *caritas* (chapter I.ii, Clagett 1968: 170).

56. Cf. discussion in Clagett (1968a: 206ff). Part of the discussion will be found in Oresme's *Quaestiones* to the *Elements* (ed. Busard 1961; cf. Murdoch 1964 and Zoubov 1968). This commentary is indeed quite different from those of earlier times, and raises questions concerned with Oresme's own theories toward the Euclidean material, on which the *Quaestiones* have sometimes little bearing.

57. Ed. Curtze (1868). A partial translation accompanied by a not fully empathetic commentary will be found in Grant (1965).

58. It should be observed that the ratios in question need not be rational—the domain as defined explicitly by Oresme encompasses all ratios that can be written as "part or parts" of a rational ratio, in modern language all rational powers of rational ratios (ed. Curtze 1868: 12).

59. We could even claim with Olaf Pedersen (1956: 99) that Oresme cannot be understood as a "mathematician" but only as a philosopher of nature.

60. Grabmann (1930: 77f.) ("intelligibidem" in question 14 corrected to "intelligibilem"). The first 26 questions were formulated by Sebastian of Aragonia, the last four by Theobald of Anchora.

61. Clagett (1964: 40). Works illustrating the field of interest are, e.g., Bradwardine's *Tractatus de continuo* (described in Stamm 1936) and Buridan's (c. 1295–c.

1358) *Quaestio de puncto* (ed. Zoubov 1961). See also Zoubov (1959); Clagett (1962); and articles by Murdoch, Sylla, and Normore in Kretzman (1982).

62. Molland (1978) is a discussion of the work based on an unpublished critical edition. I used the 1530 edition.

63. Brief, mutually supplementing descriptions in G. l'Huillier (1980: 194f.), and Poulle, ''John of Murs,'' 129; a number of excerpts were given by Karpinski (1912).

64. Briefly discussed in Victor (1979: 49–51), and in Poulle, ''John of Murs,'' 129. Busard (1974) contains a more thorough discussion of a single section, whereas Clagett (1978: 19–44) reproduces and discusses a number of Archimedean passages.

65. See below. G. l'Huillier (1980) describes the annotations made in the manuscript by Regiomontanus.

66. Firmly enough, in fact, to propose a crusade to the Pope, the success of which seemed guaranteed by a favorable conjunction–see Poulle, ''John of Murs,'' 131.

67. E.g., a *Libro d'abaco* from Lucca, published by Arrighi (1973), with second-degree algebra (pp. 108–14) and geometrical calculations (pp. 114–21); the Florentine Paolo dell'Abbaco's (c. 1281–1374) *Trattato d'arithmetica* (ed. Arrighi 1964, with geometrical calculation pp. 104–38 and second-degree algebra pp. 143–48; and the mid-fourteenth century *Rascionei d'Algorsmo* from Cortona (ed. Vogel 1977, with geometrical calculation pp. 134–41 and 149f. but no algebra).

68. See Folkerts (1971). One short collection is dated to the thirteenth century, and a manuscript containing a single problem to 1292. (Cf. Chapter 2 on the relation between practitioners' mathematics and its recreational outgrowth.)

69. Examples of such connections could perhaps be dug up at the crossroads of Joachimism, alchemy, and astrology. Arnaldo di Villanova (c. 1240–1311) might be worthwhile investigating.

70. In Kleb's list (1938) of scientific books printed before 1500, Euclid is represented by two editions of the *Elements*. Bradwardine is represented by three editions (the *Geometria speculativa*, the *Arithmetica speculativa*, and the *Tractatus proportionum*); Albertus Saxonus' *De proportionibus* was printed nine times before 1500, and Richard Swineshead's *Calculationes* twice, on a par with the pseudo-Oresmian *De latitudinibus formarum* and Peckham's *Perspectiva communis;* Sacrobosco's *De sphaera* reached 31 editions.

71. ''Der ein kunst nit allein versteet und weist, sonder auch derselbigen kunst durch stete übung vorteiligen brauch überkomen hat, wirt von den lateinern Practicus genent. Dieweil nun die Wellisch rechnung nicht anders ist dann ein geschwinder außzug in die Regel de Tri gegründet, wirt sie auch derhalben practica gesprochen'' (Rudolff 1540). The *Practica* is thus precisely the genre that we have already met as *Trattati d'abaco*.

72. On the fringes of the teaching of commercial arithmetic, however, one remarkable development took place, *viz.*, through its connection to pictorial art. Cf. below.

It could be mentioned that already Paolo dell'Abbaco's treatise (cf. note 67) contains a large number of drawings, demonstrating the artistic affinities of the environment.

73. In the University Library of Copenhagen I stumbled upon an *Algorithmus linealis proiectilium* by one Magister Johannes Cusanus, printed by Hermann Busch in Vienna in 1514, containing four pages on the basic arithmetical operations (half of which discuss progressions) and 1½ page on commercial calculation; and upon a more extensive, anonymous *Algorismus novus de integris. De minutiis vulgaribus. De minutiis physicis. Addita regula proportionum tam de integris quam de fractis, quae vulgo mercatorum regula dicitur,* printed by Sigismund Grimm in Vienna in 1520.

74. This formulation may sound like eschatology produced by hindsight. It can be given a more concrete meaning by the observation that the persons representing the "new" trends happened to be in a situation that was exceptional in the fifteenth century but was to become more widespread in the sixteenth: connected to Italian or Italianizing courtly culture, and at the same time, by profession or because of personal inclinations, interested in mathematics.

75. So, the definition of *the point* is orthodox, "quod nullas queat in partes dividi"; but already the following definition, that of *the line*, suggests the way the line is painted and is, as far as I know, quite original: "Lineam fieri dicunt puncto in oblongo deducto. Erit igitur lineae prolixitas divisibilis, latitudo autem omnino erit indivisibilis" (Ed. Grayson 1973: 115).

The language in itself is remarkable. Alberti's Latin is the Latin of a humanist; even when using Euclid's ideas he phrases them himself, instead of quoting the translations at hand.

76. I have not investigated Leonardo da Vinci's (1452–1519) works myself, but from all I know from the secondary literature his points of view and his style belong in the same orbit (cf., for instance, Keele, "Leonardo da Vinci. Life, Scientific Methods, and Anatomical Works," 195–98, and Marinoni, "Leonardo da Vinci. Mathematics," 234–43). What little I have read from Dürer's (1471–1528) hand shows that *he* at least does (so his *Underweysung der Messung*, ed. Strauss 1977; cf. also Steck, "Dürer," 258–61). To both, as to Luca, *proportion* was the all-important mathematical subject (cf. also below).

It should be observed that although Luca's main writings deal with applied mathematics, he would also occasionally lecture on purely theoretical subjects.

77. On Piero's treatises, see also the descriptions in M. D. Davis (1977). On Luca Pacioli's *De divina proportione* (written 1496–97, published 1509), see Jayawardene, "Pacioli," 269.

78. "*Conciosia* che dicte mathematici sienno fondamento e scala de peruenire a la notitia de ciascun altra scientia per esser loro nel primo grado de la certeza af-

fermondolo il philosopho cosi dicendo. Mathematice enim scientie sunt in primo gradu certitudinis et naturales sequuntur eas. Sonno commo edicto le scientie e mathematici discipline nel primo grado de la certezza loro sequitano tutte le naturali. E senza lor notitia fia impossible alcunaltra bene intendere e nella sapientia ancora e scripto. quod omnia consistunt in numero pondere et mensura cioe che tutto cio che per lo vniuerso inferiore e superiore si squaterna quello de necessita al numero peso e mensura fia soctoposto. E in queste tre cose laurelio Augustino in de ciuitate dei dici el summo opefici summamente esser laudato perche in ella fecit stare ea que non erant'' (ed. Winterberg 1896: 36). Cf. also Chapter 2 in general, pp. 35–40. The description of Archimedes as an ''ingegnoso geometra e dignissimo architetto'' is found on p. 36; the claim that Luca's book is necessary for ''ciascun studioso di Philosophia, Perspectiua, Pictura, Sculptura, Architectura, Musica e altre Mathematiche suauissima sottile e admirabile doctrina'' is taken from the title page, p. 18.

79. It is worthwhile remembering that Luca was, precisely when this was written, a colleague of Leonardo da Vinci at the Sforza court in Milan; he did not forget to praise him in Chapter 1, nor to refer to him as a compatriot.

80. Arithmetic, geometry, and proportion is the claim of Chapter 2 (ed. Winterberg 1896: 35–40), the modified quadrivial scheme that of Chapter 3 (pp. 40–42).

81. ''[ . . . ] ex quo evenit mundi machinam perire non posse'' (ed. Wilpert 1967: I, 67f. [ed. Argent. I, 50]).

82. It is one thing to explain Trinity numerologically, quite another to prove (through analogy with the basic role of triangulation in surveying coupled with the identity of maximal and minimal entities) that Quaternity or further extensions of the Divine are *impossible,* as done in *De docta ignorantia* I,xx, (ed. Wilpert 1967: I, 59f. [ed. Argent. I, 20–22]).

83. *De docta ignorantia* I,xi, ed. Wilpert (1967: I,13 [ed. Argent. I,11]). Cf. Volkmann-Schluck (1968: 25–35). This position is, if one wants to put a label on it, closer to Plato than to neoplatonism—and the same chapter *does* represent it as Platonic.

84. In this respect, then, his approach foreshadows the whole *analysis infinitorum* of the early Modern age that, in its disrespect for Archimedean rigor, created something effective and *potentially rigorous.* The mathematicians from Cavalieri to Newton and Leibniz (to name but a few) could do this because they dealt carefully with the infinitely great and the infinitely small. They might well have accepted Cusanus's claim that a circular arc approaches gradually to a straight line as its radius approaches infinity—but they would have parted company with him when discovering that he referred not to an arc of fixed, limited length but to a quarter of the full circle (e.g., *Aurea propositio in mathematicis,* trans. Hofmann 1952; 180; cf. *De docta ignorantia* I,xiii (ed. Wilpert 1967: I, 15f. [ed. Argent I, 13–14]).

Cusanus's failure can be explained through a lack of trained mathematical intuition, or as the result of an interest directed by his philosophical aims and convictions rather than by those practical norms which are acquired by the working mathematician. Conversely, the success of the seventeenth–eighteenth century mathematicians of the infinite can be ascribed to a well-trained intuition, and to dominance of the outspoken and tacit standards of the discipline over those imposed externally by philosophical principles.

85. See Toscanelli's letter and the commentary in Hofmann (1952: 128–35, 233–35).

86. The two algorithms mentioned in note 73 are excellent examples of this Aristotelianism by facile habit and convention. The first quotes the *Topica* to explain the ratio 10 between successive lines on the line abacus; the second introduces the subject in a beautiful mix-up of references to Boethius, Augustine, and Aristotle (in descending order of importance).

87. See Vogel, "John of Gmunden," 117–22; cf. also Benjamin (1954) on Johann's very close reliance upon Campanus's *Theorica planetarum*.

88. In the sense, at least, that intimate technical knowledge of astronomy would almost inevitably entail skepticism toward all sorts of marketplace astrology— be it even the well-paying marketplace of the court.

89. All these works are reprinted (together with Regiomontanus' *De triangulis* etc.) in Schmeidler (1972).

90. Lecturing *inter alia* on the *Aeneid* and on Juvenal—see Vogel, "Peurbach," 474.

91. The advertising circular is reprinted in Schmeidler (1972: 532).

92. The difference is highlighted by a comparison between the first definitions of Euclid's *Data* and Regiomontanus's *De triangulis*, which run, respectively.

"Surfaces, lines and angles to which we can procure equals are said to be *given in magnitude*"—a question of theoretically possible construction (ed. Menge 1896: 2; emphasis added), *and*

"A quantity is called *known* when it is measured, either by a well-known or by an arbitrarily fixed measure, according to a known number" (ed. Schmeidler 1972: 283; emphasis added). Obviously, the intricacies of irrational ratios are of less concern than the actual process of constructing a table of numerical values.

93. The advertising circular contains, besides the classical astrological works of Ptolemy and Firmicus Maternus, only one astrological author, with the possibility to include fragments of another—and possibly still other authors of predictions "if they are seen to be worthy." However, if the prefatory letters to a horoscope of the later Emperor Maximilian I are to be relied upon (which could of course be problematic, since the addressees were the Emperor and the Empress), Regiomontanus believed precise astrological prediction possible but difficult in practice, as great knowledge was required (extensive quotations in Zinner 1968: 51f.). More unambiguous evidence is constituted by numerous quite private notes, aiming, e.g., at building up an astrological meteorology (Zinner 1968: 54f., and *passim*). In a lecture

held in Padua in 1463/64 (introducing a series of lectures on al-Farghānī) it is also stated once more that astrological prediction is possible but requires much more knowledge than what is acquired from Sacrobosco's *Sphere* and similar compendia (ed. Schmeidler 1972: 52f.)

Evidence of another kind of Regiomontanus's high esteem for astrology comes from a stylistic analysis of the Padua lecture. Whereas the praise of mathematics proper (identified with arithmetic and geometry) is held in relatively sober terms, and astronomy is only celebrated in a somewhat more elevated style, the opening of the final laudation of astrology (Zinner 1968: 51f.) is nothing less than an exalted hymn in prose to "the divine numen of Astrology, without doubt the most faithful herald of the immortal God who, interpreting his secrets, displays the Law according to which the Almighty resolved that the Heavens be made, on which he sprinkled the starry fires, testimonials of the Future." Astrology itself is an "angelical doctrine" that "makes us no less the kindred of God than we are separated from the beasts by the other arts." In the last phrase we notice an echo of Isidore—whereas the whole passage verges on heresy in its apotheosis of astrological naturalism.

94. This is made explicit in the Padua lecture (ed. Schmeidler 1972: 50). Precisely as did Luca, Regiomontanus claims for mathematics the role of a *first philosophy*, that of a foundation on which philosophy can build (Cf. the quotations in Chapter 7, Section 2).

95. "[ . . . ] a guisa di giuoco, per pigliarse alcuna ricreatione & solazzo." From Ferrari's *Primo cartello*, ed. Masotti 1974: 5.

96. The history of this and other equivalent constructions from Leonardo Fibonacci's *Practica geometriae* onward is presented in H. l'Huillier (1979: 54–56).

97. Also to be found as a theoretical astronomer in *Practica generalis*, chapter 46, which shows him to be identical with the "Gebar hispanus" of Regiomontanus's Padua lecture, and hence with Jābir ibn Aflah, from whom Regiomontanus had borrowed freely for his *De triangulis* (cf. Lorch, "Jābir ibn Aflaḥ," 39).

98. Both works were published by Ioh. Petreius in Nürnberg, who also published Cardano's *Ars magna* and *De subtilitate* and Copernicus' *De revolutionibus*. (Cantor 1900: 431–43 contains a rather detailed description of the *Arithmetica integra*, which is a quite rare book).

99. A good example of this combination is offered by Agrippa von Nettesheym's *De occulta philosophia* (ed. Nowotny 1976), Book II of which deals according to its heading with mathematics, but brings nothing more than anecdotes about automata and other "mathematical" technologies, cabala, and numerology in the manner of Augustine together with similar reflections on the magical properties of geometric figures. Interest in the harmonic proportions of the human body and its inscription in circle and square shows affinity with Leonardo, Dürer, and others, but presupposes no more mathematical insight than the rest.

100. Cf. Rose (1975: 26–75) (the chapter "Patrons, Collectors and Translators [ . . . ]") and *passim*.

101. *De expetendis et fugiendis rebus,* see Rose (1976: 300f). Extensive information on Valla, including the catalogue of his library, is given by Heiberg (1896).

102. Fifty of these were published in various volumes of Boncompagni's *Bulletino*—most by Steinschneider (1872) and Narducci (1886); a complete list is given in vol. 20 (1887), p. 731.

103. This is, in fact, the supposed "Platonism" dominating the source quotations in Crombie (1977); cf. Chapter 7, Section 1.

The status of mathematics as (practically) *first* philosophy is curiously seen in Bessarion's defense of Plato. Whereas the High Middle Ages would argue for the value of mathematics from the words of the philosophers, Bessarion uses mathematicians as authorities and mathematics as an argument for his favorite philosopher (see Rose 1975: 44f.). Others, when arguing from philosophical writings in favor of mathematics, would do so not so much from the philosopher's authoritative words as from a claim that mathematics was the real foundation on which philosophers had built (cf. Possevino as quoted in Crombie 1977: 70), and hence a necessary prerequisite for understanding their words.

104. "E stato Archimede il principe de'matematici; onde con molta ragione diceua il Commandino, a pena potersi chiamare matematico chi con diligenza non haueua studiato l'opere d'Archimede"—ed. Narducci (1886: 453).

105. "Satis prope [ . . . ] ad disciplinarum culmen perducere"—fol. $21^r$ of the volume containing first Bradwardine's *Geometria speculativa* and next the second edition of the *Elementale geometricum* (1530). According to Cantor (1900: 409), the 1546 edition prefaced by Melanchton was virtually unchanged.

106. Lefèvre's preface to the Jordanus edition (1514, but probably unaltered since the first edition from 1496, which I have not seen) refers to the public utility of mathematics, to the *Prisca theologia*, to Pythagoras (quoted for the opinion that nothing can be known without numbers), to Plato's inscription over the entrance to the Academy and to Book VIII of his *Republic*, and to Theon of Smyrna. For once Aristotle and Boethius are not referred to, and the Isidorean *tolle numerus* only turns up indirectly and in a way that shows that the traditional formulation is consciously cast in new form.

107. Prominent among whom were, it is true, philosophers such as Ficino, Pico, and Ermolao Barbaro rather than mathematicians (cf. Randall 1962: 92).

108. Moreover, as Cantor (1900: 364) observes, it was Jordanus's *Arithmetica* that was printed and not the much more "modern" *De numeris datis,* Jordanus's theoretical algebra, which did not fit into a simple quadrivial scheme.

109. Information on its publishing history can be found in Geldsetzer (1973: vii–ix). Reisch's participation in the Cusanus venture is mentioned in Wilpert (1967: viii).

The following account of the work is based upon the (unpaginated) Straßburg edition from 1512 and on the Basal edition from 1517 as reprinted in Geldsetzer (1973) (all features which are mentioned subsequently are found in both editions). The woodcuts are also found in the first edition.

110. A separate edition of the book on geometry appeared as late as 1549 in Paris (Reisch 1549).

111. Thus, when Isidore says that "Pythagoras is supposed to be first among the Greeks to have written about the discipline of numbers, but then it was ordered more broadly by Nicomachos and was transferred to the Latins first by Apuleius, and then by Boethius" (*Etymologiae* III,ii), Reisch simply claims that Apuleius and Boethius are said to have translated Pythagoras. When Isidore derives the term *arithmetica* from *arithmós*, Reisch gives a double derivation from *arithmós* and *aretē*, an explanation which is also given in Hugue's *Didascalicon* II,viii.

112. In the agrimensor tradition, this term designated the side opposing the base of a quadrangle; possibly because he misunderstands the "similarly directed" used in his sources (this is Gerbert's formulation, ed. Bubnov 1899: 77), Reisch tells in the following that this line is equal to the base, and his diagram shows a square.

113. Characterization quoted from Poulle, "Fine," 156; the biography as a whole gives ample support for the harsh words.

114. A brief biography is Rosen, "Schöner," 199–200.

115. Schöner (1534). The algorism itself is a mathematically stringent presentation of the subject written in the mid-thirteenth century by an anonymous follower of Jordanus. In the preface Schöner quotes Greek authors and Virgil, and on the title page he declares that the book "will put the mathematical demonstrations of that calculating art which is popularly called algorism, and thereby its source and origin as well as its reason and certitude, clearly (as it is usual in all branches of mathematics) to the eyes" of the reader.

An appendix "De proportionibus" demonstrates the tendency of humanist mathematics to merge different levels and approaches, presenting the theory of arithmetical, geometrical, and harmonic means, and the 18 Ptolemean "rules of six."

116. Cf. Chapter 4 on the similar correlation between early Islamic non-institutionalized religious fundamentalism and the character of early Islamic science (especially mathematics).

117. The expression used by Garin (1965: 96) to characterize Ficino's Platonic theology.

118. Particularly, it goes by itself, by those who anyhow made a living of mathematics, or were engaged in mathematics *con amore*. But the outcome of Lefevre d'Étaples' intercourse with humanists such as Ficino, Pico, and Barbaro demonstrates that the idea was both accepted and propagated by others who had no such personal reasons to share it.

119. Explicitly stated, e.g., in Cassiodorus, *Institutiones* II.v.2 (trans. L. W. Jones 1946: 190).

120. My conceptualization of the field is vague on purpose. Occultism is most often concerned with *the hidden forces of Nature,* and hence to be characterized as naturalism. During the sixteenth century it came to support itself increasingly on the *Corpus hermeticum,* which has led many scholars to speak indiscriminately of Hermeticism, but which might still be regarded as naturalism, albeit turned specifically toward neoplatonism. When we come to Cabala, numerology, and the magical use of geometric figures, however, a relevant concept of nature is not easily found, even if these interests were often mixed up with naturalist occultism (*vide,* e.g., Pico and Agrippa von Nettescheim). A useful survey is Shumaker (1972); sound caution on "Hermeticism" is recommended by Westman and McGuire (1977)—cf. Schmitt (1978).

121. It is characteristic that neither Yates (1979) nor Shumaker (1972) mentions Archimedes in their respective indexes. The two references in the index of Yates (1964) are immaterial, and the single reference in Yates (1972) almost so. No less characteristic is Archimedes' absence from Agrippa of Nettesheym's *Occulta philosophia.*

122. A number of secondary figures are mentioned in Feingold (1984). Stifel, whose private cabalism appeared as an exception when discussed in the context of Early Renaissance mathematics, could be seen as only premature.

123. See Yates (1964: 173), and Westman (1977: 42ff.).

124. See French (1972, *passim*). Some extra information on the English *Elements* is drawn from Easton, "Dee," 5. The Arabic translation and the collaboration with Commandino is discussed in Rose (1972).

125. See Kirschvogel, "Faulhaber," 549–53. The width of his engineering mathematics as depicted in copperplate in the two editions of his *Ingenieurs-Schul* is reproduced in Scriba (1985: 50f.). In the second edition from 1637, the four quadrivial disciplines have been increased to no fewer than 18.

126. So *inter alia* fol. b.iii–c.i.$^r$, quoting in full six theorems of hydrostatics; fol. c.i$^v$, telling that Archimedes sought and found the squaring of the circle, that "great Secret: of him, by great trauaile of minde"; fol. c.iiii$^v$ and d.i.$^r$, presenting him as the engineer *sans pareil.*

127. This, at least, is my impression gained from the secondary literature; I have seen none of his works in original.

128. Cf. the list of his titles in Kirschvogel, "Faulhaber," 552.

129. The difference in attitude is illustrated by Kepler's criticism of standard astrology. He dismissed zodiacal astrology, the division of the zodiac being purely human and conventional, devoid of physical reality, and hence unable to influence

anything. Planetary *aspects,* on the other hand, depending on the harmonic proportion between the angle between planets and the full circle, had physical reality; the validity of aspectual astrology was therefore not only a theoretical possibility but almost necessary in a universe governed by harmony and proportion ("Since God the Creator derived the structure of the corporeal world from the form of body [ . . . ] it is reasonable to suppose that the positions, the spacing and the bulk of bodies should bear to one another the proportions that arise from the regular solid figures"—*De fundamentis astrologiae certioribus* XXXVII, trans. Field 1984: 250). Cf. also Simon (1979: 36–48).

130. Cf. the survey in Keller (1972). The construction of automata (also discussed extensively by Keller), especially of planetary clocks, as a new branch of courtly science bringing higher artisanate and mathematics into contact is the subject of Moran (1977).

131. I used the original, anonymously published Paris edition from 1560, the existence of which appears to be virtually unknown. Neither Cantor (1900: 612, 641) nor Mahoney ("Ramus," 289) know anything but an edition by Lazarus Schoner from 1591, which they regard as dubious. At least two copies have the author's name written carefully on the title page in ink, in a way that is intended to be mistaken for print, and which suggests systematic repair of a printer's omission (Christ Church College, Oxford, see Ong 1958, #564; and the University Library of Copenhagen, Section II, where Warren van Egmond revealed that I had indeed been fooled).

132. "Algebra est pars Arithmeticae, quae é figuratis continué proportionalibus numerationem quandam propriam instituit."

133. Ed. Hofmann (1970). A German translation (together with selections from other algebraic writings) is given in Reich and Gericke (1973). Witmer's recent English translation (1983) omits the dedicatory letter, from which the present quotations are taken. Since it also replaces Viète's notation with modern algebraic symbolism, it is unfit to show on its own what Viète really did *to* algebra, whereas it may assist those who want to know what he was able do *in* algebra, and serve as a useful interpretative commentary to the Latin text.

134. "Ecce ars quam profero nova est, aut demùm ita vetusta, et à barbaris defoedata et conspurcata, ut novam omninò formam ei inducere [ . . . ] excogitare necesse habuerim [ . . . ] At sub suâ, quam praedicabant, et magnam artem vocabant, Algebrâ vel Almucabulâ, incomparabile latere aurum omnes agnoscebant Mathematici, inveniebant verò minimè" (ed. Hofmann 1970: XI).

135. Not even by Diophantos, the prominent "algebraist" of Greek antiquity, although he supplies a number of problems for Viète's *Zeteticorum libri quinque*—see the tabulation in Reich and Gericke 1973: 93–96. Diophantos is characterized as "subtilissime" in the "zetetic art" (the derivation of the equation expressing a problem); but even though his *working* is algebraic, his *presentation* is numerical, by which his subtlety and skill is made even more admirable, but which also loads the field with unnecessary abstruseness (*Isagoge* V,14, ed. Hofmann 1970: 10).

136. Both are also Archimedists in a more literal fashion. In Book I of the *Scholae mathematicae* (1569), where Ramus sets forth a general view of mathematics, Archimedes is discussed over eight consecutive pages; no other author is mentioned on more than two pages. Viète is more parsimonious in his references to authorities; but whereas Euclid and Apollonios are mentioned seven times each in the collected mathematical works (and Plato four times and Aristotle twice), Archimedes turns up 13 times (thus according to the index—Hofmann 1970: XXXII*-XXXIX*).

137. I checked my prophecy in Randall (1962) and Copleston (1963). It proved correct, apart from the absence of Hooke from Copleston.

138. Exceptions can of course be found, such as Della Porta's optics.

139. Cf. also Vickers (1984a) on the "rejection of occult symbolism, 1580–1680" and on the differentiation of the experimental from the late occult tradition on account of their distinction or non-distinction between *word* and *thing*, between *signifier* and *signified*.

140. See, e.g., Gilbert's *De magnete* from 1600 (ed., trans. S. P. Thompson 1900).

141. The "geometric method" could, of course, be identified with Euclidean as well as with Archimedean method; I prefer a continued reference to Archimedes, because *he* continued to be regarded (with Apollonios) as the supreme geometer, the *Elements* being primarily *prolegomena*. Cf. also Chapter 7, Section 5.

142. "Objections and Replies," trans. Haldane and Ross (1931: II, 48 and 52–59). Remarkably, Descartes organized his *Geometrie* (ed., trans. Smith and Latham 1954) discursively and **not** *more geometrico*.

143. Ed. Chevalier (1954: 575–604, quotation from p. 576). Precisely the same two qualities were emphasized by Descartes when he characterized the method (*Objections and Replies*, trans. Haldane and Ross 1931: II, 48).

Once more, the *weight, number, and measure* of Wisd. 11:21 turn up in Pascal's argument (p. 583)—but in a way that shows him to be closer to d'Alembert and Kant than to Augustine (not to speak of Isidore).

### 6. Jordanus de Nemore: A Case Study on Thirteenth-Century Mathematical Innovation and Failure in Cultural Context

1. This question is the subject of Booß and Høyrup (1979), which investigates a series of concrete exemplifications from the physical as well as the social sciences.

2. Excellent and recommendable expositions of this theme are Werner (1976; 1980).

3. All these decrees will be found in Denifle and Chatelain (1889).

4. The prohibition is found in Denifle (1885: 222, cf. pp. 188–191).

5. *Physica* I,I,i. In order to guard the proportions of the intellectual change of climate one should keep in mind that not all members of the Dominican order had attained the same favorable attitude to philosophy by 1250—cf. van Steenberghen (1966: 276f.).

6. Both passages continue to be quoted in elementary mathematical treatises down to the thirteenth century, and even beyond.

7. Often, the device goes under the name of "Gerbert abacus." The first references, however, antedate Gerbert's mathematical activity by a few years; most probably it is a result of the first tender Arabo-Christian scholarly contacts in the Lorraine.

8. And yet, Héloïse and Abelard, the greatest of dialecticians, baptized their little son Astralabius.

9. E.g., in the twelfth-century Castilian epos *Poema de mío Cid* (I,31), the hero decides to keep his prisoners as servants "because we cannot sell them, and we shall gain nothing from decapitating them."

10. I follow Gherardo's translation (with a single instance of recourse to the Arabic text).

11. Original, "short" set: *Opus numerorum* on integers and *Tractatus minutiarum* on fractions. Revised, "long" set: *Demonstratio Jordani de algorismo* and *Demonstratio de minutiis*. Descriptions and excerpts in Eneström (1906) (*Demonstratio de algorismo*); Eneström (1907) (*Opus numerorum*); and Eneström (1913) (both treatises on fractions).

12. By *ab*, Jordanus means the sum of *a* and *b*. This is his only touch of algebraic formalism. His letters serve the purpose to represent unspecified numbers, and not symbolic abbreviation.

13. The taciturnity amounts to deliberate deceit. In fact, in *one* place a method is referred to "the Arabs." This regards an alternative method for a problem of "purchase of a horse" (borrowed perhaps from Leonardo, and stripped of its concrete dress)—as if the whole subject had not been borrowed from "the Arabs"!

14. The arguments for this are utterly complex; they involve among other things access to the same manuscripts; handwritings and other clues to the origin of manuscripts; and the quality of the information to which various authors had access. Since it is hopeless to make a brief but still concrete sketch of the argumentation, I shall only refer to the unabridged version of the study.

15. Many of the Albertian ascriptions are spurious, it is true; still, as long as we seek to estimate the fame of various authors this plays no role. We might even

claim that spurious ascriptions ought to count doubly, fame having in this case to compensate for the lack of true evidence.

16. I borrow the quoted terminology from Donald T. Campbell's work (1981 *inter alia*).

### 7. Platonism or Archimedism: On the Ideology and Self-Imposed Model of Renaissance Mathematicians (1400 to 1600)

1. I use this term in the strict (and original) sense, as that interpretation of history, *in casu* the history of science, where everything from Roger Bacon or Siger of Brabant onward leads naturally to the culmination in and justification through nineteenth century liberalism and inductivism. Cf. Mayr 1990 on the rampant misuse of the expression as a stigma put on *any* interpretative history.

2. Made, e.g., or rather intimated as a self-evident causal connection, by Marie Boas (1970: 23) in a renowned standard history of Renaissance science: "the hunanist [ . . . ] praised both Plato the Socratic rejecter of the material world, and Plato the cosmologer, who had insisted on geometry as a preliminary to the study of higher things. Indeed, Plato's precepts were followed; for wherever humanist schools were set up, mathematics, pure and applied, was always associated with the purely literary study of Latin and Greek." And again on p. 185:

> The Platonic tradition was of enormous consequence for Renaissance mathematics. Most obviously, it encouraged the study of pure mathematics and the search for previously neglected Greek mathematical texts. It stimulated the founding of chairs of mathematics in the new humanist schools like the Collège Royale [sic] in France, though these were intended as linguistic centres. It helped the revival of professorships of mathematics in the established universities, though it did not raise the professors' salaries. It suggested that mathematics was better training for the mind than dialectic. If offered a number of useful varieties of mathematics, suitable for non-academic education: fortification for the gentleman-soldier, surveying for the landed proprietor, practical astronomy and some knowledge of the use of maps for all. On a less rational plane, Platonism and neo-Platonism encouraged so much number mysticism and astrology that to the layman "mathematicus" and "astrologer" were identical. (Indeed, they were when the mathematician was a Cardan or Dee, though the latter protested that he only dealt with "marvellous Acts and Feats, Naturally, Mathematically and Mechanically wrought," and it was unfair to call him a conjurer for that.)

3. "It would be useful, too, to look at a sphere with its motions, as Archimedes once did, and which recently a certain Florentine by the name of Lorenzo has made. Not just to look at it, but to reflect on it in the soul" (trans. Boer 1980: 153)—one of the two passages where a mathematician different from Pythagoras and Ptolemy is mentioned (the other regards Archytas's automata).

4. Chapter I,xi—ed. Wilpert et al. 1964: I, 40. In the following chapter, the merely symbolic role of mathematical objects is highlighted by the observation that they are finite and thus fundamentally different from that Divine reality to which they lead. Ultimate reality is not mathematical, and cannot be described analytically by mathematics, we might say.

5. Cf. Chapter 5, Section 2, according to which thirteenth-century mathematics was mostly not Aristotelian *stricto sensu*, even though certain philosophical writings *about* mathematics were.

6. "[ . . . ] belli pensiero circa la doctrina di Platone, e di Aristotele, per essere versatissimo, in ambi due questi autori"—Crombie (1977: 78 note 6).

7. This role as a philosophical *sine qua non* is different from and much more ambitious than that of being a propaedeutic preparation of the mind, i.e., the role ascribed to mathematics by Plato.

8. Not as a matter of definition, which would make part of the following argument circular, but as it follows from a survey of the mathematical authors of the time—cf. Chapter 5, Section 5.

9. It is essential for understanding humanist utilitarianism to notice that it is *civic*, i.e., concerned with *public utility* as understood by the culturally and politically leading strata.

10. None of them, in fact, is mentioned anywhere in the mathematical works—except for the claim in *De pictura* II,27 (ed. Grayson 1973: 48) that Plato, together with Socrates and others, "furono in pittura conosciuti." The absence of Euclid is all the more remarkable since one of Alberti's mathematical works in the *Elementa picturae*, of obvious Euclidean inspiration.

The *De re aedificatoria* (ed., trans. Theuer 1912) contains 19 references to Aristotle and 24 to Plato. Insofar as these are not purely anecdotic, however, they refer to Plato the city planner and the social planner, and to Aristotle the natural historian or the city planner. Neither of them stand out as authorities; both are authors with whom one may agree or—if needed—disagree.

Archimedes is mentioned only a few times in the latter work: three times in connection with the movement of very large weights, and once as a representative of that sophistication in the treatment of angles and lines to which the architect should *not* aspire (Theuer 1912: 519).

11. "Io solea maravigliarmi insieme e dolermi che tante ottime e divine arti e scienze, quali per loro opere e per le istorie veggiamo copiose erano in que' vertuosissimi passati antiqui, ora così siano mancate e quasi in tutto perdute: pittori, scultori, architetti, musici, ieometri, rhetorici, auguri e simili nobilissimi e maravigliosi inteletti oggi si truovano rarissimi e poco da lodarli"—ed. Grayson (1973: 7).

12. Ed. Winterberg (1896: 18, 36). Italian text in Chapter 5, note 78.

13. "[ . . . ] per quotidiana experientia a Vostra Ducal celsitudine non e ascosto, [ . . . ] che la deffensione de le grandi e piccole republiche per altro nome arte

militare apellata non e possibili senza la notitia de Geometria Arithmetica e Proportione egregiamente poterse con honore e vtile exercitare" (ed. Winterberg 1896: 37).

14. "Haec de patre omnium Geometrarum Euclide, cui succedunt Archimedes Siracusanus ciuis, et Apollonius Pergaeus ob ingenij altitudinem diuinus uocari solitus, quorum uter alteri praeferendu sit, non facile dixero. Nam etsi Apollonius elementa conica in octo libris, quos nondum uidit latinitas, subtilissime conscripserit, Archimedi tamen Siculo uarietas rerum editarum principatum contulisse uidetur, quem sub Nicolao quinto Pontifice Iacobus quidam Cremonensis Latinum ex Graeco reddidit, duos de sphaera, et chylindro libros composuit, de Conoidalibus et Sphaeroidalibus duos, totidem de aequeponderantibus, scripsit item de lineis spiralibus, ubi circumferentiae circuli aequalem rectam designare conatur, quatenus circulum quadrare liceat, quod quidem plerisque uetustissimis philosophis quaesitum est, ad tempora usque Aristotelis autem à nemine compertum, cuius rei gloriam nonnulli nostra tempestate uiri clarissimi praestolantur. Archimede insuper mensurationem circuli accepimus, quadraturam, parabolae et arenae numerum. Sunt qui scripsisse eum asserant Moechanicam, ubi electissima ad uarios usus colligit ingenia, de ponderibus, de aqueductibus, et caeteris quae usquehac uidere non licuit" (ed. Schmeidler 1972: 45; not fully literal translation).

15. "Quod de nostris disciplinis nemo nisi insanus praedicare ausit, quandoquidem neque aetas neque hominum mores sibi quicquam detrahere possunt: Theoremata Euclidis eandem hodie quam ante mille annos habent certitudinem. Inuenta Archimedis post mille secula uenturis hominibus non minorem inducent admirationem, quam legentibus nobis iucunditatem" (ed. Schmeidler 1972: 51).

16. Less important than the value of mathemathics for philosophy and liberal studies but still briefly mentioned (under the guise of a list of topics to be omitted) is the utility of mathematics for "mechanical" (we would rather say "manual" or "practical") purposes. Their order is informative: At bottom we find the services rendered to smiths and masons; somewhat higher is the status of surveyors, who are followed by merchants, soldiers, and builders of warlike machines; just below the liberal studies we find the building of musical instruments. The beauty of courtly music and warfare are evidently more "civic" than trade and surveying (abacus school topics), which on their part rank higher than genuine productive crafts.

17. Even Regiomontanus, it is true, is more interested in his *results* than in his ingenious and sophisticated *methods*. As expressed by Clagett (1978: 383) it "is regrettable that Regiomontanus did not live long enough to exploit his very promising beginning in the study of Archimedes. If he had lived, possessing as he did the techniques and abilities in language and mathematics, he might have been able to anticipate the mastery of the Archimedean corpus first achieved in the sixteenth century by Maurolico and Commandino. As it was, by the time of his death, Regiomontanus had not developed any interest in the higher geometrical problems susceptible to the method of exhaustion"—and given his primary interest in astronomy and the mathematics of astronomy it is perhaps to be doubted whether a longer life would have changed the focus of his mathematical interest.

18. "[ . . . ] even if [Petrarch's] Archimedes is a long way from the rigorous and certain geometer extolled by later mathematicians, nevertheless the humanistic Archimedes is one foundation of the Archimedean tradition of the Renaissance," as formulated by P. L. Rose (1975: 9).

19. Never fulfilled, one will remember, because of Regiomontanus's early and sudden death. But his publishing plans can be consulted in a circular printed in facsimile by Schmeidler (1972: 533).

20. "Le grand Archimède de nostre age"—quoted from Westman (1977: 42).

21. "Sig. Luca Valerio, nuovo Archimede dell'età nostra," and "Luca Valerio, altro Archimede secondo dell'età nostra"—ed. Favaro (1890: VIII, 76 and 184).

22. See Strømholm, "Valerio."

23. Even though Marie Boas's conflation of his occult and his mathematical interests (see note 2) certainly does not correspond to his major mathematical works.

24. In Cardano (1663: 440–45). The quotations are taken from p. 443.

25. One of these should be noted apropos of mathematical Platonism, viz., that Cardano finds it appropriate to supplement Proclos's praise of Plato the mathematician by the information that Plato did not follow up his sympathy for mathematics with actual work on the subject.

26. "[ . . . ] omnes Graeci, Graecéque scripserunt, sed quorum tamen monumenta interierint, aliorumque tantum testimonium viuant. Verùm omnes hos vincit Archimedes Syracusius, cuius fermè omnia inuenta habemus: vir summo ingenio, et qui circuli periferiam proximius ostenderit, et solida demonstratione duabus lineis duas interponere continua proportione docuerit: sed hoc periit."

27. In Cardano (1663a: 352–672). Quotations from p. 607f.

28. "Archimedes primus sit, non solùm ob monumenta illius nunc vulgata, sed ob mechanica, quibus vt Plutarchus auctor est, vires Romanorum saepius fregit [ . . . ]."

29. "Sextus locus Apollonio Pergeo debetur, qui fermè aetate aequalis fuit Archimedi."

30. "Omnis enim veritas diuina est"—De subtilitate, p. 607.

31. Ed. Narducci (1887).

32. "In tutte le facoltà ui sono stati alcuni, che, ariuati al colmo dell'eccellenza, hanno mostrato quanto in quella possa auanzarsi l'intelletto humano. Tale senza alcun dubbio fù Archimede fra'Matematici, poichè ad esso ragioneuolmente si conuiene il primo luogo; onde, douende scriuere de lui, mi dolerò di più cose, cioè dell'ingegno non proportionato a'meriti del suggetto, dell'antichità, che non me las-

cia giungere alla cognitione di tutte le sue cose degne d'historia, e l'altra la penuria
dei libri, e il luogo oue mi trouo, non solo lontano dalle librarie famose, ma'anco
dalle minime" (ed. Narducci 1886: 388).

33. Ed. Narducci (1886: 453). Italian text in Chapter 5, note 104.

34. A discussion of the whole wave, from the mid-fifteenth century beginnings
to its disappearance with Estienne de la Roche in the 1520s, is in van Egmond (1988).

35. There is no reason to substantiate these claims here. But see Chapter 5,
Section 5.

36. Or perhaps it was precisely the utterly elementary level which he found in
his (selective) reading of the ancients. *Numerandi et metiendi scientia* has a similar
ring to the "arts of *arithmetikē* and *metrētikē*" of Plato's *Philebus* 55 E (ed. Fowler
and Lamb 1975: 358f.)—but if this is more than mere coincidence it is all the more
striking that the ensuing distinction (56 D) between the utilitarian and the "philosophical" kinds of these sciences seems to have escaped More completely. *If* More
learned from Plato regarding mathematics, what he learned will have been to *avoid*
the realm of theory.

37. Erasmus, it is true, knew that "references to music, geometry, arithmetic,
astronomy, medicine, and if you like to add it, magic too, occur not infrequently in
poetry"; this, on the other hand, is his only reason to argue for some teaching of
"arithmetic, music, and astronomy." Still, these subjects "need only be sampled,"
i.e., be treated on the level of Isidore of Sevilla (*De recta pronuntiatione,* trans. M.
Pope, *in* Craig R. Thompson et al. (eds.) 1985: IV, 371, 387). Apart from that, arithmetic and geometry are (if the indices are to be trusted) only mentioned once in the
six volumes of literary and educational writings, *viz.,* in an endorsement of Jerome's
and Augustine's support for study of "the liberal disciplines, logic, rhetoric, physics,
arithmetic, geometry, music, finally histories and the knowledge of antiquity," but
finding it at the same time "scarecely profitable to recall" all the "complications" of
Augustine's discussion of the mathematical disciplines (*The Antibarbarians,* I, 95f.,
trans. M. M. Phillips). Archimedes is never mentioned, and Euclid only once in a
satirical passage in the *Praise of Folly* (V, 116, trans. B. Radice), where one instance
in a long list of fellows deceived by self-love is the one "who has only to trace three
arcs with a compass to imagine himself Euclid."
Not much in almost 2000 pages concerned with education and teaching.

38. Lefèvre d'Étaples is thus also an exception to the rule that Ficinian neoplatonism was not conducive to mathematical activity.

39. In his *Protomathesis* (1532) a reference to Archimedes *will* be found, and
even an Archimedean proof: fols 85$^v$ to 89$^r$ contain an Archimedean proof that the
ratio between the circular periphery and the diameter lies between $3^{10}/_{71}$ and $3\frac{1}{7}$, and
tell that the proof was found by Archimedes "according to common assumption"
(*iuxta vulgatum*). Alas, this correct proof (yet very medieval approach) is followed by
an "exact" construction of Fine's own making (fols. 89$^r$ to 91$^v$).

40. "Voluit deus in omnibus artibus aliquam velut ideam singularem esse, quam omnes ejus disciplinae studiosi ad imitandum sibi proponerent: ut in eloquentia, Demosthenem et Ciceronem: in medicina Hippocratem et Galenum: sic in mathematicis Archimedem" (Ramus 1569: 26).

41. "Nam si mathematicae institutionis compositio et conformatio spectetur, στοιχειῶται Hippocrates, Leo, Theudius, Hermotimus, Euclides, Theon principem fructum laudis ferent, si nobilitas mathematicae scholae et amplitudo perpendatur, mathematum authoritas ad Pythagoram, Platonem, Aristotelem pertinebit: si, quod summum est, mathematum non solum scolastica veritas et é libris demonstratio, sed popularis usus atque utilitas aestimatur, Archytas, Eudoxus, Eratosthenes, sed maximé et altissimé supra omnes unus Archimedes in caelum ferendus erit" (Ramus 1569: 40).

42. Ramus 1560: A ij; Latin text in Chapter 5, note 132.

43. In a more literal way, too, Viète can be counted an "Archimedist," in the sense that he appears to have regarded Archimedes as the most accomplished of mathematicians (though, like most of the "Archimedists" who were listed, he was not a follower of Archimedes in his own mathematical style and techniques). Admittedly, he is more parsimonious than Ramus in the discussion of authors; but if references can count as praise it can be observed in the index to the collected works that Archimedes turns up 13 times, whereas Euclid and Apollonios must content themselves with 7 references each (and Plato with 4 and Aristotle with 2). See Hofmann (1970: XXXII*–XXXIX*).

44 Ed. Hofmann (1970: XI); Latin text in Chapter 5, note 134.

45. So in Richard de Fournival's *Biblionomia*, where Jordanus de Nemore's *apodixeis* (i.e., works based on rigorous mathematical proof) were opposed to purely empirical *experimenta* on algebra or progressions—e.g., in N[os] 45 and 48 (ed. Birkenmajer 1970: 166f.).

46. Letter to Mersenne, 11 October 1638, trans. Crombie, "Descartes," p. 53.

47. Tycho, on the other hand, in his *De disciplinis mathematicis oratio* in 1574, appears to have subscribed to the usual Archimedist tenets, but mainly as commonplaces that were unnecessary to expand upon (according to Jardine [1984: 262]; I have not seen the work).

48. Still, Books I–II of Ramus's *Scholae mathematicae* contain "an exhortation to the mathematical arts to Catharina of Medici, Queen and Mother of the King" (Ramus 1560: α 1ᵛ), whereas Pacioli's *De divina proportione* was written to Ludovico Sforza of Milano. To a large extent, humanist high culture *was* courtly culture, in particular from the mid-fifteenth century onward.

49. Evidently, this conjecture does not hold without qualification. Neoplatonism was an important strain in Christian, not least Augustinian, thought, and was important throughout the Middle Ages. Similarly, Hermetic material was not unknown

to medieval natural philosophers—see, e.g., the pseudo-Hermetic *Liber Hermetis mercurii Triplicis de vi rerum principiis* (ed. Silverstein 1955) and N^os 85 and 103 in Richard de Fournival's *Biblionomia* (ed. Birkenmajer 1970: 186, 191; N° 85 also contains Plato's *Timaeus* and some Apuleian treatises). Nor was Archimedes unknown to Richard and his contemporaries, however, and they rewrote him just as Hermes was rewritten. This could be done precisely because neither author had yet become sacred symbols.

50. In itself, one may say, this constitutes indirect support for the "materialist" interpretation (although it is also compatible with a "Zeitgeist" interpretation). Social structures and social life can more reasonably be claimed to reflect themselves directly in a general cultural climate that then functions as a hotbed for the formulation of specific sets of ideas and particular philosophies than to be the direct cause of such formulated ideas and philosophies. A strictly "idealist" interpretation, on the other hand, will almost be definition start from ideas rather than from a cultural climate.

51. In a way it was ideologically obliged to be so by the tradition; but contradictions could easily be removed by definitions. As observed by Biagioli (1989: 53), in "the many sixteenth century debates on the status of different arts (medicine, law, visual arts), 'abstractness' or 'mechanicity' became little more than rhetorical tropes which told more about the relative power of a profession that called itself 'abstract' and the other 'mechanical,' than about the actual 'abstraction' or 'mechanicity' of the so-judged discipline." See also p. 61, on the particular case of aristocratic and non-aristocratic military engineers, and cf. notes 9 and 16 of the present chapter, on the concept of "civic utility" and its contents.

52. The plausible effect of such a change of balance might have been a steady growth in empirically based techniques and enhanced neglect of theoretical investigation. This would have been fully compatible with the technological needs of early capitalism—after all, European *technological* progress was, until well into the nineteenth century, almost exclusively based on empirical trial and error and ingenuity, and on "theory" very close to common-sense empirical studies.

A development of this kind would have made the European situation similar to that of ancient China, to which Zilsel points indeed (1942: 560) as a puzzling counterexample.

## 8. On Mathematics and War: An Essay on the Implications, Past and Present, of the Military Involvement of the Mathematical Sciences for Their Development and Potentials

1. As illustrations of these uncertainties, as they look at this very moment of writing (March 20, 1992): It is discussed in the European press whether President Bush is most likely to attack Libya or Iraq in a predicted attempt to improve his opinion poll ratings, and the optimists among us discuss whether he is really going to at-

tack; elsewhere on the globe it is a serious question whether Ukraine and Russia, formally members of a confederation, are actually moving toward a terror balance on atomic weapons.

When this book appears, the reader will certainly know the answers. Undoubtedly, new questions will have arisen.

2. When mathematics is spoken about in the following, mathematics *as a science* is thus intended, i.e., the creation of new theoretical knowledge. Mathematics as a subject for teaching and as a means for applications will mostly be considered only in their connection to mathematical science or to the community of mathematical scientists.

3. So was, e.g., Bernal's opinion: "Science and warfare have always been most closely linked; in fact, except for a certain portion of the nineteenth century, it may be fairly claimed that the majority of significant technical and scientific advances owe their origin directly to military or naval requirements" (1939: 165).

4. See the introductory paragraphs of his *Metrica* and *Dioptra*. The latter introduction (chapter II) first refers to the utility of the dioptra in surveying in general, but then passes on to the specific case where part of the terrain is occupied by an enemy or inaccessible for other reasons. Finally, Hero points to the specific utility of the device for ensuring that siege towers and ladders are built sufficiently high.

5. Cf. Gross (1978: 246–48). Fuller discussions are Waters (1976) and Beaujouan (1966).

6. Gross (1978: 248). Already now we may point out the general importance of these observations for the discussion of contemporary military research.

7. John Dee, the mathematician-magician who published prolificly on "occult" secrecies, kept "his treatises on navigation and navigational instruments deliberately[ . . . ]in manuscript" (Easton, "Dee," p. 5).

8. Thus, in 1727, the French Académie des Sciences asked for the most efficient arrangement of masts on a ship—a problem that the sixteen-year-old Euler answered, though without receiving more than *accessit* (see A. P. Youschkevitch, "Euler," p. 468).

9. This observation does not detract from the global importance of total societal practice as a governing condition and source of energy for mathematics as a specific practice. This question is discussed in some depth in Section 5.

10. See Schneider (1970), and the fuller presentation of English mathematical practice in Taylor (1954, 1966).

11. Diderot's *Encyclopédie raisonnée* [ . . . ] had asserted that *war is unreasonable,* "a fruit of the depravation of men; it is a convulsive and violent illness in the body politic" (ed. Soboul 1976: 176). Clausewitz, who was anything but a sympathizer of the Revolution, had learned from Napoleon's strategy *that war*

*must be waged with as much reason as possible.* This is the cardinal message of his *Vom Kriege.* In Max Weber's terminology, we could speak of a shift from *Wert-* to *Zweckrationalität.*

12. Conceptual differentiation, combined with an attempt to preserve institutional unity, is reflected in the titles of the first specialized mathematical journals—cf. Box V.

13. Still, Horst-Eckart Gross observes that mathematicians "participated to a considerable extent" in the development of aerodynamics (used for the construction of airplanes), which may have "inspired the development of mathematics" (private communication). Cf. Box VII.

14. At least in 1940, he did not suspect the potentialities of his favorite discipline—see Hardy (1967: 140). But he was no convinced opponent (nor, for sure, a determined protagonist) of the application of science in war, if war was inevitable (Hardy 1967: 141f.).

15. The gradual dismantlement applies primarily to England (see Rose and Rose 1970: 40ff.) and the U.S. (see Kevles 1978: 148–54). Defeat and blockade called forth a special situation in Germany, where the break was abrupt. The *Notgemeinschaft der deutschen Wissenschaften,* founded 1920, managed to bring German fundamental science successfully through the twenties, with the intention to save scientific culture and thus preserve the basis for later technical development, but did (and aimed at) nothing more (cf. A. Hermann 1982: 116–25). Approximately the same situation seems to be reflected in the gradual expansion of grants for and positions in applied mathematics during the twenties (cf. Bernhardt 1980 and Tobies 1984).

16. Actually, the conflict between private enterprise and publicly initiated research had made itself felt even during the war. Physicists had thus been excluded from the earliest U.S. sonar research in 1917 because their presence would complicate the patent situation, as explained by Admiral Griffin (cited in Kevles 1978: 120).

17. See, e.g., Jungk (1958: 112–23; Kevles (1978: 324–33); Greenberg (1969: 117–25); and *A History of Technology* VI, 226–76, *passim.*

18. The same simultaneity can be observed in the elaboration of ideas from mathematical statistics, leading to a similar conclusion.

19. This is no mere speculation by analogy, but corresponds to formulated mathematical ideals. Cf. the eulogy praising Fefferman for having shown that "in many problems complications cannot be avoided."

20. Cf. Conway (1980). Actually, the way from the group theory of the nineteenth century, which aimed at simplification of the seemingly complex and irregular, to the modern "monsters" has passed, *inter alia,* over the irreducible groups of quantum physics. Even in the case of algebra, real complexities and not only the cultural problem of complexity have played a role for the orientation of research.

21. At the same time we observe that "real-world problems" only become productive when mediated adequately, through mathematized sciences or through other mathematical disciplines. On the level of systems theory abstractions, the function of these intermediaries is analogous to that of seventeenth- to eighteenth-century scientific academies.

22. As with any rash generalization, prominent exceptions can be found: catastrophe theory, and the qualitative theory of turbulence.

23. This latter assertion may sound overly Machiavellian—but see the description of the Mathematics Research Center, U.S. Army in note 42 and appurtenant text.

24. We shall presently discuss various negative consequences of military influences on scientific development, summed up in the concepts of "overloading" and "punctualization." They are not conspicuous in mathematics proper—mostly because no mathematics is produced if they are too strong. One example of this was discussed already: Portuguese navigational research (cf., however, note 36).

25. Continuing, once again, prewar ideas—see Wallis (1980: 326).

26. The Statistical Research Group (Columbia, 1942–1945). The description was formulated by its head, W. Allen Wallis (1980: 322). According to Wallis, the group was not only unprecedented: Even afterward, this "model[ . . . ]of an efficient statistical research group" has not been equaled.

27. Anticipating later discussions we may consequently observe that as a *socially concrete* practice war *is* something specific: When not submitted to the extreme pressure of total war, capitalist society is (or was at least until 1945) not able to launch this sort of rational organization, and hence not to make full and well-considered use of the potentialities of science.

28. According to Wallis (1980: 326, quoting his own private report from 1950), he first got Wald interested (and that only with difficulty) in the problem of sequential tests by arguments referring to the intrinsic mathematical interest of the problem and not at all linked to the war.

29. H. B. Hansen, private communication. See also S. Rosen (1967: 13), who ascribes the declaration unspecifically to the "United States government." In practice, *equivalence* could only be demonstrated if the language *was* COBOL.

30. Wernicke (1982: 9). As an electronics engineer and a former military researcher (now technical advisor of the West German *Grünen*) Wernicke knows better than most of what he speaks.

There is one important difference between military electronics and the superlanguage. Whereas the overloading of a tank or a fighter-bomber is at least to some extent balanced by the extreme standards for single components, there are good reasons to believe that the global complexity of Ada and the necessary interconnectedness of single procedures will make even the single "elements" of the high order language less reliable than in less complex languages. In civil application, this will be unpleasant and generate ineffectiveness; in the military domain things grow worse: "in applications where reliability is critical, i.e., nuclear power stations, cruise missiles, early warning systems, anti-ballistic missile defense systems," nobody will be able to intervene when things have gone astray because of "an unreliable programming language generating unreliable programs." "The next rocket to go astray as a result of a programming language error may not be an exploratory space rocket on a harmless trip to Venus: It may be a nuclear warhead exploding over one of our own cities," as C. A. R. Hoare warned the community of American Computer Scientists in his Turing Award Lecture (1981: 83).

Cf. also note 36 concerning the specific question of overloaded computer software.

31. The SDI project is often cited as an exception to this rule, or even as a portend of new trends. First, however, it should be asked whether this luxurious red herring is more than a wriggle on the curve—giving exorbitant power to ignorant zealots surrounded by ill-willed advisors may break any institutional rationality for some time. Second, the proclamations of those enjoying the profits function as publicity and may hide more than they reveal; their truth value will only be known years later. Cf. *Scientific American* 255(5) (November 1986): 54–56 (European pagination), "SDI Boom—or Bust?"

32. Cf. also Wernicke (1982) as cited in note 30.

33. Especially that of contemporary monopolistic capitalism, where the weapon is (real or fake) product development and cost reduction, and not price reduction. Cf. note 40.

34. For one ugly example among many, see R. J. Smith (1982), on the involvement of scientific experts in the cover-up of the atomic test fall-down in Utah in 1953.

35. H. Werner (1982: 67), in the President's Speech to the Deutsche Mathematikervereinigung.

36. Obviously, this line of argument concerns "mathematics proper." In the less "proper" but, from a global perspective, no less important field of *computer software*, the need for an "inclusive understanding" is as great as if physical phenomena are concerned, and a practical problem may find no solution before a fundamental breakthrough has taken place. The creation of COBOL and FORTRAN (and thus, we might claim, of software as a separate concern) can be quoted as examples—COBOL especially is striking because the COBOL Committee set out to find short-time solutions based on the existing state of the art and found out that only the creation of a new language would do (see S. Rosen 1967: 12f.).

In later years, there seems to be a tendency that military influence has indeed not only called forth specific broad developments but even permeated the global way to think about software development—and has done so negatively. Enormous resources have been thrown into projects overloaded with advance specification of every imaginable feature, which has effectively prevented any dialogue between user and manufacturer and any feedback during the programming process (see Abrahams

1988). According to Paul Abrahams, this hampers not only the development of military software but software engineering in general, because it is an ''unvoiced assumption that the software is being built to military specifications,'' and because ''the reasonableness of those specifications'' is accepted tacitly (p. 481).

The trend reminds strikingly of the ERA/UNIVAC experience (see Box X), though repeated on an immensely larger scale. There are thus historical parallels to Abraham's claim that the military orientation is hampering and ineffective, producing much ''less knowledge per dollar'' than would be the case in a feedback-oriented organization of development. But ''costs being not an appropriate'' criterion (!; explicit DoD statement quoted by Abrahams), diminished effectiveness is balanced by abundance of dollars. ''Pentagon-style'' software hence becomes quantitively dominating to such a degree that it also acquires structural dominion. We may guess that the ensuing curtailing of engineering and computer effectiveness will in the long run be conquered by competition from civilly based development—but as long as competition is distorted by the availability of unbounded resources on the military side, current trends are likely to continue. The self-defeating character of military influence in the field stops being self-defeating when society as a whole is completely militarized—not because the negative effects are vanquished but because they are carried by society as a whole in the form of inadequate technology and curtailed welfare.

37. It may, however, be of importance in this connection that Pontrjagin the topologist set himself the task. Similarly, many of those advances in early computer science which are often credited to military influence can be traced directly to the person of John von Neumann (cf. Goldstine 1972: 329–32 and *passim*).

38. Quoted from Serge Lang (1971: 73). That institutions and not only individuals are (still) bought will be seen in the list of 10 ''grand benefactors'' to the 1986 International Congress of Mathematicians held in Berkeley. Among these we find Department of the Air Force, Air Force Office of Scientific Research; Department of the Army, Army Research Office; and Department of the Navy, Office of Naval Research (*Proceedings . . . 1986*, p. xvii, cf. ii). Nobody can expect to get results of specific military value from the arrangement of a mammoth congress; only the loyalty of the profession as a broad national (and even international) average can be in question.

39. In summer 1988, after the agreement on land-based intermediate range missiles, the anxiety of the Reagan administration to avoid the inclusion of sea-based missiles in further agreements underscores that these considerations have conserved their full validity in spite of apparent changes of policy.

40. Private estimate from comparative international R&D statistics. The mechanism is that firms will have no advantage from marginal sales price reductions under monopolistic (in the technical language of Anglo-Saxon economists: oligopolistic) conditions. Instead, competition will concern costs (involving, among other things, process development); product development and differentiation (if only sham development through changes in the design of ''this year's model''); public image; political protection (important not least in the armaments industry); etc.

41. Zuckerman 1982: 102–6, quotation from p. 103. In order to drive home his point more convincingly, Zuckerman takes care only to mention the ''soldier or sailor

or airman," i.e., the military professionals, as alternatives to "the man in the laboratory" when assigning responsibility.

42. This three-level structure of the military entanglement is clearly visible in the case of the Mathematics Research Center—United States Army, of the University of Wisconsin. A meticulous investigation of the work and workings of this institution will be found in *The AMRC Papers* (1973).

43. This passage is left as it was written in 1983/84. Since then counter-experts have, indeed, substantiated much of the criticism so thoroughly that even the military planners have been forced to take up, e.g., the nuclear-winter phenomenon.

44. One may get funny reactions. "Some don't approve of sexual intercourse, some are opposed to card-playing, and JH cannot accept what I do," as one professor commented in the weekly of the Danish Technical High School when his work on antennas for use in the Vietnam war was questioned.

45. See *Scientific American* 254(1) (January 1986): 48 (European pagination), "Signing off."

46. This comfortable complacency may have much to do with the outcome of a survey of scholars' political attitudes conducted in 1969, at the height of the Vietnam War (Ladd & Lipset 1972). Mathematicians were found not only to the right of average faculty in all fields (as were all scientists except physicists); they were also perceptibly more right-wing than scientists in general. As in all other fields, "achievers" (faculty at elite universities having published 10 or more professional works during the last two years) were more left-wing than average for the field; but even they were less so than achievers in all other sciences (engineering apart), and whereas "achieving" physicists would rather be "very liberal" (33 percent) than "liberal" (27 percent), the numbers for mathematicians were 9 percent and 44 percent, respectively.

# Abbreviations and Bibliography

*A History of Technology*. Edited by Charles Singer, E. J. Holmyard, A. R. Hall and Trevor I. Williams. 7 vols. Oxford: Clarendon Press, 1954–78.

Abert, "Sakadas aus Argos." Paulys *Realencyclopädie der Classischen Altertumswissenschaften*. Neue Bearbeitung begonnen von Georg Wissowa. Zweite Reihe, Zweiter Halbband 1768f.

Abrahams, Paul. 1988. "President's Letter: Specifications and Illusions." *Communications of the ACM* 31:480–481.

Adams, Robert McC. 1974. "Anthropological Perspectives on Ancient Trade." *Current Anthropology* 15:239–49, discussion 249–59.

———— 1982. "Die Rolle des Bewässerungsbodenbaus bei der Entwicklung von Institutionen in der mesopotamischen Gesellschaft." In Hermann and Sellnow (eds.) 1982: 119–40.

Adams, Robert McC., and Hans J. Nissen. 1972. *The Uruk Countryside. The Natural Setting of Urban Societies*. Chicago: University of Chicago Press.

Addington, Larry H. 1984. *The Patterns of War Since the Eighteenth Century*. London: Croom Helm.

Aeschylos. *Works*. Ed., trans. Smyth 1973.

Agrippa von Nettesheym. *De occulta philosophia*. Ed. Nowotny 1967.

Ahrens, W. 1916. "Studien über die 'magischen Quadrate' der Araber." *Islam* 7: 186–250.

Al-Fārābī. *Catálogo de las ciencias*. Ed., trans. Palencia 1953.

Albers, Donald J., G. L. Alexanderson and Constance Reid. 1987. *International Mathematical Congresses. An Illustrated History, 1893–1986*. Revised edition. New York: Springer.

Albertus Magnus. *Physicorum* libri VIII. In Beati Alberti Magni *Operum* tomus secundus.

*Algorismus novus de integris, De minutiis vulgaribus, De minutiis physicis. Addita regula proportionum tam de integris quam de fractis, quae vulgo mercatorum regula dicitur*. Vienna: Sigismund Grimm, 1520.

Alster, Bendt. 1975. *Studies in Sumerian Proverbs.* (Mesopotamia: Copenhagen Studies in Assyriology, 3.) Copenhagen: Akademisk Forlag.

Amir-Moéz, Ali R. 1959. *"Discussion of Difficulties in Euclid,* by Omar Ibn Abrahim al-Khayyami (Omar Khayyam), translated." *Scripta Mathematica* 24:275–303.

*The AMRC Papers.* By Science for the People, Madison Wisconsin Collective. Madison: Science for the People, 1973.

Anawati, G. C., and A. Z. Iskandar. "Ḥunayn ibn Ishāq." *DSB XV,* 230–49.

Anbouba, Adel. 1978. "Acquisition de l'algèbre par les Arabes et premiers développements. Aperçu général." *Journal for the History of Arabic Science* 2:66–100.

———— "Al-Samaw'al." *DSB XII,* 91–95.

———— 1979. "Un traité d'Abū Ja'far [al-Khazin] sur les triangles rectangles numériques." *Journal for the History of Arabic Science* 3:134–56 (Arabic pp. 178–57).

Anderson, J. Edward. 1981. "First Strike: Myth or Reality." *Bulletin of the Atomic Scientists* 37(9) (1981):6–11.

*Anthologia graeca* [The arithmetical epigrams]. Ed., trans. Paton 1979.

Aristotle. *Analytica posteriora.* Eds., trans. Tredennick and Forster 1960.

———— *Analytica priora.* Eds., trans. Cook and Tredennick 1962.

———— *De sophistis elenchis.* Eds., trans. Forster and Furley 1955.

———— *Metaphysica.* Ed., trans. Tredennick 1980.

———— *Physica.* Eds., trans. Wicksteed and Cornford 1970.

———— *Politica.* Ed., trans. Rackham 1977.

———— *Topica.* Eds., trans. Tredennick and Forster 1960.

Arrighi, Gino (ed.). 1964. Paolo Dell'Abaco, *Trattato d'arithmetica.* Pisa: Domus Galileana.

———— (ed.). 1970. Piero della Francesca, *Trattato d'abaco.* Dal codice ashburnhamiano 280 (359*–291*) della Biblioteca Medicea Laurenziana di Firenze. A cura e con introduzione di Gino Arrighi. (Testimonianze di storia della scienza, 6). Pisa: Domus Galileana.

———— (ed.). 1973. *Libro d'abaco.* Dal Codice 1754 (sec. XIV) della Biblioteca Statale di Lucca. Lucca: Cassa di Risparmio di Lucca.

*Arts libéraux et philosophie au Moyen Age. Actes du Quatrième Congrès International de Philosophie Médiévale, 27 août–2 septembre 1967.* Montréal: Institut d'Études Médiévales/Paris: Vrin, 1969.

Ascher, Marcia. 1988. "Graphs in Cultures: A Study in Ethnomathematics." *Historia Mathematica* 15:201–27.

Ascher, Marcia, and Robert Ascher. 1986. "Ethnomathematics." *History of Science* 24(2):125–44.

Augustine. *The City of God against the Pagans.* Ed., trans. McCracken 1966.

Aujac, Germaine (ed., trans.) 1975. Géminos, *Introduction aux phénomènes.* Paris: "Les Belles Lettres".

———— 1984. "Autolycos de Pitané, prédécesseur d'Euclide." *Cahiers du Séminaire d'Histoire des Mathématiques* 5:1–12.

———— 1984a. "Le langage formulaire dans la géométrie grecque." *Revue d'Histoire des Sciences* 37(2):97–109.

Baillet, J. 1892. *Le Papyrus mathématique d'Akhmîm.* (Mission Archéologique Française au Caire, Mémoires 9,1). Paris: Leroux.

Baines, John. 1988. "Literacy, Social Organization, and the Archaeological Record: The Case of Early Egypt." In Gledhill et al. (eds.) 1988: 192–214,

Barth, G. 1982. "Die Programmiersprache Ada." *Jahrbuch Überblicke Mathematik* 1982, 141–57.

Bauer, Josef. 1975. "Darlehensurkunden aus Girsu." *Journal of the Economic and Social History of the Orient* 18:189–217.

Baur, Ludwig (ed.). 1912. *Die philosophischen Werke des Robert Grosseteste, Bischofs von Lincoln.* (Beiträge zur Geschichte der Philosophie des Mittelalters. Texte und Untersuchungen, 9). Münster: Aschendorffsche Verlagsbuchhandlung.

Beale, Thomas W. 1978. "Bevelled Rim Bowls and Their Implication for Change and Economic Organization in the Later Fourth Millenium B.C." *Journal for Near Eastern Studies* 37:289–313.

Beale, Thomas W., and S. M. Carter. 1983. "On the Track of the Yahya Large Kuš: Evidence for Architectural Planning in the Period IVC Complex at Tepe Yahya." *Paléorient* 9(1):81–88.

Beati Alberti Magni. *Operum* tomus secundus. Lyon 1651.

Beaujouan, Guy. 1954. "L'enseignement de l'arithmétique élémentaire à l'Université de Paris aux XIIIᵉ et XIVᵉ siècles. De l'abaque à l'algorisme." Pp. 93–124 in *Homenaje a Millás-Vallicrosa,* vol. 1. Barcelona: Consejo Superior de Investigaciones Científicas.

———— 1957. *L'interdépendance entre la science scolastique et les sciences utilitaires.* (Les Conférences du Palais de la Découverte, série D Nº 46). Paris: Université de Paris.

———— 1966. "Science livresque et art nautique au XVᵉ siècle." In Mollat and Adam 1966: 61–89.

Becker, Oskar. 1936. "Die Lehre vom Geraden und Ungeraden im Neunten Buch der Euklidischen Elemente." *Quellen und Studien zur Geschichte der Mathematik, Astronomie und Physik.* Abteilung B: *Studien* 3 (1934–36):533–53.

Benjamin, Francis S. 1954. "John of Gmunden and Campanus of Novara." *Osiris* 11:221–46.

Benjamin, Francis S., and G. J. Toomer (eds., trans.). 1971. *Campanus of Novara and Medieval Planetary Theory. Theorica planetarum.* Madison: University of Wisconsin Press.

Berger, M. S. 1977. *Nonlinearity and Functional Analysis. Lectures on Nonlinear Problems in Mathematical Analysis.* New York: Academic Press.

Bergsträsser, G. 1923. "Zu den magischen Quadraten." *Islam* 13:227–35.

Bernal, John Desmond. 1939. *The Social Function of Science.* London: Routledge & Kegan Paul.

Bernhardt, H. 1980. "Zur Institutionalisierung der angewandten Mathematik an der Berliner Universität 1920–1933." *NTM—Schriftenreihe für Geschichte der Naturwissenschaften, Technik und Medizin* 17(1):23–31.

Besthorn, R. O., and J. L. Heiberg (eds.). 1893. *Codex Leidensis 399, 1. Euclidis Elementa ex interpretatione al-Hadschdschadschii cum commentariis al-Narizii.* 3 vols. Copenhagen: Gyldendalske Boghandel, 1893–1932.

Biagioli, Mario. 1989. "The Social Status of Italian Mathematicians, 1450–1600." *History of Science* 27:41–95.

Birkenmajer, A. 1930. "Le rôle joué par les médecins et les naturalistes dans la réception d'Aristote au XII-e et XIII-e siècles." Pp. 1–15 in *La Pologne au VIᵉ Congrès International des Sciences Historiques, Oslo 1928.* Warszawa: Société Polonaise d'historie.

———— 1970. *Études d'histoire des sciences et de la philosophie du Moyen Age.* (Studia Copernicana, 1). Wroclaw: Zaklad Narodowy Imienia Ossolinskich.

Blasucci, Luigi (ed.). 1965. Dante Alighieri, *Tutte le opere.* Florence: Sansoni.

Bloor, David, 1976. *Knowledge and Social Imagery.* London: Routledge & Kegan Paul.

Boas, Marie. 1970. *The Scientific Renaissance 1450–1630.* London: Collins.

Boer, Charles (trans.). 1980. Marsilio Ficino, *The Book of Life.* A Translation of *Liber de vita* (or *De vita triplici*). Irving, TX: Spring Publications.

Boncompagni, Baldassare. 1851. "Della vita e delle opere di Leonardo Pisano matematico del secolo decimoterzo." *Atti dell' Accademia pontificia de' Nuovi Lincei 5* (1851–52):5–9, 208–46.

———— 1851a. "Della vita e delle opere di Gherardo cremonese, traduttore des secolo duodecimo, e di Gherardo da Sabbionetta astronomo del secolo decimoterzo." *Atti dell' Accademia pontificia de' Nuovi Lincei 4* (1850–51):387–493.

———— (ed.). 1857. *Scritti* di Leonardo Pisano matematico del secolo decimoterzo. I. Il *Liber abaci* di Leonardo Pisano. Rome: Tipografia delle Scienze Matematiche e Fisiche.

———— 1881. "Intorno ad uno scritto inedito di Adelardo di Bath intitolato 'Regule abaci.' " [followed by Adelard] "Regule Abaci." *Bulletino de Bibliografia e di Storia delle Scienze matematiche e fisiche* 14:1–90, 91–134.

Booß, Bernhelm, and Jens Høyrup. 1979. "From Kepler's Planetary Motion to the 'Quark Confinement' in Elementary Particle Physics. Empirical Investigations in the Problem of Model Transfer and Analogy. Fable with a Moral." Presented to the meeting on "Mathematical Economics", Oberwolfach, January 1979. *Materialien des Universitätsschwerpunkt Mathematisierung der Einzelwissenschaften, Universität Bielefeld* 14:211–42.

Booß, Bernhelm, and Mogens Niss (eds.). 1979. *Mathematics and the Real World.* Proceedings of an International Workshop, Roskilde University Centre (Denmark), 1978. (Interdisciplinary Systems Research, 68). Basel: Birkhäuser Verlag.

Booß, Bernhelm, and Rasmus Ole Rasmussen. 1979. "Challenging Problems in Science—Technology for a Better Use of the Earth's Resources." Paper contributed to the Symposium "Problems of the Conversion from War to Peace Production" at the International Institute for Peace in Vienna, March 30–April 1, 1979. *Roskilde Universitetscenter, Institute for Geography, Socio-Economic Analysis and Computer Science, Working Paper* 6/1979.

Boyaval, B. 1971. "Le P. Ifao 88: Problèmes de conversion monétaire." *Zeitschrift für Papyrologie und Epigraphik* 7:165–68, Tafel VI.

Boyer, Carl B. 1968. *A History of Mathematics.* New York: Wiley.

Bradwardine, Thomas. 1530. *Geometria speculativa;* & Johann Vögelin, *Elementale geometricum ex Euclidis geometria [ . . . ] decerptum.* Paris: Réginald Chauldière.

Brentjes, Burchard. 1966. "Einige Quellen zur Geschichte der Klassenkämpfe im Alten Orient." *Klio* 46 (Berlin):27–44.

Brentjes, Sonja, and Manfred Müller. 1982. "Eine neue Interpretation der ersten Aufgabe des altbabylonischen Textes AO 6770." *NTM. Schriftenreihe für Geschichte der Naturwissenschaften, Technik und Medizin* 19(2):21–26.

Brewer, J. S. (ed.). 1859. Fr. Rogeri Bacon *Opera* quaedam hactenus inedita. Vol. I containing *Opus tertium. Opus minus. Compendium philosophiae.* (Rerum Britannicorum Medii Aevi Scriptores). London: Longman, Green, Longman, and Roberts.

Brown, Joseph Edward. 1967. "The *Scientia de ponderibus* in the Later Middle Ages." Dissertation, The University of Wisconsin.

———. 1967a. "The *Scientia de ponderibus* in the Later Middle Ages." *Dissertation Abstracts International* A 28 (1967–68):3097.

Brown, Martin (ed.). 1971. *The Social Responsibility of the Scientist.* New York: The Free Press/London: Collier-Macmillan.

Brown, Peter. 1971. *The World of Late Antiquity.* London: Thames and Hudson.

Brunner, Hellmut. 1957. *Altägyptische Erziehung.* Wiesbaden: Otto Harrassowitz.

Bubnov, Nicolaus (ed.). 1899. Gerberti postea Silvestri II papae *Opera mathematica* (972–1003). Berlin: Friedländer.

Bukharin, N. I. 1931. "Theory and Practice from the Standpoint of Dialectical Materialism." In *Science at the Cross Roads.* Pp. 11–33.

Bulliet, Richard W. 1979. *Conversion to Islam in the Medieval Period: An Essay in Quantitative History.* Cambridge, MA: Harvard University Press.

Bulmer-Thomas, Ivor. "Hippias of Elis." *DSB* VI, 405–10.

———. "Oenopides of Chios." *DSB* X, 179–82.

Burhop, E. H. S. 1980. "Die Kernenergie und ihre Perspektive." *Wissenschaftliche Welt* 24(1):2–3.

Burkert, Walter. 1962. *Weisheit und Wissenschaft. Studien zu Pythagoras, Philolaos und Platon.* (Erlanger Beiträge zur Sprach- und Kunstwissenschaft, 10). Nürnberg: Hans Carl.

Bury, R. G. (ed., trans.). 1967. Plato, *Laws.* 2 vols. (Loeb Classical Library). London: Heinemann/Cambridge, MA: Harvard University Press. First ed. 1926.

Busard, H. L. L. (ed.). 1961. Nicole Oresme, *Quaestiones super Geometriam Euclidis.* (Janus, suppléments, 3). Leiden: Brill.

———. 1968. "L'algèbre au moyen âge: Le 'Liber mensurationum' d'Abû Bekr." *Journal des Savants,* Avril–Juin 1968, 65–125.

———. 1972. "The Translation of the *Elements* of Euclid from the Arabic into Latin by Hermann of Carinthia (?), Books VII, VIII and IX." *Janus* 59:125–87.

———. 1974. "The Second Part of Chapter 5 of the *De arte mensurandi* by Johannes de Muris." In Cohen et al. 1974: 147–64.

————. (ed.). 1991. Jordanus de Nemore, *De elementis arithmetice artis. A Medieval Treatise on Number Theory.* Part I: *Text and Paraphrase.* Part II: *Conspectus siglorum and Critical Apparatus.* (Boethius, 22,1–2). Stuttgart: Franz Steiner, 1991.

Caillois, Roland, Madeleine Francès and Robert Misrahi (eds., trans.). 1954. Spinoza, *Oeuvres complètes.* (Bibliothèque de la Pléiade, 108). Paris: Gallimard.

Cammann, Schuyler. 1969. "Islamic and Indian Magic Squares." *History of Religions* 8 (1968–69):181–209, 271–99.

Campbell, Donald T. 1981. "A Tribal Model of the Social System Vehicle Carrying Scientific Knowledge." *Knowledge: Creation, Diffusion, Utilization* 1:181–201.

Cantor, Moritz. 1875. *Die römischen Agrimensoren und ihre Stellung in der Geschichte der Feldmesskunst. Eine historisch-mathematische Untersuchung.* Leipzig: Teubner.

————. 1900. *Vorlesungen über Geschichte der Mathematik.* Zweiter Band, *von 1200–1668.* Zweite Auflage. Leipzig: Teubner.

————. 1907. *Vorlesungen über Geschichte der Mathematik.* Erster Band, *von den ältesten Zeiten bis zum Jahre 1200 n. Chr.* Dritte Auflage. Leipzig: Teubner.

————. 1908. *Vorlesungen über Geschichte der Mathematik.* Vierter Band. *Von 1759 bis 1799.* Leipzig: Teubner.

Cardano, Girolamo. 1550. *De subtilitate libri XXI.* Nürnberg: Ioh. Petreius.

————. 1663. Hieronymo Cardani Mediolanensis Philosophi ac Medici celeberrimi *Operum* tomus quartus; quo continentur *Arithmetica, Geometrica, Musica.* Lyon: Jean Antoine Huguetan & Marc Antoine Raguad.

————. 1663a. Hieronymo Cardani Mediolanensis Philosophi ac Medici celeberrimi *Operum* tomus tertius; quo continentur *Physica.* Lyon: Jean Antoine Huguetan & Marc Antoine Ragaud.

Carneiro, Robert L. 1970. "A Theory of the Origin of the State," *Science* 169: 733–38.

————. 1981. "The Chiefdom: Precursor of the State." In G. D. Jones and R. R. Kautz 1981: 37–79.

Cassirer, Ernst, Paul Oskar Kristeller and John Herman Randall, Jr. (eds.). 1956. *The Renaissance Philosophy of Man.* Selections in Translation. Chicago: University of Chicago Press.

Castoriadis, Cornelius. 1986. *Domaines de l'homme.* (*Les carrefours du labyrinthe* II). Paris: Seuil.

Chace, Arnhold Buffum. Ludlow Bull and Henry Parker Manning. 1929. *The Rhind Mathematical Papyrus.* II. Photographs, Transcription, Transliteration, Literal Translation. *Bibliography of Egyptian and Babylonian Mathematics* (Supplement), by R. C. Archibald. *The Mathematical Leather Roll in the British Museum,* by S. R. K. Glanville. Oberlin, OH: Mathematical Association of America.

Channell, David F. 1982. "The Harmony of Theory and Practice: The Engineering Science of W. J. M. Rankine." *Technology and Culture* 23:39–52.

Charbonnier, M. P. 1927. "Essais sur l'histoire de la balistique." *Mémorial de l'Artillerie Francqise* 6:955–1251.

Chenu, Marie-Dominique, O. P. 1966. *La théologie au douzième siècle.* Second edition. Paris: Vrin.

――――. 1974. "Civilisation urbaine et théologie. L'école de Saint-Victor au XII siècle." *Annales. Économies, sociétés, civilisations* 29:1253–63.

――――. 1977. "Praxis historique et relation Église-société." *La Pensée,* N° 192, 5–11.

Chevalier, Jacques (ed.). 1954. Pascal, *Oeuvres complètes.* (Bibliothèque de la Pléiade, 34). Paris: Gallimard.

Childe, V. Gordon. 1950. "The Urban Revolution." *Town Planning Review* 21:3–17.

Cicero. *Tusculanae disputationes.* Ed., trans. King 1971.

Claessen, Henri J. M., and Peter Skalník (eds.). 1978. *The Early State.* (New Babylon. Studies in the Social Sciences, 32). Hague: Mouton.

Clagett, Marshall. "Adelard of Bath." *DSB* I, 61–64.

――――. 1962. "The Use of Points in Medieval Natural Philosophy and Most Particularly in the *Questiones de spera* of Nicole Oresme." Pp. 215–21 in *Actes du Symposium International R. J. Boškovič 1961.* Beograd. Reprinted in Clagett 1979.

――――. 1964. "Archimedes and Scholastic Geometry." Pp. 40–60 in *L'aventure de la science. Mélanges Alexandre Koyré,* vol. 1. Paris. Reprinted in Clagett 1979.

――――. 1968. *Nicole Oresme and the Medieval Geometry of Qualities and Motions.* A Treatise on the Uniformity and Difformity of Intensities Known as *Tractatus de configurationibus qualitatum et motuum.* Madison: University of Wisconsin Press.

――――. 1968a. "Some Novel Trends in the Science of the Fourteenth Century." In Singleton 1968: 275–303. Reprinted in Clagett 1979.

————. 1976. *Archimedes in the Middle Ages.* Vol. 2, *The Translations from the Greek by William of Moerbeke.* (Memoirs of the American Philosopical Society, 117 A + B). Philadelphia: The American Philosophical Society.

————. 1978. *Archimedes in the Middle Ages.* Vol. 3, *The Fate of the Medieval Archimedes 1300–1565.* (Memoirs of the American Philosophical Society, 125 A + B + C). Philadelphia: The American Philosophical Society.

————. 1979. *Studies in Medieval Physics and Mathematics.* London: Variorum Reprints.

————. 1984. *Archimedes in the Middle Ages.* Vol. 5. *Quasi-Archimedean Geometry in the Thirteenth Century.* (Memoirs of the American Philosophical Society, 157 A + B). Philadelphia: The American Philosophical Society.

Clough, Cecil H. (ed.). 1976. *Cultural Aspects of the Italian Renaissance.* Essays in Honour of Paul Oskar Kristeller. Manchester: Manchester University Press/New York: Zambelli.

Cock, A. G. 1983. "Chauvinism and Internationalism in Science: The International Research Council, 1919–1926." *Notes and Records of the Royal Society of London* 37 (1982–83):249–88.

Cohen, H. J. 1970. "The Economic Background and the Secular Occupations of Muslim Jurisprudents and Traditionists in the Classical Period of Islam (until the Middle of the Eleventh Century)." *Journal of the Economic and Social History of the Orient* 13:16–59.

Cohen, R. S., J. J. Stachel and M. W. Wartofsky (eds.). 1974. *For Dirk Struik. Historical and Political Essays* in Honor of Dirk Struik. (Boston Studies in the Philosophy of Science, 15). Dordrecht: Reidel.

Colebrooke, H. T. (ed., trans.). 1817. *Algebra, with Arithmetic and Mensuration from the Sanscrit of Brahmagupta and Bhascara.* London: John Murray. Reprint Wiesbaden: Martin Sändig, 1973.

Conway, J. H. 1980. "Monsters and Moonshine." *The Mathematical Intelligencer* 2(4):165–71.

Cook, Harold P., and Hugh Tredennick (eds., trans.). 1962. Aristotle, *The Categories. On Interpretation. Prior Analytics.* (Loeb Classical Library). London: Heinemann/Cambridge, MA: Harvard University Press. 1st ed. 1938.

Copleston, Frederick. 1963. *A History of Philosophy.* Vol. 3, *Late Mediaeval and Renaissance Philosophy.* Garden City, NY: Doubleday.

Crew, Henry, and Alfonso de Salvio (trans.). 1914. Galileo Galilei, *Dialogues Concerning Two New Sciences.* With an Introduction by Antonio Favaro, New York: Macmillan.

Crombie, Alistair C. "Descartes." *DSB* IV, 51–55.

——. "Grosseteste, Robert." *DSB* V, 548–54.

——. (ed.). 1963. *Scientific Change. Historical Studies in the Intellectual, Social and Technical Conditions for Scientific Discovery and Technical Invention, from Antiquity to the Present.* Symposium on the History of Science, University of Oxford 9–15 July 1961. London: Heinemann.

——. 1977. "Mathematics and Platonism in the Sixteenth-Century Italian Universities and in Jesuit Educational Policy." In Maeyama and Saltzer 1977:63–94.

Crosby, H. Lamar (ed., trans.). 1955. *Thomas of Bradwardine His* Tractatus de Proportionibus. *Its Significance for the Development of Mathematical Physics.* Madison: University of Wisconsin Press.

Curtze, Maximilian (ed.). 1868. Der *Algorismus Proportionum* des Nicolaus Oresme zum ersten Male nach der Lesart der Handschrift R. 4°.2. der königlichen Gymnasial-Bibliothek zu Thorn herausgegeben. Berlin: Calvary & Co.

—— (ed.). 1902. *Urkunden zur Geschichte der Mathematik im Mittelalter und der Renaissance.* (Abhandlungen zur Geschichte der mathematischen Wissenschaften, 12–13). Leipzig: Teubner.

Czwalina, Arthur (ed., trans.). 1931. Autolykos, *Rotierende Kugel* und *Aufgang und Untergang der Gestirne.* Theodosios von Tripolis, *Sphaerik.* (Ostwalds Klassiker der eksakten Naturwissenschaften, 232). Leipzig: Akademische Verlagsgesellschaft.

Dadić, Žarko (ed.). 1968. Marini Ghetaldi *Opera omnia.* Zagreb: Institut za Povijest Prirodnih, Matematičkih i Medicinskih Nauka.

Damerow, Peter, and Robert K. Englund. 1987. "Die Zahlzeichensysteme der Archaischen Texte aus Uruk." Kapitel 3 (pp. 117–66) in Green and Nissen, *Zeichenliste der Archaischen Texte aus Uruk,* Band II (ATU 2). Berlin: Gebr. Mann.

——. 1989. *The Proto-Elamite Texts from Tepe Yahya.* (The American School of Prehistoric Research, Bulletin 39). Cambridge, MA.: Peabody Museum of Archaeology and Ethnology/Harvard University Press.

Damerow, Peter, Robert K. Englund and Hans J. Nissen. 1988. "Die Entstehung der Schrift." *Spektrum der Wissenschaften,* Februar 1988, 74–85.

——. 1988a. "Die ersten Zahldarstellungen und die Entwicklung des Zahlbegriffs." *Spektrum der Wissenschaften,* März 1988, 46–55.

Damerow, Peter, and Wolfgang Lefèvre (eds.). 1981. *Rechenstein, Experiment, Sprache. Historische Fallstudien zur Entstehung der exakten Wissenschaften.* Stuttgart: Klett-Cotta.

Dandamajev, Mohammed A. 1971. "Die Rolle des *tamkārum* in Babylonien im 2. und 1. Jahrtausend v.u.Z." In Klengel (ed.) 1971: 69–78.

Datta, Bibhutibhusan, and Avadhesh Narayan Singh. 1962. *History of Hindu Mathematics. A Source Book.* Parts I and II. Bombay: Asia Publishing House. First ed. 1935–38.

Dauben, Joseph W. 1980. "Mathematicians and World War I: The International Diplomacy of G. H. Hardy and Gösta Mittag-Leffler as Reflected in Their Personal Correspondence." *Historia Mathematica* 7:261–88.

Davis, Chandler. 1974. "Materialist Mathematics." In Cohen et al. (eds.) 1974: 37–66.

Davis, Charles T. 1965. "Education in Dante's Florence." *Speculum* 40:415–35.

Davis, Margaret Daly. 1977. *Piero della Francesca's Mathematical Treatises. The "Trattato d'abaco" and "Libellus de quinque corporibus regularibus."* Ravenna: Longo Editore.

Debus, Allen G. (ed.). 1972. *Science, Medicine and Society in the Renaissance.* 2 vols. London: Heinemann.

———— (ed.). 1975. John Dee, *The Mathematicall Praeface to the Elements of Geometrie of Euclid of Megara* (1570), with an Introduction. New York: Science History Publications.

Defoe, Daniel. 1927. *The Life and Strange Surprising Adventures of Robinson Crusoe of York, Mariner.* 3 vols. (The Shakespeare Head Edition of the Novels and Selected Writings of Daniel Defoe). Oxford: Blackwell.

Deimel, Anton. 1923. *Die Inschriften von Fara. II, Schultexte aus Fara.* In Umschrift herausgegeben und bearbeitet. (Wissenschaftliche Veröffentlichungen der Deutschen Orient-Gesellschaft, 43). Leipzig: J. C. Hinrichs'sche Buchhandlung.

Delisle, Léopold. 1868. *Le Cabinet des manuscrits de la Bibliothèque nationale.* 3 vols. (Histoire générale de Paris. Collection de documents). Paris: Imprimérie impériale/nationale, 1868, 1874, 1881.

Denifle, Heinrich. 1885. "Die Constitutionen des Prediger-Ordens vom Jahre 1228." *Archiv für Literatur- und Kirchengeschichte des Mittelalters* 1:165–227.

Denifle, H., and É. Chatelain (eds.). 1889. *Chartularium Universitatis Parisiensis.* 4 vols. Paris: Frères Delalain, 1889–97.

Diakanoff, Igor M. (ed.). 1969. *Ancient Mesopotamia, Socio-Economic History. A Collection of Studies by Soviet Scholars.* Moscow: "Nauka", Central Department of Oriental Literature.

————. 1969a. "The Rise of the Despotic State in Ancient Mesopotamia." In Diakonoff (ed.) 1969: 173–202.

————. 1971. ''On the Structure of Old Babylonian Society.'' In Klengel (ed.) 1971: 15–31.

————. 1975. ''The Rural Community in the Ancient Near East.'' *Journal of the Economic and Social History of the Orient* 18:121–33.

Diels, Hermann. 1951. *Die Fragmente der Vorsokratiker, Griechisch und Deutsch.* Herausgegeben von Walther Kranz. 3 vols. 6. Auflage. Berlin: Weidmann, 1951–52.

Djebbar, A. 1981. *Enseignement et recherche mathématiques dans le Maghreb des XIII<sup>e</sup>–XIV<sup>e</sup> siècles (étude partielle).* (Publications mathématiques d'Orsay, 81-02). Orsay: Université de Paris-Sud.

Dodge, Bayard (ed., trans.). 1970. The *Fihrist* of al-Nadīm. *A Tenth-Century Survey of Muslim Culture.* 2 vols. (Records of Civilization. Sources and Studies, N° 43). New York: Columbia University Press.

Drake, Stillman (ed., trans.). 1957. *Discoveries and Opinions of Galileo.* New York: Doubleday.

Drake, Stillman, and I. E. Drabkin (eds., trans.). 1969. *Mechanics in Sixteenth-Century Italy. Selections from Tartaglia, Benedetti, Guido Ubaldo, & Galileo.* Madison: University of Wisconsin Press.

*DSB: Dictionary of Scientific Biography.* Charles Coulston Gillispie, Editor-in-Chief. 16 vols. New York: Charles Scribner's Sons, 1970–80.

Dupuis, J. (ed., trans.). 1892. Théon de Smyrne, philosophe platonicien, *Exposition des connaissances mathématiques utiles pour la lecture de Platon.* Paris: Hachette.

Easton, Joy B. ''Dee.'' *DSB IV*, 5–6.

Eco, Umberto. 1979. *Lector in Fabula. La cooperazione interpretativa nei testi narrativi.* Milano: Bompiani.

Edzard, Dietz Otto. 1969. ''Eine altsumerische Rechentafel (OIP 14, 70).'' In Röllig (ed.) 1969: 101–4.

————. 1974. '' 'Soziale Reformen' im Zweistromland bis nach 1600 v. Chr.: Realität oder literarischer Topos?'' *Acta Antiqua Academiae Scientiarum Hungaricae* 22:145–56.

————. 1980. ''Keilschrift.'' *Reallexikon der Assyriologie und Vorderasiatischen Archäologie* V, 544–68. Berlin: de Gruyter.

Endesfelder, Erika. 1988. ''Zur Herausbildung von Klassen und Staat im alten Ägypten.'' In Herrmann and Köhn (eds.) 1988: 372–77.

Eneström, Georg. 1906. "Über die 'Demonstratio Jordani de algorismo.' " *Bibliotheca Mathematica*, 3. Folge 7 (1906–7):24–37.

———. 1907. "Über eine dem Nemorarius zugeschriebene kurze Algorismusschrift." *Bibliotheca Mathematica*, 3. Folge 8 (1907–8):135–53.

———. 1913. "Das Bruchrechnen des Nemorarius." *Bibliotheca Mathematica*, 3. Folge 14 (1913–14):41–54.

Englund, Robert K. 1988. "Administrative Timekeeping in Ancient Mesopotamia." *Journal of the Economic and Social History of the Orient* 31:121–85.

———. 1990. *Organisation und Verwaltung der Ur III-Fischerei*. (Berliner Beiträge zum Vorderen Orient, 10). Berlin: Dietrich Reimer.

*Études de civilisation médiévale (IX<sup>e</sup>–XII<sup>e</sup> siècles)*. Mélanges offerts à Edmond-René Labande. Poitiers; C. E. S. C. M., 1974.

Euclid. *Elements*. Ed., trans. J. L. Heiberg 1883.

*Euclidis Megarensis mathematici clarissimi Elementorum libri XV. Cum expositione Theonis in Priores XIII à Bartholomeo Veneto Latinitate donata, Campani in omnes & Hypsiclis Alexandrini in duos postremos. His adiecta sunt Phaenomena, Catoptrica & Optica, deinde Protheoria Marini et Data. Postremum verò, Opusculum de Levi & Ponderoso, hactenus non visum, eiusdem autoris*. Basel: Johannes Hervagius, 1546.

Evans, Gillian R. 1977. "From Abacus to Algorism: Theory and Practice in Medieval Arithmetic." *British Journal for the History of Science* 10:114–31.

———. 1978. "Introductions to Boethius's 'Arithmetica' of the Tenth to the Fourteenth Century." *History of Science* 16:22–41.

Fakhry, Majid. 1969. "The Liberal Arts in the Medieval Arabic Tradition from the Seventh to the Twelfth Centuries." In *Arts Libéraux et philosophie . . .*, pp. 91–97.

Falkenstein, Adam. 1953. "Die babylonische Schule." *Saeculum* 4:125–37.

Fanfani, Amintore. 1951. "La préparation intellectuelle à l'activité économique, en Italie, du XIV<sup>e</sup> au XVI<sup>e</sup> siècle." *Le Moyen Age* 57:327–46.

Farmer, Henry George. 1931. *The Organ of the Ancients from Eastern Sources (Hebrew, Syriac and Arabic)*. London: Reeves.

Farrington, Benjamin. 1938. "Prometheus Bound: Government and Science in Classical Antiquity." *Science and Society* 2 (1937–38):435–47.

———. 1965. *Science and Politics in the Ancient World*. Second edition. London: Allen & Unwinn. First ed. 1939.

————. 1969. *Greek Science*. Harmondsworth: Penguin. First ed. 1944–49.

Favaro, Antonio (ed.). 1890. *Le Opere* di Galileo Galilei. Edizione nazionale. 20 vols. Florence: G. Barbéra, 1890–1909.

Feingold, Mordechai. 1984. "The Occult Tradition in the English Universities of the Renaissance: A Reassessment." In Vickers 1984: 73–94.

Field, Judith V. 1984. "A Lutheran Astrologer: Johannes Kepler." [Includes a translation of Kepler's *De fundamentis atrologiae certioribus*]. Archive for History of Exact Sciences 31:189–272.

Fine, Oronce. 1532. *Protomathesis*. Paris.

Finkbeiner, U., and W. Röllig (eds.). 1986. Ǧamdat Nasr—Period or Regional Style? (Beihefte zum Tübinger Atlas des Vorderen Orients, Reihe B [Geisteswissenschaften], 62). Wiesbaden: Ludwig Reichert.

Finkelstein, J. J. 1969. "The Laws of Ur-Nammu." *Journal of Cuneiform Studies* 22:66–82.

Folkerts, Menso. 1970. *"Boethius" Geometrie II. Ein mathematisches Lehrbuch des Mittelalters*. Wiesbaden: Franz Steiner.

————. 1971. "Mathematische Aufgabensammlungen aus dem ausgehenden Mittelalter. Ein Beitrag zur Klostermathematik des 14. und 15. Jahrhunderts." *Sudhoffs Archiv* 55:58–71.

————. 1978. "Die älteste mathematische Aufgabensammlung in lateinischer Sprache: Die Alkuin zugeschriebenen *Propositiones ad acuendos iuvenes." Österreichische Akademie der Wissenschaften, Mathematisch-Naturwissenschaftliche Klasse. Denkschriften*, 116(6) (Vienna).

————. 1980. "Probleme der Euklidinterpretation und ihre Bedeutung für die Entwicklung der Mathematik." *Centaurus* 23:185–215.

Forman, Paul. 1971. "Weimar Culture, Causality, and Quantum Theory, 1918–1927: Adaptation by German Physicists and Mathematicians to a Hostile Intellectual Environment." *Historical Studies in the Physical Sciences* 3:1–115.

Forster, E. S., and D. J. Furley (eds., trans.). 1955. Aristotle, *On Sophistical Refutations; On Coming-to-Be and Passing-Away,* and *On the Cosmos*. (Loeb Classical Library). Cambridge, MA: Harvard University Press/London: Heinemann.

Foster, Benjamin. 1981. "A New Look at the Sumerian Temple State." *Journal of the Economic and Social History of the Orient* 24:225–81.

————. 1982. "Archives and Record-Keeping in Sargonic Mesopotamia." *Zeitschrift für Assyriologie und Vorderasiatische Archäologie* 72:1–27.

Fowler, Harold North (ed., trans.). 1977. Plato, *Theaetetus, Sophist.* (Loeb Classical Library). Cambridge, MA: Harvard University Press/London: Heinemann. First ed. 1921.

———— (ed., trans.). 1977a. Plato, *Euthyphro, Apology, Crito, Phaedo, Phaedrus.* (Loeb Classical Library). Cambridge, MA: Harvard University Press/London: Heinemann. First ed. 1914.

Fowler, Harold North, and W. R. M. Lamb (eds., trans.). 1975. Plato, *The Statesman, Philebus, Ion.* (Loeb Classical Library). Cambridge, MA: Harvard University Press/London: Heinemann. First ed. 1925.

Fox, Robert (ed.). 1978. Sadi Carnot, *Réflexions sur la puissance motrice du feu.* (Collection des Travaux de l'Académie Internationale d'Histoire des Sciences, n⁰ 26). Paris: Vrin.

Freeman, Kathleen. 1953. *The Pre-Socratic Philosophers. A Companion to Diels, "Fragmente der Vorsokratiker."* Third edition. Oxford: Blackwell.

French, Peter. 1972. *John Dee: The World of an Elisabethan Magus.* London: Routledge & Kegan Paul.

Friberg, Jöran. 1978. "The Third Millennium Roots of Babylonian Mathematics. I. A Method for the Decipherment, through Mathematical and Metrological Analysis, of Proto-Sumerian and proto-Elamite Semi-Pictographic Inscriptions." *Department of Mathematics, Chalmers University of Technology and the University of Göteborg* No. 1978-9.

————. 1979. "The Early Roots of Babylonian Mathematics. II: Metrological Relations in a Group of Semi-Pictographic Tablets of the Jemdet Nasr Type, Probably from Uruk-Warka." *Department of Mathematics, Chalmers University of Technology and the University of Göteborg* No. 1979-15.

————. 1986. "The Early Roots of Babylonian Mathematics. III: Three Remarkable Texts from Ancient Ebla." *Vicino Oriente* 6:3–25.

————. 1990. "Mathematik." *Reallexikon der Assyriologie und Vorderasiatischen Archäologie* VII, 531–85. Berlin: De Gruyter.

Fried, Morton. 1967. *The Evolution of Political Society. An Essay in Political Anthropology.* New York: Random House.

Friedlein, Gottfried (ed.). 1867. Anicii Manlii Torquati Severini Boetii *De institutione arithmetica* libri duo. *De Institutione musica* libri quinque. Accedit *Geometria* quae fertur Boetii. Leipzig: Teubner.

———— (ed.). 1873. Procli Diadochi *In primum Euclidis Elementorum librum commentarii.* Leipzig: Teubner.

Friedländer, M. (ed., trans.). 1904. Moses Maimonides, *The Guide for the Perplexed.* Second, Revised edition. London: Routledge & Kegal Paul. Reprint New York: Dover, 1956.

Gadd, C. J. 1956. *Teachers and Students in the Oldest Schools.* An Inaugural Lecture Delivered on 6 March 1956. London: School of Oriental and African Studies, University of London.

*GAL:* Carl Brockelmann. *Geschichte der arabischen Literatur.* Band 1–2, Supplementband 1–3. Berlin: Emil Fischer, 1898, 1902; Leiden: Brill, 1937, 1938, 1942.

Gandz, Solomon (ed., trans.). 1932. *The Mishnat ha-Middot, the First Hebrew Geometry of about 150 C.E., and the Geometry of Muchammad ibn Musa al-Khowarizmi, the First Arabic Geometry ⟨c. 820⟩, Representing the Arabic Version of the Mishnat ha Middot.* (Quellen und Studien zur Geschichte der Mathematik, Astronomie und Physik. Abteilung A: *Quellen* 2). Berlin: Springer.

Gardiner, Alan H. 1911. *Egyptian Hieratic Texts.* Series I: *Literary Texts from the New Kingdom.* Part I: *The Papyrus Anastasi I* and the *Papyrus Koller,* together with Parallel Texts. Leipzig: J. C. Hinrichs'sche Buchhandlung.

Garin, Eugenio. 1965. *Italian Humanism. Philosophy and Civic Life in the Renaissance.* Oxford: Blackwell/New York: Harper & Row.

*GAS:* Fuat Sezgin. *Geschichte des arabischen Schrifttums.* 9 vols. Leiden: Brill, 1967–84.

Gelb, Ignace J. 1965. "The Ancient Mesopotamian Ration System." *Journal of Near Eastern Studies* 24:230–43.

Geldsetzer, Lutz (ed.). 1973. Gregor Reisch, *Margarita philosophica.* Reprint of Basel edition, 1517. Düsseldorf.

Gerjuoy, Edward. 1982. "Embargo on Ideas: The Reagan Isolationism." *Bulletin of the Atomic Scientists* 38(9):31–37.

Gille, Bertrand. 1980. *Les mécániciens grecs. La naissance de la technologie.* Paris: Seuil.

Ginzburg, Carlo. 1980. *The Cheese and the Worms. The Cosmos of a Sixteenth-Century Miller.* London: Routledge & Kegan Paul.

Gledhill, John, Barbara Bender and Mogens Trolle Larsen (eds.). 1988. *State and Society. The Emergence and Development of Social Hierarchy and Political Centralization.* (One World Archaeology, 4). London: Unwin Hyman.

Godley, A. D. (ed., trans.). 1975. Herodotus. (Loeb Classical Library). Cambridge, MA: Harvard University Press/London: Heinemann. First ed. 1920–25.

Goldstine, Herman H. 1972. *The Computer from Pascal to von Neumann*. Princeton, NJ: Princeton University Press.

Goldthwaite, Richard A. 1972. "Schools and Teachers of Commercial Arithmetic in Renaissance Florence." *Journal of European Economic History* 1:418–33.

Goldziher, Ignácz. 1915. "Die Stellung der alten islamischen Orthodoxie zu den antiken Wissenschaften." *Abhandlungen der Preußischen Akademie der Wissenschaften. Philosophisch-historische Klasse* 1915 Nr. 8 (Berlin, 1916).

Grabiner, Judith V. 1974. "Is Mathematical Truth Time-Dependent?" *American Mathematical Monthly* 81:354–65.

Grabmann, Martin. 1930. "Mitteilungen aus Münchner Handschriften über bisher unbekannte Philosophen der Artistenfakultät (Codd.lat. 14246 und 14383)." Pp. 73–83 in *Festschrift für Georg Leidinger zum 60. Geburtstag am 30. December 1930*. Munich: Hugo Schmidt.

————. 1934. "Eine für Examinazwecke abgefaßte Quaestionensammlung der Pariser Artistenfakultät aus der ersten Hälfte des XIII. Jahrhunderts." *Revue néoscolastique de philosophie* 36:211–29.

————. 1941. *I divieti ecclesiastici di Aristotele sotto Innocenzo III e Gregorio IX* (Miscellanea Historiae Pontificiae, 5,1). Rome: Saler.

Granger, Frank (ed., trans.). 1970. Vitruvius, *De Architectura*. 2 vols. (Loeb Classical Library). London: Heinemann/Cambridge, MA: Harvard University Press. First ed. 1931–34.

Grant, Edward (ed., trans.). 1965. "Part I of Nicole Oresme's *Algorismus Proportionum*." *Isis* 56:327–41.

———— (ed., trans.). 1966. Nicole Oresme, *De proportionibus proportionum* and *Ad pauca respicientes*. Madison: University of Wisconsin Press.

———— (ed., trans.). 1971. *Nicole Oresme and the Kinematics of Circular Motion. Tractatus de commensurabilitate vel incommensurabilitate motuum celi*. Madison: University of Wisconsin Press.

Grayson, Cecil (ed.). 1973. Leon Battista Alberti, *Opere volgari*. Volume terzo. *Trattati d'arte, Ludi rerum mathematicarum, grammatica della lingua toscana, Opuscoli amatori, Lettere*. (Scrittori d'Italia, 254). Bari: Laterza.

Green, M. W. 1981. "The Construction and Implementation of the Cuneiform Writing System." *Visible Language* 15:345–72.

Greenberg, Daniel S. 1969. *The Politics of American Science*. Harmondsworth: Penguin. First published as *The Politics of Pure Science*. New American Library, 1967.

Gregory, Tullio. 1975. "La nouvelle idée de nature et de savoir scientifique au XIIe siècle." In Murdoch and Sylla (eds.) 1975: 193–218.

Grigor'jan, A. T., and A. P. Juškevič (eds.). 1966. *Fiziko-matematičeskie nauki v stranax vostoka*. Sbornik statej i publikacij. Vypusk I (IV). Moscow: "Nauka".

Gross, Horst-Eckart. 1978. "Das sich wandelnde Verhältnis von Mathematik und Produktion." In Plath and Sandkühler 1978: 226–69.

Größing, Helmuth. 1980. "Der Humanist Regiomontanus und sein Verhältnis zu Georg von Peuerbach." In Schmitz and Krafft 1980: 69–82.

Gundlach, Karl-Bernhard, and Wolfram von Soden. 1963. "Einige altbabylonische Texte zur Lösung 'quadratischer Gleichungen.' " *Abhandlungen aus dem mathematischen Seminar der Universität Hamburg* 26:248–63.

Günther, Siegmund. 1887. *Geschichte des mathematischen Unterrichts im deutschen Mittelalter*. (Monumenta Germaniae Paedagogica, 3). Berlin.

Hahnloser, H. R. (ed.). 1935. *Villard de Honnecourt. Kritische Gesamtausgabe des Bauhüttenbuches ms. fr 19093 der Pariser Nationalbibliothek*. Vienna: Anton Schroll.

Haldane, Elisabeth S. and G. R. T. Ross (eds., trans.). 1931. The *Philosophical Works* of Descartes Rendered into English. 2 vols. Second corrected edition. Cambridge: Cambridge University Press. First ed. 1911.

Hallo, William W. 1976. "Toward a History of Sumerian Literature." In *Sumerological Studies . . . ,* pp. 181–203.

Hardy, Godfrey Harold. 1967. *A Mathematician's Apology*. With a Foreword by C. P. Snow. Cambridge: Cambridge University Press. First ed. 1940.

Harvey, E. Ruth. "Qusṭā ibn Lūqā al-Baʿlabakkī." *DSB* XI, 244–46.

Haskins, Charles Homer. 1924. *Studies in the History of Mediaeval Science*. Cambridge, Mass: Harvard University Press.

———. 1929. *Studies in Mediaeval Culture*. Oxford: Clarendon Press.

Hassan, Ahmad Y. al-, Ghada Karmi, and Nizar Namnum (eds.). 1978. *Proceedings of the First International Symposium for the History of Arabic Science, April 5–12, 1976*. Vol. 2, Papers in European Languages. Aleppo: Institute for the History of Arabic Science, Aleppo University.

Hay, Cynthia (ed.). 1988. *Mathematics from Manuscript to Print, 1300–1600*. (Oxford Scientific Publications). New York: Oxford University Press.

Haydn, Hiram. 1950. *The Counter-Renaissance*. New York: Scribner.

Heath, Thomas L. 1921. *A History of Greek Mathematics*. 2 vols. Oxford: Clarendon Press.

Heiberg, J. L. (ed., trans.). 1883. Euclidis *Elementa*. 5 vols. (Euclidis Opera omnia, vol. I-V). Leipzig: Teubner, 1883–1888.

Heiberg, J. L. 1896. "Beiträge zur Geschichte Georg Valla's und seiner Bibliothek." *Centralblatt für Bibliothekswesen, Beihefte* 6 (1896–97):353–481 (= XVI. Beiheft).

Heinen, Anton M. 1978. "Mutakallimūn and Mathematicians." *Islam* 55:57–73.

Heinrich, Ernst. 1938. "Grabungen im Gebiet des Anu-Antum-Tempels." Pp. 19ff in Nöldeke et al., *Neunter vorläufiger Bericht über die von der deutschen Forschungsgemeinschaft in Uruk-Warka unternommenen Ausgrabungen*. Berlin: Verlag der Akademie der Wissenschaften.

———. 1982. *Die Tempel und Heiligtümer im Alten Mesopotamien*. 2 vols. (Denkmäler Antiker Architektur, 14). Berlin: De Gruyter.

Hermann, Armin. 1982. *Wie die Wissenschaft ihre Unschuld verlor. Macht und Mißbrauch der Forscher*. Stuttgart: Deutsche Verlagsanstalt.

Hermelink, Heinrich. 1958. "Die ältesten magischen Quadrate höherer Ordnung und ihre Bildung." *Sudhoffs Archiv* 42:199–217.

———. 1978. "Arabic Recreational Mathematics as a Mirror of Age-Old Cultural Relations Between Eastern and Western Civilizations." In Hassan et al. 1978: 44–52.

Herodotos. *Histories*. Ed., trans. Godley 1975.

Hero. *Dioptra*. Ed., trans. Schöne 1903.

———. *Metrika*. Ed., trans. Schöne 1903.

Herrmann, Joachim, and Jens Köhn (eds.). 1988. *Familie, Staat und Gesellschaftsformation. Grundprobleme vorkapitalistischer Epochen einhundert Jahre nach Friedrich Engels' Werk "Der Ursprung der Familie, des Privateigentums und des Staats."* Berlin: Akademie-Verlag.

Herrmann, Joachim, and Irmgard Sellnow (eds.). 1982. *Produktivkräfte und Gesellschaftsformationen in vorkapitalistischer Zeit*. (Veröffentlichungen des Zentralinstituts für Alte Geschichte und Archäologie, 12) Berlin: Akademie-Verlag.

Hesiodos. *Works and Days*. Ed., trans. Mazon 1979.

Hillard, Denise, and Emmanuel Poulle. 1971. "Bibliographie des travaux d'O. Fine." *Bibliothèque d'humanisme et renaissance* 33:335–51.

Hoare, Charles Antony Richard. 1981. "The Emperor's Old Clothes." *Communications of the ACM* 24:75–84.

Hochheim, Adolph (ed., trans.). 1878. *Kafî fîl Hisâb (Genügendes über Arithmetik)* des Abu Bekr Muhammed ben Alhusein Alkarkhi. 3 Hefte. Halle: Louis Nebert.

Hofmann, Joseph Ehrenfried (ed.). 1942. "Ramon Lulls Kreisquadratur. [Followed by] Raimundus Lullus, *De quadratura et triangulaturo circuli.*" (Die Quellen der Cusanischen Mathematik, 1; Cusanus-Studien, 7). *Sitzungsberichte der Heidelberger Akademie der Wissenschaften: Philosophisch-historische Klasse* 1941/42 Nr. 4.

——— (ed.). 1952. Nikolaus von Cues, *Die mathematischen Schriften.* Übersetzt von Joseph Hofmann, mit einer Einführung und Anmerkungen versehen. (Schriften von Nikolaus von Cues in deutscher Übersetzung, Heft 11). Hamburg: Felix Meiner.

——— (ed.). 1970. François Viète, *Opera mathematica* recognita à Francisci à Schooten. Hildesheim: Georg Olms Verlag.

Hogendijk, Jan P. 1985. "Thābit ibn Qurra and the Pair of Amicable Numbers 17296, 18416." *Historia Mathematica* 12:269–73.

——— (ed.). 1986. Abū Kāmil Shujāᶜ ibn Aslam (Second half ninth century A.D.), *The Book of Algebra. Kitāb al-Jabr wa l-muqābala.* (Publications of the Institute for the History of Arabic-Islamic Science, Series C: Facsimile Editions, 24). Frankfurt am Main: Institute for the History of Arabic-Islamic Science.

Holt, P. M., Ann K. S. Lambton and Bernard Lewis (eds.). 1970. *The Cambridge History of Islam.* 2 vols. Cambridge: Cambridge University Press.

Homer, *The Odyssey.* Ed., trans. Murray 1966.

———. *The Iliad.* Ed., trans. Murray 1978.

Horowitz, Ellis (ed.). 1983. *Programming Languages: A Grand Tour.* A Collection of Papers. Berlin-Heidelberg: Springer.

Høyrup, Jens. 1980. "Influences of Institutionalized Mathematics Teaching on the Development and Organization of Mathematical Thought in the Pre-Modern Period. Investigations into an Aspect of the Anthropology of Mathematics." *Materialien und Studien. Institut für Didaktik der Mathematik der Universität Bielefeld* 20:7–137.

———. 1982. "Investigations of an Early Sumerian Division Problem, c. 2500 B.C." *Historia Mathematica* 9:19–36.

———. 1985. *Babylonian Algebra from the View-Point of Geometrical Heuristics. An Investigation of Terminology, Methods, and Patterns of Thought.* Second, slightly corrected printing. Roskilde: Roskilde University Centre, Institute of Educational Research, Media Studies and Theory of Science.

———. 1986. "Al-Khwârizmî, Ibn Turk, and the Liber Mensurationum: On the Origins of Islamic Algebra." *Erdem* 2 (Ankara):445–84.

———. 1988. "Jordanus de Nemore, 13th Century Mathematical Innovator: An Essay on Intellectual Context, Achievement, and Failure." *Archive for History of Exact Sciences* 38:307–63.

———. 1989. "Zur Frühgeschichte algebraischer Denkweisen." *Mathematische Semesterberichte* 36:1–46.

———. 1990. "Algebra and Naive Geometry. An Investigation of Some Basic Aspects of Old Babylonian Mathematical Thought." *Altorientalische Forschungen* 17:27–69, 262–354.

———. 1990a. "*Dýnamis,* the Babylonians, and Theaetetus 147c7—148d7." *Historia Mathematica* 17:201–22.

———. 1990b. "On Parts of Parts and Ascending Continued Fractions." *Centaurus* 33:293–324.

———. 1992. "Geometrical Patterns in the Pre-Classical Greek Area: Prospecting the Borderland between Decoration, Art, and Structural Inquiry." Contribution to the Symposium "Early Greek Mathematics", Athens, August 17–21, 1992. *Mimeo,* Roskilde University Centre, Institute of Languages and Culture.

———. 1993. "Sumerian: The Descendant of a Proto-Historical Creole? An Alternative Approach to 'the Sumerian Problem.' " Revised contribution to the Thirteenth Scandinavian Conference of Linguistics, University of Roskilde, January 9–11, 1992. *ROLIG-Papir* N° 51 (Roskilde University, 1993).

Hrabanus Maurus. *De clericorum institutione.* In *PL* 107.

Hruška, Blahoslav. 1973. "Die innere Struktur der Reformtexte Urukaginas von Lagaš." *Archiv Orientální* 41:4–13, 104–32.

Hughes, Barnabas B., O.F.M. (ed., trans.). 1981. Jordanus de Nemore, *De numeris datis.* Berkeley: University of California Press.

——— (ed.). 1985. "John Pecham, *De numeris misticis.*" *Archivum Franciscanum Historicum* 78:3–28, 333–83.

——— (ed.). 1986. "Gerard of Cremona's Translation of al-Khwārizmī's *Al-Jabr.*" *Mediaeval Studies* 48:211–63.

Hugue de Saint-Victor. *Didascalicon.* In *PL* 176.

Hultsch, Friedrich (ed., trans.). 1876. Pappi Alexandrini *Collectionis* quae supersunt. E libris manu scriptis edidit et commentariis instruxit Fridericus Hultsch. 3 vols. Berlin: Weidmann, 1876, 1877, 1878.

*Humanismus und Menschenbild im Orient und in der Antike.* Konferenzvorträge. Herausgegeben von der Sektion Orient- und Altertumswissenschaften der Martin-Luther-Universität Halle-Wittenberg. (Martin-Luther-Universität Halle-Wittenberg, Wissenschaftliche Beiträge 1977/28 [I 2]). Halle (Saale).

Hume, David. *Enquiries Concerning Human Understanding and Concerning the Principles of Morals.* Reprinted from the Posthumous Edition of 1777 and Edited with Introduction [ . . . ] by L. A. Selby-Bigge. Third edition with text revised and notes by P. H. Nidditch. Oxford: Oxford University Press, 1975.

Isidore of Seville. *Etymologiae.* In *PL* 82.

Jackson, D. E. P. 1980. "Toward a Resolution of the Problem of τὰ ἑνὶ διαστήματι γραφόμενα in Pappus' Collection Book VIII." *The Classical Quarterly,* n.s. 30:523–33.

Jacobsen, Thorkild. 1943. "Primitive Democracy in Mesopotamia." *Journal of Near Eastern Studies* 2:159–82.

———. 1957. "Early Political Development in Mesopotamia." *Zeitschrift für Assyriologie und Vorderasiatische Archäologie,* Neue Folge 18:91–140.

———. 1976. *The Treasures of Darkness. A History of Mesopotamian Religion.* New Haven: Yale University Press.

Jakobson, Vladimir A. 1971. "Some Problems Connected with the Rise of Landed Property (Old Babylonian Period)." In Klengel (ed.) 1971: 33–37.

Jardine, Nicholas (ed., trans.). 1984. *The Birth of History and Philosophy of Science: Kepler's "A Defence of Tycho against Ursus",* With Essays on its Provenance and Significance. Cambridge: Cambridge University Press.

Jarvad, Ib Martin. 1981. "Friedliche Koexistenz—Utopie oder Wirklichkeit?" Manuscript, Roskilde Universitetscenter.

Jasim, Sabah Abboud, and Joan Oates. 1986. "Early Tokens and Tablets in Mesopotamia: New Information from Tell Abada and Tell Brak." *World Archaeology* 17:348–62.

Jayawardene, S. A. "Pacioli." *DSB* X, 269–72.

———. 1976. "The 'Trattato d'abaco' of Piero della Francesca." In Clough (ed.) 1976: 229–43.

Jestin, Raymond. 1937. *Tablettes sumériennes de Šuruppak au Musée de Stamboul.* (Mémoires de l'Institut Français d'Archéologie de Stamboul, III). Paris: Boccard.

Johannes Cusanus. 1514. *Algorithmus linealis proiectilium. De integris perpulchris Arithmetrice artis regulis: earundemque probationis claris exornatus: Studiosis admodum utilis et necessarius.* Vienna: Hermann Busch.

Johnson, Gregory A. 1975. "Locational Analysis and the Investigation of Uruk Local Exchange Systems." In Sabloff and Lamberg-Karlovsky 1975: 285–339.

Jolivet, Jean. 1974. "Les *Quaestiones naturales* d'Adelard de Bath ou la nature sans livre." In *Études de civilisation médiévale . . .* , pp. 437–45.

Jones, Charles W. (ed.). 1943. Bedae *Opera de temporibus*. (The Mediaeval Academy of America, Publication N° 41). Cambridge, Massachusetts: The Mediaeval Academy of America.

———. 1963. "An Early Medieval Licensing Examination." *History of Education Quarterly* 3:19–29.

Jones, Grant D., and Robert R. Kautz (eds.). 1981. *The Transition to Statehood in the New World*. (New Directions in Archaeology). Cambridge: Cambridge University Press.

Jones, Leslie Webber (ed., trans.). 1946. Cassiodorus Senator, *An Introduction to Divine and Human Readings*. (Records of Civilization). New York: Columbia University Press.

Jordanus de Nemore. *Arithmetica*. Ed. Busard 1991.

———. *De numeris datis*. Ed., trans. Hughes 1981.

———. *De plana spera*. Ed., trans. Thomson 1978.

———. *De ratione ponderis*. Eds. Moody and Clagett 1952.

———. *Elementa super demonstrationem ponderum*. Eds. Moody and Clagett 1952.

———. *Liber philotegni*. Ed. Clagett 1984.

Jungk, Robert. 1958. *Brighter Than A Thousand Suns. A Personal History of the Atomic Scientists*. New York: Harcourt Brace Jovanovich.

Juschkewitsch, A. P. 1964. *Geschichte der Mathematik im Mittelalter*. Leipzig: Teubner.

Kaldor, Mary. 1982. *The Baroque Arsenal*. London: André Deutsch.

Karpinski, Louis Charles. 1912. "The 'Quadripartitum numerorum' of John of Meurs." *Bibliotheca Mathematica*, 3. Folge 13 (1912–13):99–114.

——— (ed., trans.). 1915. *Robert of Chester's Latin Translation of the Algebra of al-Khowarizmi*. (University of Michigan Studies, Humanistic Series, 11). New York. Reprinted in L. C. Karpinski and J. G. Winter, *Contributions to the History of Science*. Ann Arbor: University of Michigan, 1930.

Kasir, Daoud S. (ed., trans.). 1931. *The Algebra of Omar Khayyam*. Dissertation, Faculty of Philosophy, Columbia University. New York: Bureau of Publications, Teachers College, Columbia University.

Keele, Kenneth D. "Leonardo da Vinci. Life, Scientific Methods, and Anatomical Works." *DSB* VIII, 193–206.

Keller, Alex. 1972. "Mathematical Technologies and the Growth of the Idea of Technical Progress in the Sixteenth Century." In Debus 1972: I, 11–27.

Kennedy, E. S. 1968. "The Lunar Visibility Theory of Ya'qûb ibn Ṭâriq." *Journal of Near Eastern Studies* 27:126–32.

Kennedy, E. S., and David Pingree. 1971. *The Astrological History of Māshā'allāh.* Cambridge, MA: Harvard University Press.

Kerferd, G. B. "Antiphon." *DSB* I, 170–72.

Kevles, Daniel. 1978. *The Physicists.* New York: Knopf.

Khaldûn, ibn. *Muqaddimah.* Ed., trans. Rosenthal 1958.

King, J. E. (ed., trans.). 1971. Cicero, *Tusculan Disputations.* (Loeb Classical Library). Cambridge, MA: Harvard University Press/London: Heinemann. First ed. 1927.

Kirk, G. S., J. E. Raven and M. Schofield. 1983. *The Presocratic Philosophers. A Critical History with a Selection of Texts.* Second edition. Cambridge: Cambridge University Press.

Kirschvogel, Paul A. "Faulhaber." *DSB* IV, 449–553.

Klebs, Arnold C. 1938. "Incunabula scientifica et medica." *Osiris* 4:1–359.

Klein, Felix. 1926. *Vorlesungen über die Geschichte der Mathematik im 19. Jahrhundert.* 2 vols. (Die grundlehren der mathematischen Wissenschaften, 24–25). Berlin: Julius Springer, 1926–27.

Klein, Jacob. 1981. *Three Šulgi Hymns. Sumerian Hymns Glorifying King Šulgi of Ur.* Ramat-Gan, Israel: Bar-Ilan University Press.

Klengel, Horst (ed.). 1971. *Beiträge zur sozialen Struktur des alten Vorderasien.* (Schriften zur Geschichte und Kultur des Alten Orients, 1). Berlin: Akademie-Verlag.

―――. 1974. "Einige Bemerkungen zur sozialökonomischen Entwicklung in der altbabylonischen Zeit." *Acta Antiqua Academiae Scientiarum Hungaricae* 22:249–57.

―――. 1977. "Zur Rolle der Persönlichkeit in der altbabylonischen Gesellschaft." In *Humanismus und Menschenbild . . . ,* pp. 109–17.

―――. 1980. *Hammurapi von Babylon und seine Zeit.* 4., überarbeitete Auflage. Berlin: Deutscher Verlag der Wissenschaften.

———— (ed.). 1982. *Gesellschaft und Kultur im alten Vorderasien.* (Schriften zur Geschichte und Kultur des Alten Orients, 15). Berlin: Akademie-Verlag.

Kline, Morris. 1972. *Mathematical Thought from Ancient to Modern Times.* New York: Oxford University Press.

Knorr, Wilbur R. 1975. *The Evolution of the Euclidean Elements. A Study of the Theory of Incommensurable Magnitudes and Its Significance for Early Greek Geometry.* (Synthese Historical Library, 15). Dordrecht: Reidel.

————. 1983. " 'La croix des mathématiciens': The Euclidean Theory of Irrational Lines." *Bulletin of the American Mathematical Society,* n.s. 9(1):41–69.

Kokian, P. Sahak (ed., trans.). 1919. "Des Anania von Schirak arithmetische Aufgaben." *Zeitschrift für die deutsch-österreichischen Gymnasien* 69 (1919–20):112–17.

Komoróczy, Géza. 1978. "Landed Property in Ancient Mesopotamia and the Theory of the So-Called Asiatic Mode of Production." *Oikumene* 2 (Budapest):9–26.

————. 1979. "Zu den Eigentumsverhältnissen in der altbabylonischen Zeit: Das Problem der Privatwirtschaft." In Lipinski (ed.) 1979: 411–42.

Kraeling, Carl, and Robert McC. Adams (eds.). 1960. *City Invincible.* A Symposium on Urbanization and Cultural Development in the Ancient Near East, held at the Oriental Institute of the University of Chicago. December 4–7, 1958. Chicago: University of Chicago Press.

Kramer, Samuel Noah. 1949. "Schooldays: A Sumerian Composition Relating to the Education of a Scribe." *Journal of American Oriental Studies* 69:199–215.

————. 1983. "The Ur-Nammu Law Code: Who Was its Author?" *Orientalia,* NS 52:453–56.

Krasnova, S. A. (ed. trans.). 1966. "Abu-l-Vafa al-Buzdžani, *Kniga o tom, čto neobxodimo remeslenniku iz geometričeskix postroenij."* In Grigor'jan and Juškevič 1966: 42–140.

Kraus, F. R. 1973. *Vom mesopotamischen Menschen der altbabylonischen Zeit und seiner Welt.* (Mededelingen der Koninklijke Nederlandse Akademie van Wetenschappen, Afd. Letterkunde. Nieuwe Reeks—Deel 36—No. 6). Amsterdam: North-Holland.

Krecher, Joachim. 1973. "Neue sumerische Rechtsurkunden des 3. Jahrtausends." *Zeitschrift für Assyriologie und Vorderasiatische Archäologie* 63 (1973–74): 145–271.

————. 1974. "Die Aufgliederung des Kaufpreises nach sumerischen Kaufverträgen der Fara- und der Akkade-Zeit." *Acta Antiqua Academiae Scientiarum Hungaricae* 22:29–32.

Kretzmann, Norman (ed.). 1982. *Infinity and Continuity in Ancient and Medieval Thought*. Ithaca: Cornell University Press.

Kudrow, Viktor. 1981. "Der wissenschaftlich-technische Fortschritt und die Swächung der internationalen Stellung der USA." *Deutsche Außenpolitik* 26(6):84–102.

Kuhn, Thomas S. 1963. "The Function of Dogma in Scientific Research." In Crombie (ed.) 1963: 347–69.

l'Huillier, Ghislaine. 1980. "Regiomontanus et le *Quadripartitum numerorum* de Jean de Murs." *Revue d'Histoire des Sciences et de leurs applications* 33:193–214.

l'Huillier, Hervé (ed.). 1979. *Nicolas Chuquet, "La géométrie." Première géométrie algébrique en langue française* (1484). Paris: Vrin.

Ladd, Everett Caril, Jr., and Seymour Martin Lipset. 1972. "Politics of Academic Natural Scientists and Engineers." *Science* 176:1091–1100.

Lamb, W. R. M. (ed., trans.). 1964. Plato, *Charmides. Alcibiades* I and II. *Hipparchus. The Lovers. Theages. Minos. Epinomis.* (Loeb Classical Library). London: Heinemann/Cambridge, MA: Harvard University Press. 1st. ed. 1927.

——— (ed., trans.). 1977. Plato, *Laches. Protagoras. Meno. Euthydemus.* (Loeb Classical Library). London: Heinemann/Cambridge, MA: Harvard University Press. First ed. 1924.

Lamberg-Karlovsky, C. C. 1976. "The Economic World of Sumer." In Schmandt-Besserat (ed.) 1976: 59–68.

Lambert, Maurice. 1952. "La période présargonique." *Sumer* 8:57–77, 198–216.

Lambert, W. G., and A. R. Millard. 1969. *Atra-ḫasīs: The Babylonian Story of the Flood.* Oxford: Oxford University Press.

Landsberger, Benno. 1956. "Babylonian Scribal Craft and Its Terminology." Pp. 123–26 in Sinor (ed.), *Proceedings of the 23rd International Congress of Orientalists, Cambridge, 21–28 August 1954.* London.

———. 1960. "Scribal Concepts of Education." In Kraeling and Adams 1960: 94–102, discussion 104–23.

Lang, Serge. 1971. "A Mathematician on the DOD, Government, and Universities". In Brown 1971: 51–79.

Larsen, Mogens Trolle. 1988. "Literacy and Social Complexity." In Gledhill et al. (eds.) 1988: 173–91.

Le Brun, A., and François Vallat. 1978. "L'origine de l'écriture à Suse." *Cahiers de la Délégation Archéologique Française en Iran* 8:11–59.

Leemans, W. F. 1950. *The Old Babylonian Merchant: His Business and his Social Position*. (Studia et documental ad iura Orientis antiqui pertinentia, 3). Leiden: Brill.

Lefèvre, Wolfgang. 1981. "Rechenstein und Sprache". In Damerow and Lefèvre (eds.) 1981: 115–69.

Lefèvre d'Étaples, Jacques (ed.). 1514. *In hoc opere contenta. Arithmetica decem libris demonstrata. Musica libris demonstrata quatuor. Epitome in libros Arithmeticos divi Severini Boetii. Rithmimachie ludus qui et pugna numerorum appellatur*. Secundaria aeditio. Paris: Henricus Stephanus.

Lemay, Richard. 1962. *Abu Maʿshar and Latin Aristotelianism in the Twelfth Century. The Recovery of Aristotle's Natural Philosophy Through Arabic Astrology*. Beirut: American University of Beirut.

———. 1976. "The Teaching of Astronomy in Medieval Universities, Principally at Paris in the Fourteenth Century." *Manuscripta* 20:197–217.

Levey, Martin (ed., trans.). 1966. The *Algebra* of Abū Kāmil, *Kitāb fī al-jābr (sic) wa'l-muqābala*, in a Commentary by Mordechai Finzi. Madison: University of Wisconsin Press.

Levey, Martin, and Marvin Petruck (eds., trans.). 1965. Kūshyār ibn Labbān, *Principles of Hindu Reckoning*. A Translation with Introduction and Notes of the *Kitāb fī usūl hisāb al-hind*. Madison: University of Wisconsin Press.

*Liber de triangulis Jordani*. Ed. Clagett 1984.

Lindberg, David C. "Pecham." *DSB* X, 473–76.

———. 1971. "Lines of Influence in Thirteenth-Century Optics: Bacon, Witelo, and Pecham." *Speculum* 46:66–83.

Lipinski, Edward (ed.). 1979. *State and Temple Economy in the Ancient Near East*. 2 vols. Proceedings of the International Conference Organized by the Katholieke Universiteit Leuven from the 10th to the 14th of April 1978. (Orientalia Lovaniensa Analecta, 5–6). Leuven: Department Oriëntalistiek.

Liverani, Mario. 1988. *Antico Oriente. Storia, società, economia*. Rome: Laterza.

Lloyd, G. E. R. 1979. *Magic, Reason and Experience. Studies in the Origin and Development of Greek Science*. Cambridge: Cambridge University Press.

Lohne, J. A. "Harriot." *DSB* VI, 124–29.

Lorch, Richard P. "Jābir ibn Aflaḥ." *DSB* VII, 37–39.

Lucas, Christopher J. 1979. "The Scribal Tablet-House in Ancient Mesopotamia." *History of Education Quarterly* 19:305–32.

Luckey, Paul. 1941. "Tābit b. Qurra über den geometrischen Richtigkeitsnachweis der Auflösung der quadratischen Gleichungen." *Sächsischen Akademie der Wissenschaften zu Leipzig. Mathematisch-physische Klasse. Berichte* 93:93–114.

Luria, Aleksandr R. 1976. *Cognitive Development. Its Cultural and Social Foundations.* Edited by Michael Cole. Cambridge, MA: Harvard University Press.

Maeyama, Y., and W. G. Saltzer (eds.). 1977. *Prismata. Naturwissenschaftliche Studien.* Festschrift für Willy Hartner. Wiesbaden: Franz Steiner.

Mahoney, Michael S. "Ramus." *DSB* XI, 286–90.

————. 1971. "Babylonian Algebra: Form *vs.* Content." *Studies in History and Philosophy of Science* 1 (1970–71):369–80.

Makdisi, George. 1961. "Muslim Institutions of Higher Learning in Eleventh-Century Baghdad." *Bulletin of the School of Oriental and African Studies* 24:1–56.

————. 1971. "Law and Traditionalism in the Institutions of Learning in Medieval Islam." In von Grünebaum 1971: 75–88.

Mancini, G. (ed.). 1916. "L'opera 'De corporibus regularibus' di Pietro Franceschi detto Della Francesca usurpata da Fra Luca Pacioli". *Atti della R. Accademia dei Lincei,* anno CCCVI. Serie quinta. *Memorie della Classe di Scienze morali, storiche e filologiche,* 14 (Roma 1909–16), 446-580.

Mandonnet, Pierre, O. P. 1914. "La crise scolaire au début du XIIIe siècle et la fondation de l'ordre des Frères-Prêcheurs." *Revue d'histoire ecclésiastique* 15:35–49.

Mannheim, Karl. 1965. *Ideologie und Utopie.* 4. Auflage. Frankfurt am Main: Schulte-Bulmke.

Marinoni, Augusto. "Leonardo da Vinci. Mathematics." *DSB* VIII, 234–41.

Marquet, Yves. "Ikhwān al-Ṣafāʾ." *DSB* XV, 249–51.

Marre, Aristide (ed.). 1880. "Le Triparty en la science des nombres par Maistre Nicolas Chuquet Parisien." *Bulletino di Bibliografia e di Storia delle Scienze Matematiche e Fisiche* 13:593–659, 693–814.

Martzloff, Jean-Claude. 1988. *Historie des mathématiques chinoises.* Paris: Masson.

Mason, Stephen F. 1962. *A History of the Sciences.* Revised edition. New York: Collier.

Masotti, Arnaldo (ed.). 1974. Ludovico Ferrari e Niccolò Tartaglia, *Cartelli di sfida matematica.* Riproduzione in facsimile delle edizioni originali 1547–1548 edita con parti introduttorie. Brescia: Ateneo di Brescia.

Matvievskaya, G. P. 1981. "On Some Problems of the History of Mathematics and Astronomy of the Middle East." Pp. 21–31 in *XVIth International Congress of the History of Science. Papers by Soviet Scientists: The 1000th Anniversary of Ibn Sina's Birth (Avicenna)*. Moscow: "Nauka".

Maurer, Armand (ed., trans.). 1963. St. Thomas Aquinas, *The Division and Methods of the Sciences: Questions V and VI of his Commentary on the "De Trinitate" of Boethius*. Third revised edition. Toronto: The Pontifical Institute of Mediaeval Studies.

Mayr, Ernst. 1990. "When is Historiography Whiggish?" *Journal of the History of Ideas* 51:301–9.

Mazon, Paul (ed., trans.). 1979. Hésiode, *Théogonie—Les travaux et les jours—Le bouclier*. 10ᵉ tirage. Paris: "Les Belles Lettres".

McCracken, D. S. (ed., trans.). 1966. Saint Augustine, *The City of God Against the Pagans*. 7 vols. (Loeb Classical Library) London: Heinemann/Cambridge, Mass: Harvard University Press, 1966 etc.

McGarry, Daniel D. (ed., trans.). 1971. *The Metalogicon* of John of Salisbury. A *Twelfth-Century Defense of the Verbal and Logical Arts of the Trivium*. Gloucester, Massachusetts: Peter Smith. First ed. 1955.

*MCT*: O. Neugebauer and A. Sachs, *Mathematical Cuneiform Texts*. (American Oriental Series, 29). New Haven, CT: American Oriental Society, 1945.

Mellaart, James. 1978. "Early Urban Communities in the Early Near East, *c.* 9000–3400 BC." In Moorey 1979: 22–33.

Menge, Heinrich (ed.). 1896. Euclidis *Data* cum *Commentario* Marini et scholiis antiquis. (Euclidis Opera Omnia, 6). Leipzig: Teubner.

Menninger, Karl. 1957. *Zahlwort und Ziffer. Eine Kulturgeschichte der Zahl*. I. *Zählreihe und Zahlsprache*. II. *Zahlschrift und Rechnen*. 2. neubearbeitete und erweiterte Auflage. Göttingen: Vandenhoeck & Ruprecht, 1957–58.

Merlan, Philip. "Bryson of Heraclea." *DSB* II, 549f.

Merton, Robert K. 1938. "Science, Technology and Society in Seventeenth Century England." *Osiris* 4(2).

⸻. 1938a. "Science and the Social Order." *Philosophy of Science* 5:321–37.

⸻. 1942. "A Note on Science and Democracy." *Journal of Legal and Political Sociology* 1:115–26.

⸻. 1968. *Social Theory and Social Structure*. Enlarged edition. New York: The Free Press.

————. 1970. *Science, Technology and Society in Seventeenth Century England*. New York: Harper and Row. [Reprint of Merton 1938 with a new introduction].

Metropolis, N., J. Howlett, and Gian-Carlo Rota (eds.). 1980. *A History of Computing in the Twentieth Century*. New York: Academic Press.

*MEW:* Karl Marx and Friedrich Engels. *Werke*. 39 vols. Berlin: Dietz Verlag, 1965–72.

Minio-Paluello, Lorenzo. "Moerbeke, William of." *DSB* IX, 434–40.

*MKT:* O. Neugebauer. *Mathematische Keilschrift-Texte*. 3 vols. (Quellen und Studien zur Geschichte der Mathematik, Astronomie und Physik. Abteilung A: Quellen. 3. Band, erster–dritter Teil). Berlin: Julius Springer, 1935, 1935, 1937.

Molland, A. G. 1978. "An Examination of Bradwardine's Geometry." *Archive for History of Exact Sciences* 19:113–75.

————. 1980. "Mathematics in the Thought of Albertus Magnus." In Weisheipl 1980: 463–78.

Mollat, Michel, and Paul Adam (eds.). 1966. *Les aspects internationaux de la découverte océanique aux XV<sup>e</sup> et XVI<sup>e</sup> sicles. Actes du cinquième colloque international d'histoire maritime* (Lisbonne—14–16 septembre 1960). (Bibliothèque générale de l'École Pratique des Hautes Études, VI<sup>e</sup> section). Paris: S. E. V. P. E. N.

Moody, Ernest A., and Marshall Clagett (eds.). 1952. *The Medieval Science of Weights (Scientia de ponderibus). Treatises Ascribed to Euclid, Archimedes, Thabit ibn Qurra, Jordanus de Nemore, and Blasius of Parma*. Madison: University of Wisconsin Press.

Moorey, P. R. S. 1979. *The Origins of Civilization*. Wolfson College Lectures 1978. Oxford: Clarendon Press.

Moran, Bruce T. 1977. "Princes, Machines, and the Valuation of Precision in the Sixteenth Century." *Sudhoffs Archiv* 61:209–28.

Morse, Marston. 1943. "Mathematics and the Maximum Scientific Effort in Total War." *Scientific Monthly* 56:50–55.

Muckle, J. T., C. S. B. (ed.). 1950. "Abelard's Letter of Consolation to a Friend *(Historia Calamitatum)*." *Mediaeval Studies* 12:163–213.

Müller, Martin (ed.). 1934. *Die "Quaestiones naturales" des Adelardus von Bath, herausgegeben und untersucht*. (Beiträge zur Geschichte der Philosophie und Theologie des Mittelalters. Texte und Untersuchungen 31,2). Münster: Verlag der Aschendorffschen Verlagsbuchhandlung.

Murdoch, John E. "Euclid: Transmission of the Elements." *DSB* IV, 437–59.

————. 1962. *"Rationes mathematice"* : *Un aspect du rapport des mathématiques et de la philosophie au Moyen Age.* (Les Conférences du Palais de la Découverte. Série D N° 81). Paris: Université de Paris, Palais de la Découverte.

————. 1964. [Essay review of Busard 1961]. *Scripta Mathematica* 27:67–91.

————. 1968. "The Medieval Euclid: Salient Aspects of the Translations of the *Elements* by Adelard of Bath and Campanus of Novara." *Revue de Synthèse* 89:67–94.

————. 1969. *"Mathesis in philosophiam scholasticam introducta:* The Rise and Development of Mathematics in Fourteenth Century Philosophy and Theology." In *Arts Libéraux . . .* , pp. 215–54.

Murdoch, John Emery, and Edith Dudley Sylla (eds.). 1975. *The Cultural Context of Medieval Learning.* Proceedings of the First International Colloquium on Philosophy, Science, and Theology in the Middle Ages—September 1973. (Boston Studies in the Philosophy of Science, 26). Dordrecht: Reidel.

Murray, A. T. (ed., trans.). 1966. Homer, *The Odyssey.* 2 vols. (Loeb Classical Library). London: Heineman/Cambridge, MA: Harvard University Press. First ed. 1919.

———— (ed., trans.). 1978. Homer, *The Iliad.* 2 vols. (Loeb Classical Library). London: Heineman/Cambridge, MA: Harvard University Press. First ed. 1924–25.

Nader, A. N. 1956. *Le système philosophique des Muʿtazila (Premiers penseurs de l'Islam).* Beirut. Quoted via A. M. Heinen 1978.

Narducci, Enrico (ed.). 1886. "Vite inedite di matematici italiani scritti da Bernardino Baldi." *Bulletino di Bibliografia e di Storia delle Scienze matematiche e fisiche* 19:335–406, 437–89, 521–640.

———— (ed.). 1887. *"Vita di Pitagora,* scritti da Bernardino Baldi, tratta dall'autografo ed annotato." *Bulletino di Bibliografia e di Storia delle Scienze matematiche e fisiche* 20:197–308.

Nasr, Seyyed Hossein. "Al-Ṭūsī, [ . . . ] Naṣīr al-Dīn." *DSB* XIII, 508–14.

————. 1968. *Science and Civilization in Islam.* Cambridge, MA: Harvard University Press.

Nau, F. 1910, "Notes d'astronomie syrienne." *Journal Asiatique,* 10. série 16:209–28.

Nebbia, G. 1967. "Ibn al-Haytham nel millesimo anniversario della nascita." *Physis* 9:165–214.

Needham, Joseph. 1938. [Review of Merton 1938]. *Science and Society* 2 (1937–38):566–71.

————. 1971. "Foreword." In *Science at the Cross Roads,* pp. vii–x.

Nemet-Nejat, Karen Rhea. 1993. *Cuneiform Mathematical Texts as a Reflection of Everyday Life in Mesopotamia.* (American Oriental Series, 75). New Haven, CT: American Oriental Society.

Neuenschwander, Erwin. 1973. "Die ersten vier Bücher der Elemente Euklids. Untersuchungen über den mathematischen Aufbau, die Zitierweise und die Entstehungsgeschichte." *Archive for History of Exact Sciences* 9 (1972–73):325–80.

Neugebauer, Otto. 1933. "Babylonische 'Belagerungsrechnung.' " *Quellen und Studien zur Geschichte der Mathematik, Astronomie und Physik.* Abteilung B: *Studien* 2:305–10.

————. 1954. "Ancient Mathematics and Astronomy." In *A History of Technology* I, 785–803.

Neumann, Hans. 1987. *Handwerk in Mesopotamien. Untersuchungen zu seiner Organisation in der Zeit der III. Dynastie von Ur.* (Schriften zur Geschichte und Kultur des Alten Orients, 19). Berlin: Akademie-Verlag.

————. 1988. "Bemerkungen zu den Eigentums- und Wirtschaftsverhältnissen in Mesopotamien gegen Ende des 3. Jahrtausends v.u.Z." In Herrmann and Köhn 1988: 335–43.

————. 1989. " 'Gerechtigkeit liebe ich. . .'. Zum Strafrecht in den ältesten Gesetzen Mesopotamiens." *Das Altertum* 35(1):13–22.

Nissen, Hans J. 1974. "Zur Frage der Arbeitsorganisation in Babylonien während der Späturuk-Zeit." *Acta Antiqua Academiae Scientiarium Hungaricae* 22:5–14.

————. 1981. "Bemerkungen zur Listenlitteratur Vorderasiens im 3. Jahrtausend (gesehen von den Archaischen Texten aus Uruk)." Pp. 99–108 in Luigi Cagni (ed.), *La Lingua di Ebla.* Atti del Convegno Internazionale (Napoli, 21–23 aprile 1980). (Istituto Universitario Orientale, Seminario di Studi asiatici, series minor, 14). Naples.

————. 1982. "Die 'Tempelstadt': Regierungsform der frühdynastischen Zeit in Babylonien?" In Klengel (ed.) 1982: 195–200.

————. 1983. *Grundzüge einer Geschiche der Frühzeit des Vorderen Orients.* (Grundzüge, 52). Darmstadt: Wissenschaftliches Buchgesellschaft.

————. 1985. "The Emergence of Writing in the Ancient Near East." *Interdisciplinary Science Reviews* 10:349–61.

————. 1986. *Mesopotamia Before 5000 Years.* (Istituto di Studi del Vicino Oriente. Sussidi didattici). Rome: Istituto di Studi del Vicino Oriente.

————. 1986a. "The Archaic Texts from Uruk." *World Archaeology* 17:317–34.

————. 1986b. "The Development of Writing and of Glyptic Art." In Finkbeiner and Röllig (eds.) 1986:316–31.

Nissen, Hans J., Peter Damerow and Robert Englund. 1990. *Frühe Schrift und Techniken der Wirtschaftsverwaltung im alten Vorderen Orient. Informationsspeicherung und -verarbeitung vor 5000 Jahren.* Bad Salzdetfurth: Verlag Franzbecker.

Nowotny, Karl Anton (ed.). 1967. Agrippa von Nettesheym, *De occulta philosophia,* herausgegeben und erläutert. Graz: Akademische Druck- und Verlagsanstalt.

O'Leary, De Lacy. 1949. *How Greek Science Passed to the Arabs.* London: Routledge & Kegan Paul.

Oates, Joan. 1960. "Ur and Eridu, the Prehistory." *Iraq* 22:32–50.

Ong, Walter J., S.J. 1958. *Ramus and Talon Inventory.* Cambridge, MA; Harvard University Press.

Oppenheim, A. Leo. 1967. "A New Look at the Social Structure of Mesopotamian Society." *Journal of the Economic and Social History of the Orient* 18:1–16.

Ore, Oystein. 1953. *Cardano. The Gambling Scholar.* With a Translation from the Latin of Cardano's *Book on the Games of Chance,* by Sydney Henry Gould. Princeton, NJ: Princeton University Press.

Oschinsky, Dorothea. 1971. *Walter of Henley and Other Treatises on Estate Management and Accounting.* Oxford: Oxford University Press.

Pacioli, Luca. 1523. *Summa de Arithmetica geometria Proportioni: et proportionalita.* Novamente impressa. Toscolano.

Palencia, Angel Gonzales (ed., trans.). 1953. Al-Fārābī, *Catálogo de las ciencias.* Edición y traduccion Castellana. Segunda edición. Madrid: Consejo Superior de Investigaciones Científicas, Instituto Miguel Asín.

Parker, Richard A. 1972. *Demotic Mathematical Papyri.* Providence: Brown University Press.

Paton, W. R. (ed., trans.). 1979. *The Greek Anthology.* Vol. 5. (Loeb Classical Library). Cambridge, MA: Harvard University Press/London: Heinemann. First ed. 1918.

Pedersen, Fritz Saaby (ed.). 1983. Petri Philomena de Dacia et Petri de S. Audomaro *Opera quadrivialia.* Pars I. *Opera* Petri Philomenae. (Corpus philosophorum danicorum medii aevi, 10,1). Copenhagen: Gad.

Pedersen, Olaf. 1956. *Nicole Oresme og hans naturfilosofiske system. En undersøgelse of hans skrift "Le livre du ciel et du monde."* (Acta Historica Scientiarum Naturalium et Medicinalium, 13). Copenhagen: Munksgaard.

Peet, T. Eric (ed., trans.). 1923. *The Rhind Mathematical Papyrus, British Museum 10057 and 10058.* London: University Press of Liverpool.

Perlo, Victor. 1974. *The Unstable Economy: Booms and Recessions in the U.S. Since 1945*. Second edition. New York: International Publishers.

Pidal, Ramon Menendez (ed.). 1963. *Poema de Mio Cid*. Sexta edicion. Madrid: Aguilar.

Pines, Shlomo. 1970. "Philosophy [in Islam]." In Holt et al. 1970: II, 780–823.

Pingree, David. "Abū Maʿšar." *DSB* I, 32–39.

———. "Al-Fazārī." *DSB* IV, 555–56.

———. "Māšaʾallāh." *DSB* IX, 159–62.

———. "Umar ibn al-Farrukhān al-Tabarī." *DSB* XIII, 538–39.

———. 1963. "Astronomy and Astrology in India and Iran." *Isis* 54:229–46.

———. 1968. "The Fragments of the Works of Yaʿqûb ibn Ṭâriq." *Journal of Near Eastern Studies* 27:97–125.

———. 1970. "The Fragments of the Works of al-Fazārī." *Journal of Near Eastern Studies* 29:103–23.

———. 1973. "The Greek Influence on Early Islamic Mathematical Astronomy." *Journal of the American Oriental Society* 93:32–43.

*PL: Patrologiae cursus completus, series latina*, accurante J. P. Migne. 221 vols. Paris, 1844–64.

Plath, Peter, and Hans Jörg Sandkühler (eds.). 1978. *Theorie und Labor. Dialektik als Programm der Naturwissenschaft*. Cologne: Pahl-Rugenstein.

Plato. *Apologia*. Ed., trans. Fowler 1977a.

———. *Erastae*. Ed., trans. Lamb 1964.

———. *Leges*. Ed., trans. Bury 1967.

———. *Menon*. Ed., trans. Lamb 1977.

———. *Protagoras*. Ed., trans. Lamb 1977.

———. *Respublica*. Ed., trans. Shorey 1978.

———. *Theaetetus*. Ed., trans. Fowler 1977.

*Poema de Mio Cid*. Ed. R. M. Pidal 1963.

Poulle, Emmanuel. "Fine." *DSB* XV, 153–57.

———. "John of Murs." *DSB* VII, 128–33.

Powell, Marvin A. 1971. *Sumerian Numeration and Metrology*. Dissertation, University of Minnesota.

————. 1972. "Sumerian Area Measures and the Alleged Decimal Substratum." *Zeitschrift für Assyriologie und Vorderasiatische Archäologie* 62 (1972–73): 165–221.

————. 1976. "The Antecedents of Old Babylonian Place Notation and the Early History of Babylonian Mathematics." *Historia Mathematica* 3:417–39.

————. 1978. "Götter, Könige und 'Kapitalisten' im Mesopotamien des 3. Jahrtausend v. u. Z." *Oikumene* 2 (Budapest):127–44.

————. 1986. "Economy of the Extended Family According to Sumerian Sources." *Oikumene* 5 (Budapest):9–14.

Prager, Frank D., and Gustina Scaglia. 1972. *Mariano Taccola and His Book "De ingeneis"*. Cambridge, MA: M.I.T. Press.

Pritchard, J. B. (ed.). 1950. *Ancient Near Eastern Texts Relating to the Old Testament*. Princeton, NJ: Princeton University Press.

*Proc.* . . . *1912: Proceedings of the Fifth International Congress of Mathematicians (Cambridge, 22–28 August 1912)*. Cambridge: Cambridge University Press, 1913.

*Proc.* . . . *1920: Comptes Rendus du Congrès International des Mathématiciens (Strasbourg, 22–30 Septembre 1920)*. Toulouse, 1921.

*Proc.* . . . *1924: Proceedings of the International Mathematical Congress Held in Toronto (August 11–16, 1924)*. Toronto: The University of Toronto Press, 1928.

*Proc.* . . . *1936: Comptes Rendus du Congrès International des Mathématiciens (Oslo 1936)*. Oslo: A. W. Brøgger, 1937.

*Proc.* . . . *1950: Proceedings of the International Congress of Mathematicians (Cambridge, Massachusetts, August 30—September 6, 1950)*. American Mathematical Society, 1952.

*Proc.* . . . *1954: Proceedings of the International Congress of Mathematicians 1954 (Amsterdam, September 2—September 9)*. Groningen: Noordhoff/Amsterdam: North-Holland, 1957.

*Proc.* . . . *1958: Proceedings of the International Congress of Mathematicians (14–21 August 1958)*. Cambridge: Cambridge University Press, 1960.

*Proc.* . . . *1962: Proceedings of the International Congress of Mathematicians (15–22 August 1962)*. Djursholm, Sweden: Institut Mittag-Leffler, 1963.

*Proc.* . . . *1966: Trudy Meždunarodnogo Konressa Matematikov (Moscow—1966)*. Moscow: Izdatel'stvo "Mir", 1968.

*Proc. . . . 1970: Actes du Congrès International des Mathématiciens 1970.* Paris: Gauthier-Villars, 1971.

*Proc. . . . 1974: Proceedings of the International Congress of Mathematicians (Vancouver 1974).* Canadian Mathematical Congress, 1975.

*Proc. . . . 1978: Proceedings of the International Congress of Mathematicians (Helsinki 1978).* Helsinki: Academia Scientiarum Fennica, 1980.

*Proc. . . . 1986: Proceedings of the International Congress of Mathematicians (August 3–11, 1986, Berkeley).* American Mathematical Society, 1987.

Proclos Diadochos. *In primum Euclidis Elementorum librum commentarii.* Ed. Friedlein 1873.

Rackham, H. (ed., trans.). 1977. Aristotle, *Politics.* (Loeb Classical Library). Cambridge, MA: Harvard University Press/London: Heinemann. First ed. 1932.

[Ramus, Petrus]. 1560. *Algebra.* Paris: Andreas Wechelum.

Ramus, Petrus. 1569. *Scholarum mathematicarum libri unus et triginta.* Basel: Eusebius Episcopius.

Randall, John Herman. 1962. *The Career of Philosophy from the Middle Ages to the Enlightenment.* New York: Columbia University Press.

Rashdall, Hastings. 1936. *The Universities of Europe in the Middle Ages.* A new edition in three volumes edited by F. M. Powicke and A. B. Emden. Oxford: Clarendon Press. First ed. 1895.

Rashed, Roshdi. "Kamāl al-Dīn." *DSB* VII, 212–19.

————. 1982. "Matériaux pour l'histoire des nombres amiables et de l'analyse combinatoire." *Journal for the History of Arabic Science* 6:209–11 (French introduction), 212–78 (Arabic texts).

————. 1983. "Nombres amiables, parties aliquotes et nombres figurés aux XIII$^{ème}$ et XIV$^{ème}$ siècles." *Archive for History of Exact Sciences* 28:107–47.

———— (ed., trans.). 1984. Diophante, *Les Arithmétiques*, tômes III (*livre IV*), IV (*livres V, VI, VII*). Paris: "Les Belles Lettres".

Redman, Charles L. et al. (eds.). 1978. *Social Archeology. Beyond Subsistence and Dating.* (Studies in Archeology). New York: Academic Press.

Rees, Mina. 1980. "The Mathematical Sciences and World War II." *American Mathematical Monthly* 87:607–21.

Reich, Karin, and Helmuth Gericke (ed., trans.). 1973. François Viète, *Einführung in die Neue Algebra.* (Historiae Scientiarum Elementa, 5). Munich: Werner Fritsch.

Reid, Constance. 1979. *Richard Courant, 1888–1972. Der Mathematiker als Zeitgenosse.* Berlin: Springer.

Reisch, Gregor. 1512. *Margarita philosophica.* Straßburg.

————. 1549. *Artis metiendis seu geometriae liber.* Ex *Margarita philosophica.* Paris: Apud Guil. Morelium.

Renaud, H. P. J. 1938. "Ibn al-Bannâ' de Marrakech, ṣûfî et mathématicien (XIII$^e$-XIV$^e$ s. J.C.)." *Hespéris* 25:13–42.

————. 1944. "Sur un passage d'ibn Khaldûn relatif à l'histoire des mathématiques." *Hespéris* 31:35–47.

Renger, Johannes. 1979. "Interaction of Temple, Palace, and 'Private Enterprise' in the Old Babylonian Economy." In Lipinski (ed.) 1979:249–56.

*Report from Iron Mountain on the Possibility and Desirability of Peace.* With Introductory Material by Leonard C. Lewin. London: MacDonald, 1968.

Richard de Fournival. *Biblionomia.* In Delisle 1868: II, 518–535.

Rinaldi, Raffaele (ed.). 1980. Leon Battista Alberti, *Ludi matematici.* (Quaderni della Fenice, 66). Milan: Guanda.

Risner, Friedrich. 1572. *Opticae thesaurus. Alhazeni arabis libri septem, nunc primi editi. Eiusdem liber de crepusculis et nubium ascensionibus. Item Vitellonis Thuringopoloni libri X.* Omnes instaurati, figuris illustrati et aucti, adiectis etiam in Alhazenum commentarijs. Basel 1572. (Reprint with an introduction by David C. Lindberg, New York: Johnson Reprint Corporation, 1972).

Ritter, Hellmut. 1916. "Ein arabisches Handbuch der Handelswissenschaft." *Islam* 7:1–91.

Robinson, Joan. 1964. *Economic Philosophy.* Harmondsworth: Penguin.

Rodet, Léon. 1878. "L'algèbre d'al-Khârizmi et les méthodes indienne et grecque." *Journal Asiatique,* 7$^e$ série 11:5–98.

Röllig, W. (ed.). 1969. *Lišān mithurti. Festschrift Wolfram Freiherr von Soden zum 19.VI.1968 gewidmet von Schülern und Mitarbeitern.* (Altes Orient und Altes Testament, 1). Kevelaer: Butzon & Bercker/Neukirchen-Vluyn: Neukirchener Verlag des Erziehungsvereins.

Romaine, Suzanne. 1988. *Pidgin and Creole Languages.* London: Longman.

Römer, W. H. Ph. 1980. *Das sumerische Kurzepos "Bilgameš und Akka".* (Alter Orient und Altes Testament, 209/1. Nimwegener sumerologische Studien I. Studien zu sumerischen literarischen Texten 1). Kevelaer: Butzon & Bercker/Neukirchen-Vluyn: Neukirchener Verlag.

Rose, Hilary, and Steven Rose. 1970. *Science and Society.* Harmondsworth: Penguin.

Rose, Paul Lawrence. 1972. "Commandino, John Dee, and the *De superficierum divisionibus* of Machometus Bagdadinus." *Isis* 63:88–93.

———. 1973. "Humanist Culture and Renaissance Mathematics: The Italian Libraries of the *Quattrocento.*" *Studies in the renaissance* 20:46–105.

———. 1975. *The Italian Renaissance of Mathematics. Studies on Humanists and Mathematicians from Petrarch to Galileo.* (Travaux d'Humanisme et Renaissance, 145). Geneva: Librairie Droz.

———. 1976. "Bartolomeo Zamberti's Funeral Oration for the Humanist Encyclopaedist Giorgio Valla." In Clough 1976: 299–310.

Rosen, Edward. "Regiomontanus." *DSB* XI, 348–52.

———. "Schöner." *DSB* XII, 199–200.

Rosen, Frederic (ed., trans.). 1831. The *Algebra* of Muhammad ben Musa. London: The Oriental Translation Fund.

Rosen, Saul (ed.). 1967. *Programming Systems and Languages.* New York: McGraw-Hill.

Rosenfield, B. A., and A. T. Grigorian. "Thābit ibn Qurra." *DSB* XIII, 288–95.

Rosenthal, Franz (ed., trans.). 1958. Ibn Khaldûn, *The Muqaddimah. An Introduction to History.* 3 vols. London: Routledge & Kegan Paul.

Ross, Richard P. 1974. "Oronce Fine's Printed Works: Additions to Hillard and Poulle's Bibliography." *Bibliothèque d'humanisme et renaissance* 36:83–85.

Ross, W. D. (ed., trans.). 1928. Aristotle, *Metaphysica* (The Works of Aristotle, 8). Second edition. Oxford: Clarendon Press.

Rosser, J. Barkley. 1982. "Mathematics and Mathematicians in World War II." *Notices of the American Mathematical Society* 29(6):509–15.

Rozenfeld, Boris A. 1976. "The List of Physico-Mathematical Works of Ibn al-Haytham Written by Himself." *Historia Mathematica* 3:75–76.

Rudolff, Christoff. 1540. *Künstliche rechnung mit der ziffer und mit den zalpfenningē/ sampft der Wellischen Practica/ und allerley forteyl auff die Regel de tri. Item vergleichung mancherley Land uñ Stet/ gewicht/ Einmas/ Müntz etc.* Second edition, Vienna.

Runciman, W. G. 1982. "Origins of States: The Case of Archaic Greece." *Comparative Studies in Society and History* 24:351–77.

Sabloff, Jeremy A., and C. C. Lamberg-Karlovsky (eds.). 1975. *Ancient Civilization and Trade.* Albuquerque: University of New Mexico Press.

Sabra, A. I. 1968. "Thābit ibn Qurra on Euclid's Parallels Postulate." *Journal of the Warburg and Courtauld Institutes* 31:12–32.

———. 1969. "Simplicius' Proof of Euclid's Parallels Postulate." *Journal of the Warburg and Courtauld Institutes* 32:1–24.

Sachs, Abraham J. 1946. "Notes on Fractional Expressions in Old Babylonian Mathematical Texts." *Journal of Near Eastern Studies* 5:203–14.

Sachs, Eva. 1917. *Die fünf platonischen Körper. Zur Geschichte der Mathematik und der Elementenlehre Platons und der Pythagoreer.* (Philologische Untersuchungen, 24, Heft). Berlin: Weidmannsche Buchhandlung.

Saidan, Ahmad S. "Al-Baghdādī, Abū Manṣūr [ . . . ] ibn Ṭāhir [ . . . ]." *DSB* XV, 9–10.

———. "Al-Nasawī." *DSB* IX, 614–15.

———. "Al-Qalaṣādī." *DSB* XI, 229–30.

———. "Al-Umawī." *DSB* XIII, 539–40.

———. 1974. "The Arithmetic of Abū'l-Wafā'." *Isis* 65:367–75.

———(ed., trans.). 1978. *The "Arithmetic" of al-Uqlīdisī. The Story of Hindu-Arabic Arithmetic as Told in "Kitāb al-Fuṣūl fī al-Ḥisāb al-Hindī" by Abū al-Ḥasan Aḥmad ibn Ibrāhīm al-Uqlīdisī written in Damascus in the Year 341* (A.D. 952/53). Dordrecht: Reidel.

———. 1978a. "Number Theory and Series Summations in Two Arabic Texts." In al-Hassan 1978: 145–63.

Sajo, Géza (ed.). 1964. Boetii de Dacia *Tractatus de eternitate mundi.* (Quellen und Studien zur Geschichte der Philosophie, 4). Berlin: De Gruyter.

Sammett, Jean E. 1969. *Programming Languages: History and Fundamentals.* Englewood Cliffs, NJ: Prentice-Hall.

Sarasvati Amma, T. A. 1979. *Geometry in Ancient and Medieval India.* Delhi: Motilal Banarsidass.

Sarfatti, Gad. 1968. *Mathematical Terminology in Hebrew Scientific Literature of the Middle Ages.* [In Hebrew, English abstract]. Jerusalem: The Magnes Press/The Hebrew University.

Sarton, George. 1927. *Introduction to the History of Science.* I. *From Homer to Omar Khayyam.* (Carnegie Institution of Washington, Publication 376). Baltimore: William & Wilkins.

———. 1931. *Introduction to the History of Science.* II. *From Rabbi ben Ezra to Roger Bacon.* In two parts. (Carnegie Institution of Washington, Publication 376). Baltimore: William & Wilkins.

Sartre, Jean-Paul. 1960. *Critique de la raison dialectique* précédé de *Questions de méthode. Tôme I, Théories des ensembles pratiques*. Paris: Gallimard.

Sayılı, Aydin. 1960. *The Observatory in Islam and its Place in the General History of the Observatory*. (Publications of the Turkish Historical Society, Series VII, Nᵒ 38). Ankara: Türk Tarih Kurumu Basımevi.

———. 1962. *Abdülhamid ibn Türk'ün katışık denklemlerde mantıkî zaruretler adlı yazısı ve zamanın cebri* (Logical Necessities in Mixed Equations by ʿAbd al Ḥamîd ibn Turk and the Algebra of his Time). (Publications of the Turkish Historical Society, Series VII, Nᵒ 41). Ankara: Türk Tarih Kurumu Basımevi.

Schacht, Joseph. 1974. "Islamic Religious Law." In Schacht and Bosworth 1974: 392–405.

Schacht, Joseph, and C. E. Bosworth (eds.). 1974. *The Legacy of Islam*. Second edition. Oxford: Oxford University Press.

Schmandt-Besserat, Denise (ed.). 1976. *The Legacy of Sumer*. Invited Lectures on the Middle East at the University of Texas as Austin. (Bibliotheca Mesopotamica, 4). Malibu, CA: Undena.

———. 1977. "An Archaic Recording System and the Origin of Writing." *Syro-Mesopotamian Studies* 1(2).

———. 1978. "The Earliest Precursor of Writing." *Scientific American* 238(6) (June 1978), 38–47 (European edition).

———. 1986. "Tokens: Facts and Interpretation." *Visible Language* 20:250–72.

———. 1988. "Tokens as Funerary Offerings." *Vicino Oriente* 7:3–9, Tav. I–V.

Schmeidler, Felix (ed.). 1972. Joannis Regiomontani *Opera collectanea*. Faksimile-drucke von neun Schriften Regiomontans und einer von ihm gedruckten Schrift seines Lehrers Purbach. Zusammengestellt und mit einer Einleitung herausgegeben. (Milliaria 10,2). Osnabrück: Otto Zeller.

Schmitt, Charles B. 1978. "Reappraisals in Renaissance Science." [Essay review of Westman and McGuire 1977]. *History of Science* 16:200–14.

Schmitz, Rudolf, and Fritz Krafft (eds.). 1980. *Humanismus und Naturwissenschaften*. (Beiträge zur Humanismusforschung, 6). Boppard: Harald Boldt.

Schneider, Ivo. 1970. "Die mathematischen Praktiker im See, Vermessungs- und Wehrwesen vom 15. bis zum 19. Jahrhundert." *Technikgeschichte* 37:210–42.

———. 1979. *Archimedes: Ingenieur, Naturwissenschaftler und Mathematiker* (Reihe Erträge der Forschung, 102). Darmstadt: Wissenschaftliche Buchgesellschaft.

Schneider, Nikolaus. 1940. "Die Urkundenbehälter von Ur III und ihre archivalische Systematik." *Orientalia*, NS 9:1–16.

Schöne, Hermann (ed., trans.). 1903. Herons von Alexandria *Vermessungslehre und Dioptra*. (Heronis Alexandrini Opera quae supersunt omnia, 3). Leipzig: Teubner.

Schöner, Ioh. (ed.). 1534. *Algorithmus demonstratus*. Nürnberg: Ioh. Schöner.

Schuhl, Pierre-Maxime. 1949. *Essai sur la formation de la pensée grecque. Introduction historique à une étude de la philosophie platonicienne*. 2⁰ édition revue et augmentée. Paris: Presses Universitaires de France.

*Science and Technology. A Five-Year Outlook*. Report Prepared at the Request of the National Science Foundation. San Francisco: W. H. Freeman, 1979.

*Science at the Cross Roads*. Papers Presented to the International Congress of the History of Science and Technology [ . . . ] 1931. With a New Foreword by Joseph Needham and a New Introduction by P. G. Werskey. London: Frank Cass, 1971.

Scriba, Christoph J. 1985. "Die mathematischen Wissenschaften im mittelalterlichen Bildungskanon der Sieben Freien Künste." *Acta historica Leopoldina* 16:25–54.

Service, Elman R. 1975. *Origins of the State and Civilization. The Process of Cultural Evolution*. New York: W. W. Norton.

Sesiano, Jacques. 1976. "Un Mémoire d'Ibn al-Haytham sur un Problème arithmétique solide." *Centaurus* 20:189–215.

———. 1980. "Herstellungsverfahren magischer Quadrate aus islamischer Zeit (I)." *Sudhoffs Archiv* 64:187–96.

——— (ed., trans.). 1982. Books IV to VII of Diophantus' *Arithmetica* in the Arabic Translation Attributed to Qusṭā ibn Lūqā. (Sources in the History of Mathematics and Physical Sciences, 3). New York: Springer.

———. 1984. "Une arithmétique médiévale en langue provençale." *Centaurus* 27:26–75.

———. 1987. "Herstellungsverfahren magischer Quadrate aus islamischer Zeit (II') *Sudhoffs Archiv* 71:78–89.

Shorey, Paul (ed., trans.). 1978. Plato, *The Republic*. With an English Translation. 2 vols. (Loeb Classical Library). London: Heinemann/Cambridge, MA: Harvard University Press. First ed. 1930, 1935.

Shumaker, Wayne. 1972. *The Occult Sciences in the Renaissance. A Study in Intellectual Patterns*. Berkeley: University of California Press.

Sigerist, Henry E. 1938. "Science and Democracy." *Science and Society* 2 (1937–38):291–99.

Silverstein, Th. (ed.) 1955. *"Liber Hermetis Mercurii Triplicis de vi rerum principiis."* Archives d'histoire doctrinale et littéraire du Moyen Age 30:217–302.

Simms, D. L. 1989. "A Problem for Archimedes.'" *Technology and Culture* 30:177f.

Simon, Gérard. 1979. *Kepler astronome astrologue.* Paris: Gallimard.

Singleton, Charles S. (ed.). 1968. *Art, Science and History in the Renaissance.* Baltimore: Johns Hopkins Press.

Siraisi, Nancy G. 1973. *Arts and Sciences at Padua. The* Studium *of Padua before 1350.* Toronto: Pontifical Institute of Mediaeval Studies.

Sjöberg, Åke W. 1972. "In Praise of the Scribal Art." *Journal of Cuneiform Studies* 24 (1971–72):126–29.

———. 1973. "Der Vater und sein Mißratener Sohn." *Journal of Cuneiform Studies* 25:105–69.

———. 1975. "Der Examenstext A." *Zeitschrift für Assyriologie und vorderasiatische Archäologie* 64:137–76.

———. 1976. "The Old Babylonian Eduba." In *Sumerological Studies . . . ,* pp. 159–79.

Skabelund, Donald, and Phillip Thomas. 1969. "Walter of Odington's Mathematical Treatment of the Primary Qualities." *Isis* 60:331–50.

Smith, David Eugene, and Marcia L. Latham (eds., trans.). 1954. The *Geometry* of René Descartes. New York: Dover.

Smith, Gertrude. 1956. "More Recent Theories on the Origin and Interrelation of the First Classifications of Greek Laws." *Cahiers d'Histoire Mondiale* 3 (1956–57):173–95.

Smith, R. Jeffrey. 1982. "Scientists Involved in Atom Test Deception." *Science* 218:545–47.

Smyth, Herbert Weir. 1973. Aeschylus. I. *Suppliant Maidens. Persians. Prometheus. Seven Against Thebes.* II. *Agamemnon. Libation-Bearers. Eumenides. Fragments.* (Loeb Classical Library). Cambridge, MA: Harvard University Press/London: Heinemann. First ed. 1922–26.

Soboul, Albert (ed.). 1976. *Textes choisis de l'Encyclopédie ou Dictionnaire raisonné des Sciences, des Arts et des Métiers.* Deuxième édition revue et augmentée. (Les classiques du peuple). Paris: Éditions Sociales.

Sollberger, Edmond, and Jean-Robert Kupper. 1971. *Inscriptions royales sumériennes et akkadiennes.* (Littératures anciennes du Proche-Orient). Paris: Éditions du Cerf.

Soubeyran, Denis. 1984. "Textes mathématiques de Mari." *Revue d'Assyriologie* 78:19–48.

Souissi, Mohamed (ed., trans.). 1969. Ibn al-Bannāʾ, *Talkhīṣ aʿmāl al-ḥisāb*. Tunis: L'Université de Tunis.

———— (ed., trans.). 1988. Qalaṣādī, *Kašf al-asrār ʿan ʿilm ḥurūf al-ġubār*. Carthage: Maison Arabe du Livre.

Stamm, Edward. 1936. "Tractatus de Continuo von Thomas Bradwardina". *Isis* 26:13–32.

Steck, Max. "Dürer." *DSB* IV, 258–61.

Steinschneider, Moritz. 1865. "Die 'mittleren' Bücher de Araber und ihre Bearbeiter." *Zeitschrift für Mathematik und Physik* 10:456–98.

———— (ed.). 1872. "Vite di matematici arabi tratte da un'opera inedita di Bernardino Baldi, con note." *Bulletino di Bibliografia e di Storia delle Scienze matematiche e fisiche* 5:427–534.

Steinschneider, Moritz. 1893. *Die hebraeischen Übersetzungen des Mittelalters und die Juden als Dolmetscher*. Berlin: Kommissionsverlag des Bibliographischen Bureaus.

————. 1896. "Die arabischen Übersetzungen aus dem Griechischen. Zweiter Abschnitt: Mathematik." *Zeitschrift der Deutschen Morgenländischen Gesellschaft* 50:161–219, 337–417.

————. 1904. "Die europäischen Übersetzungen aus dem Arabischen bis Mitte des 17. Jahrhunderts." 2 Teile. *Sitzungsberichte der Kaiserlichen Akademie der Wissenschaften in Wien, philosophisch-historische Klasse* 149/4 (1904) and 151/1 (1905).

Stiefel, Tina. 1977. "The Heresy of Science: A Twelfth-Century Conceptual Revolution." *Isis* 68:347–62.

Stifel, Michael. 1544. *Arithmetica integra*. Nürnberg: Ioh. Petreius.

————. 1545. *Deutsche Arithmetica, inhaltend Die Hausrechnung, Die Deutsche Coß, Die Kirchrechnung*. Nürnberg: Ioh. Petreius.

Stone, Elisabeth C. 1982. "The Social Role of the Nadītu Women in Old Babylonian Nippur." *Journal of the Economic and Social History of the Orient* 25:50–70.

Strauss, Walter (ed., trans.). 1977. *The Painter's Manual: A Manual of Measurement of Lines, Areas and Solids by Means of Compass and Ruler Assembled by Albrecht Dürer for the Use of All Lovers of Art with Appropriate Illustrations Arranged*. New York: Abaris Books.

Struik, Dirk J. 1936. "Concerning Mathematics." *Science and Society* 1 (1936–37):81–101.

Struve, V. V. 1969. "Some New Data on the Organization of Labour and on the Social Structure of Sumer During the Reign of the Third Dynasty of Ur." In Diakonoff 1969: 127–171 [slightly abridged version of a paper from 1948].

Strømholm, Per. "Valerio". *DSB* XIII, 560f.

*Sumerological Studies* in Honor of Thorkild Jacobsen on his Seventieth Birthday, June 7, 1974. (The Oriental Institute of the University of Chicago, Assyriological Studies, 20). Chicago: University of Chicago Press, 1976.

Surtz, Edward, S. J., and J. H. Hexter (eds., trans.). 1965. Thomas More, *Utopia*. (The Yale Edition of the Complete Works of St. Thomas More, 4). New Haven: Yale University Press.

Suter, Heinrich. 1889. "Die mathematischen und naturphilosophischen Disputationen an der Universität Leipzig 1512–1526." *Bibliotheca mathematica*, 2. Folge 3:17–22.

―――. 1892. "Das Mathematikerverzeichniss im Fihrist des Ibn Abî Jaʿḳûb an-Nadîm." *Zeitschrift für Mathematik und Physik* 37 (Supplement):1–87.

―――. 1893. "Nachtrag zu meiner Ueberesetzung des Mathematikerverzeichnisses im Fihrist des Ibn Abî Jaʿḳûb an-Nadîm." *Zeitschrift für Mathematik und Physik* 38:126–27.

―――. 1900. "Die Mathematiker und Astronomen der Araber und ihre Werke." *Abhandlungen zur Geschichte der mathematischen Wissenschaften mit Einschluss ihrer Anwendungen* 10.

―――. 1901. "Das Rechenbuch des Abû Zakarîjâ el-Ḥaṣṣar." *Bibliotheca Mathematica*, 3. Folge 2 (1901–2):12–40.

―――. 1906. "Über das Rechenbuch des Alî ben Aḥmed el-Nawawî." *Bibliotheca Mathematica*, 3. Folge 7 (1906–7):113–19.

―――. 1910. "Das Buch der Seltenheiten der Rechenkunst von Abû Kāmil al-Miṣrī." *Bibliotheca Mathematica*, 3. Folge 11 (1910–11):100–20.

―――. 1910a. "Das Buch der Auffindung der Sehnen im Kreise von Abū'l-Raihān Muh. el-Bīrūnī. Übersetzt und mit Kommentar versehen." *Bibliotheca Mathematica*, 3. Folge 11 (1910–11):11–38.

Sylla, Edith. 1971. "Medieval Quantification of Qualities: The 'Merton School.' " *Archive for History of Exact Sciences* 8 (1971–72):9–39.

Szabó, Arpád. 1969. *Anfänge der griechischen Mathematik*. Munich: R. Oldenbourg/ Budapest: Akadémiai Kiadó.

Tannery, Paul (ed., trans.). 1893. Diophanti Alexandrini *Opera omnia* cum graecis commentariis. 2 vols. Leipzig: Teubner, 1893–95.

Taton, René. "Monge." *DSB* IX, 469–78.

Taylor, E. G. R. 1954. *The Mathematical Practitioners of Tudor & Stuart England.* Cambridge: Cambridge University Press.

———. 1966. *The Mathematical Practitioners of Hanoverian England, 1714–1840.* Cambridge: Cambridge University Press.

Taylor, Jerome (ed., trans.). 1961. The *Didascalicon* of Hugh of St. Victor. *A Medieval Guide to the Arts.* (Records of Civilization. Sources and Studies, N° 64). New York: Columbia University Press.

Theuer, Max (trans.). 1912. Leone Battista Alberti, *Zehn Bücher über die Baukunst.* Ins deutsche übertragen, eingeleitet und mit Anmerkungen und Zeichnungen versehen. Vienna: Hugo Heller.

Thomas, Ivor (ed., trans.). 1939. *Selections Illustrating the History of Greek Mathematics.* In two volumes. (Loeb Classical Library). London: Heinemann/Cambridge, MA: Harvard University Press, 1939, 1941.

Thompson, Craig R., J. K. Sowards and A. H. T. Levi (eds.). 1978. Erasmus, *Literary and Educational Writings.* 6 vols. (Collected Works of Erasmus, 23–28). Toronto: University of Toronto Press, 1978–86.

Thompson, James Westfall. 1929. "The Introduction of Arabic Science into Lorraine in the Tenth Century." *Isis* 12:184–93.

Thompson, Silvanus P. [anonymous editor and translator, for the Gilbert Society]. 1900. William Gilbert of Colchester, Physician of London. *On the Magnet, magnetic Bodies also, and on the Great Magnet the Earth; a new Physiology, demonstrated by many arguments and experiments.* London: The Chiswick Press. Reprinted with a foreword by Derek J. Price. New York: Basic Books, 1958.

Thompson, Stith. 1946. *The Folktale.* New York: The Dryden Press.

———. 1975. *Motif-Index of Folk-Literature. A Classification of Narrative Elements in Folktales, Ballads, Myths, Fables, Mediaeval Romances, Exempla, Fabliaux, Jest Books and Local Legends.* 6 vols. Rev. and enl. edition. London: Indiana University Press.

Thomson, Ron B. 1976. "Jordanus de Nemore: Opera." *Mediaeval Studies* 38: 97–144.

——— (ed., trans.). 1978. *Jordanus de Nemore and the Mathematics of Astrolabes: De plana spera.* Toronto: Pontifical Institute of Mediaeval Studies.

Thorndike, Lynn. 1944. *University Records and Life in the Middle Ages.* (Records of Civilization). New York: Columbia University Press.

————. 1954. "Computus." *Speculum* 29:223–38.

————. 1955. "The True Place of Astrology in the History of Science." *Isis* 46: 273–78.

Thureau-Dangin, François. 1897. "Un cadastre chaldéen." *Revue d'Assyriologie* 4:13–27.

————. 1907. *Die sumerischen und akkadischen Königsinschriften.* (Vorderasiatische Bibliothek, I. Band Abteilung 1). Leipzig: J. C. Hinrichs'sche Buchhandlung.

————. 1938. *Textes mathématiques babyloniens.* (Ex Oriente Lux, Deel 1). Leiden: Brill.

*TMS:* E. M. Bruins and M. Rutten. *Textes mathématiques de Suse.* (Mémoires de la Mission Archéologique en Iran, XXXIV). Paris: Paul Geuthner, 1961.

Tobies, Renate. 1984. "Untersuchungen zur Rolle der Carl-Zeiß-Stiftung für die Entwicklung der Mathematik an der Universität Jena." *NTM—Schriftenreihe für Geschichte der Naturwissenschaften, Technik und Medizin* 21(1):33–43.

Toomer, G. J. "Al-Khwārizmī." *DSB* VII, 358–65.

————. 1984. "Lost Greek Mathematical Works in Arabic Translation." *The Mathematical Intelligencer* 6(2):32–38.

Tredennick, Hugh (ed., trans.). 1980. Aristotle, *The Metaphysics.* 2 vols. (Loeb Classical Library). Cambridge, MA: Harvard University Press/London: Heinemann. First ed. 1933–35.

Tredennick, Hugh, and E. S. Forster (eds., trans.). 1960. Aristotle, *Posterior Analytics* and *Topica.* (Loeb Classical Library) Cambridge, MA: Harvard University Press/London: Heinemann.

Tritton, A. S. 1957. *Materials on Muslim Education in the Middle Ages.* London: Luzac.

Tropfke, J./Vogel, Kurt, et al. 1980. *Geschichte der Elementarmathematik,* 4. Auflage. Band 1: *Arithmetik und Algebra.* Vollständig neu bearbeitet von Kurt Vogel, Karin Reich, Helmuth Gericke. Berlin: De Gruyter.

Tummers, Paul M. J. E. 1980. "The Commentary of Albert on Euclid's Elements of Geometry." In J. A. Weisheipl 1980: 479–99.

Tyumenev, A. I. 1969. "The State Economy in Ancient Sumer." [Summary of the Book *State Economy in Ancient Sumer,* Moscow-Leningrad 1956]. In Diakonoff (ed.) 1969: 70–87.

————. 1969a. "The Working Personnel on the Estate of the Temple of Ba-$U_2$ in Lagaš during the Period of Lugalanda and Urukagina (25th–24th cent. B.C.)"

[Slightly abbreviated translation of pp. 136–69 in *State Economy in Ancient Sumer*, Moscow-Leningrad 1956]. In Diakonoff (ed.) 1969:88–126.

Unguru, Sabetai (ed., trans.). 1977. Witelonis *Perspectivae liber primus*. Book I of Witelo's *Perspectiva*. (Studia Copernicana, 15). Wroclaw: Ossolineum.

Vaiman, A. A. 1974. "Über die protosumerische Schrift." *Acta Antiqua Academiae Scientiarum Hungaricae* 22:15–27.

van der Vyver, A. 1936. "Les plus anciennes Traductions latines médiévales (X$^e$-XI$^e$ siècles) de Traités d'Astronomie et d'Astrologie." *Osiris* 11:658–91.

van der Waerden, B. L. 1962. *Science Awakening*. Second edition. Groningen: Noordhoff.

———. 1976. [Review of Knorr 1975]. *Historia Mathematica* 3:497–99.

———. 1979. *Die Pythagoreer. Religiöse Bruderschaft und Schule der Wissenschaft*. Zürich: Artemis Verlag.

———. 1983. *Geometry and Algebra in Ancient Civilizations*. Berlin: Springer Verlag.

van Dijk, J. J. A. 1953. *La sagesse suméro-accadienne. Recherches sur les genres littéraires des textes sapientiaux*. Leiden: Brill.

van Egmond, Warren. 1988. "How Algebra Came to France." In Hay 1988: 127–44.

van Steenberghen, Fernand. 1955. *The Philosophical Movement in the Thirteenth Century*. Lectures Given under the Auspices of the Department of Scholastic Philosophy, The Queen's University, Belfast. Edinburgh: Nelson.

———. 1966. *La Philosophie au XIII$^e$ siècle*. (Philosophes médiévaux, 9). Louvain: Publications Universitaires/Paris: Béatrice-Nauwelaerts.

Vanstiphout, H. L. J. 1978. "Lipit-Eštar's Praise in the Edubba." *Journal of Cuneiform Studies* 30:33–61.

Vernant, Jean-Pierre. 1982. *The Origins of Greek Thought*. Ithaca, NY: Cornell University Press.

Vernet, J. "Ibn al-Bannā'." *DSB* I, 437–38.

Vickers, Brian (ed.). 1984. *Occult and Scientific Mentalities in the Renaissance*. Cambridge: Cambridge University Press.

———. 1984a. "Analogy versus Identity: The Rejection of Occult Symbolism, 1580–1680." In Vickers 1984: 95–163.

Victor, Stephen K. 1979. *Practical Geometry in the Middle Ages. Artis cuiuslibet consummatio* and the *Pratike de geometrie*. Edited with Translations and Commentary. (Memoirs of the American Philosophical Society, 134). Philadelphia: The American Philosophical Society.

Vogel, Kurt. "John of Gmunden." *DSB* VII, 117–22.

———. "Peurbach." *DSB* XV, 473–79.

———. "Stifel," *DSB* XIII, 58–62.

———. 1936. "Beiträge zur griechischen Logistik." Erster Theil. *Sitzungsberichte der mathematisch-naturwissenschaftlichen Abteilung der Bayerischen Akademie der Wissenschaften zu München* 1936, 357–472.

———. (ed.). 1963. Mohammed ibn Musa Alchwarizmi's *Algorismus. Das früheste Lehrbuch zum Rechnen mit indischen Ziffern.* Nach den einzigen (lateinischen) Handschrift (Cambridge Un.Lib.Ms.Ii.6.5) in Faksimile mit Transkription und Kommentar herausgegeben. Aalen: Otto Zeller.

——— (ed., trans.). 1968. *Chiu chang suan shu. Neun Bücher arithmetischer Technik. Ein chinesisches Rechenbuch für den praktischen Gebrauch aus der frühen Hanzeit (202 v. Chr. bis 9 n. Chr.).* (Ostwalds Klassiker der Exakten Wissenschaften. Neue Folge, 4). Braunschweig: Friedrich Vieweg & Sohn.

———. 1977. *Ein italienisches Rechenbuch aus dem 14. Jahrhundert (Columbia X 511 A13).* (Veröffentlichungen des Deutschen Museums für die Geschichte der Wissenschaften und der Technik. Reihe C, Quellentexte und Übersetzungen, 33). Munich.

Vogel, Kurt, and H. Gerstinger. 1932. "Eine stereometrische Aufgabensammlung im Pap.Gr. Vind. 19996." *Mitteilungen aus der Papyrussammlung der Nationalbibliothek in Wien,* neue Serie 1, *Griech. literar. Papyri* 1:11–76.

Vögelin, Johann. 1530. *Elementale geometricum ex euclidis geometria ad omnium mathematices studiosorum utilitatem decerptum.* See Bradwardine 1530.

Volkmann-Schluck, K. H. 1968. *Nicolaus Cusanus. Die Philosophie im Übergang vom Mittelalter zur Neuzeit,* zweite, durchgesehene Auflage. Frankfurt am Main: Vittorio Klostermann.

von Fritz, Kurt. "Philolaos of Crotona." *DSB* X, 589–91.

———. 1971. *Grundprobleme der Geschichte der antiken Wissenschaft.* Berlin: De Gruyter.

von Grünebaum, G. E. (ed.). 1971. *Theology and Law in Islam.* Wiesbaden: Harrassowitz.

von Soden, Wolfram. 1936. "Leistung und Grenze sumerischer und babylonischer Wissenschaft." *Die Welt als Geschichte* 2:411–64, 507–57.

Wallace, William A. "Dietrich von Freiberg." *DSB* IV, 92–95.

Wallis, W. A. 1980. "The Statistical Research Group, 1942–1945." *Journal of the American Statistical Association* 75:320–35.

Waltz, James. 1981. [Review of Bulliet 1979]. *Speculum* 56:360–62.

Waters, D. W. 1976. *Science and the Techniques of Navigation in the Renaissance.* (Maritime Monographs and Reports, No. 19—1976). Greenwich: National Maritime Museum.

Watt, W. Montgomery. 1973. "L'influence de l'Islam sur l'Europe médiévale." *Revue des études islamiques* 41:129–56.

Watt, W. Montgomery, and Alford T. Welch. 1980. *Der Islam. I. Mohammed und die Frühzeit—Islamisches Recht—Religiöses Leben.* (Die Religionen der Menschheit, 25,1). Stuttgart: W. Kohlhammer.

Weisheipl, James A., O.P. 1966. "Development of the Arts Curriculum at Oxford in the Early Fourteenth Century." *Mediaeval Studies* 28:151–75.

———. 1975. *Friar Thomas d'Aquino. His Life, Thought, and Works.* Oxford: Blackwell.

——— (ed.). 1980. *Albertus Magnus and the Sciences: Commemorative Essays 1980.* (Studies and Texts, 49). Toronto: The Pontifical Institute of Mediaeval Studies.

Weiss, Harvey. 1977. "Periodization, Population, and Early State Formation in Khuzistan." *Bibliotheca Mesopotamica* 7:347–69.

Werner, Ernst. 1976. "Stadtluft macht Frei: Frühscholastik und bürgerliche Emancipation in der ersten Hälfte des 12. Jahrhunderts." *Sitzungsberichte der Sächsischen Akademie der Wissenschaften zu Leipzig, Philologisch-historische Klasse* 118(5)

———. 1980. *Stadt und Geistesleben im Hochmittelalter.* (Forschungen zur Mittelalterlichen Geschichte). Berlin: Hermann Böhlau.

Werner, H. 1982. [Grußwort des Vorsitzenden der Deutschen Mathematiker-Vereinigung zur Eröffnung der DMV-Tagung 1982]. *Mitteilungen der Deutschen Mathematiker-Vereinigung* 1982 Nr. 4, 66–69.

Wernicke, J. 1982. "Zur Forschungsförderung im Wehrtechnischen Bereich der Elektronik." Beitrag zum Symposium. "Ossietzky-Tage '82: Militärische Einflüsse auf die Wissenschaft und militärische Anwendung ihrer Ergebnisse", Oldenburg, 3.–4. Mai 1982. *Mimeo.*

Werskey, Gary. 1979. *The Visible College. The Collective Biography of British Scientific Socialists of the 1930s.* New York: Hold, Rinehart and Winston.

Westman, Robert S. 1977. "Magical Reform and Astronomical Reform: The Yates Thesis Reconsidered." In Westman and McGuire 1977: 2–91.

————. 1984. "Jung, Pauli, and the Kepler-Fludd-Debate." In Vickers 1984: 177–229.

Westman, Robert S. and J. E. McGuire. 1977. *Hermeticism and the Scientific Revolution.* Papers read at a Clark Library Seminar, March 9, 1974. Los Angeles: University of California.

Whitehead, Alfred North. 1925. *Science and the Modern World.* Lowell Lectures, 1925. Cambridge: Cambridge University Press.

Whiting, Robert M. 1984. "More Evidence for Sexagesimal Calculations in the Third Millennium B.C." *Zeitschrift für Assyriologie und Vorderasiatische Archäologie* 74:59–66.

Wicksteed, Philip H., and Francis M. Cornford (eds., trans.). 1970. Aristotle, *The Physics.* 2 vols. (Loeb Classical Library). Cambridge, MA: Harvard University Press/London: Heinemann. First ed. 1929–34.

Wiedemann, Eilhard. 1970. *Aufsätze zur arabischen Wissenschaftsgeschichte.* 2 vols. Mit einem Vorwort und Indices herausgegeben von Wolfdietrich Fischer. (Collectanea 6/1–2). Hildesheim: Georg Olm.

Wiener, Norbert. 1964. *I am a Mathematician. The Later Life of a Prodigy.* Cambridge, MA: M.I.T. Press.

Willner, Hans (ed.). 1903. *Des Adelard von Bath Traktat "De eodem et diverso",* zum ersten male herausgegeben und historisch-kritisch untersucht. (Beiträge zur Geschichte der Philosophie des Mittelalters. Texte und Untersuchungen 4,1). Münster: Verlag der Aschendorffschen Buchhandlung.

Wilpert, Paul (ed.). 1967. Nikolaus von Kues, *Werke.* Neuausgabe des Straßburger Drucks von 1488. 2 vols. (Quellen und Studien zur Geschichte der Philosophie, 5–6). Berlin: De Gruyter.

Wilpert, Paul, Raymond Klibansky, and Hans Gerhard Senger (eds., trans.). 1964. Nikolaus von Kues, *Die belehrte Unwissenheit.* 3 Hefte. (Schriften des Nikolaus von Kues in deutscher Übersetzung, Heft 15a, 15b, 15c). Hamburg: Felix Meiner, 1964, 1967, 1977.

Wilson, Curtis. 1953. "Pomponazzi's Criticism of Calculator." *Isis* 44:355–62.

Winterberg, Constantin (ed., trans.). 1896. Fra Luca Pacioli, *Divina proportione, Die Lehre vom goldenen Schnitt.* Nach der Venezianischen Ausgabe vom Jahre 1509 neu herausgegeben, übersetzt und erläutert. Vienna: Carl Graeser.

Witelo. *Perspectiva,* Book I. Ed., trans. Unguru 1977.

Witmer, T. Richard (ed., trans.). 1968. Girolamo Cardano, *The Great Art or The Rules of Algebra.* Cambridge, MA: M.I.T. Press.

———— (ed., trans.). 1983. François Viète, *The Analytic Art: Nine Studies in Algebra, Geometry, and Trigonometry from the Opus restitutae mathematicae analyseos, seu algebrâ novâ*. Kent, OH: Kent State University Press.

Wittfogel, Karl. 1957. *Oriental Despotism: A Comparative Study of Total Power*. New Haven, CT.: Yale University Press.

Woepcke, Franz (ed., trans.). 1851. *L'Algèbre* d'Omar Alkhayyâmî, publiée, traduite et accompagnée d'extraits de manuscrits inédits. Paris: Benjamin Duprat.

————. 1852. "Notice sur une théorie ajoutée par Thâbit ben Korrah à l'arithmétique spéculative des Grecs." *Journal Asiatique*, 4ᵉ série 20:420–29.

————. 1853. *Extrait du "Fakhrî", traité d'algèbre par Aboû Bekr Mohammed ben Alhaçan Alkarkhî; précédé d'un mémoire sur l'algèbre indéterminé chez les A-rabes*. Paris: L'Imprimerie Impériale.

————. 1854. "Recherches sur l'histoire des sciences mathématiques chez les Orientaux, d'après des traités inédits arabes et persans. Premier article. Notice sur des notations algébriques employées par les Arabes." *Journal Asiatique*, 5ᵉ série 4:348–84.

————. 1861. "Traduction d'un fragment anonyme sur la formation des triangles rectangles en nombres entiers, et d'un traité sur le même sujet par Aboû Dja'far Mohammed Ben Alhoçaïn." *Atti dell'Accademia Pontificia de' Nuovi Lincei* 14 (1860–61):221–27, 241–69, 301–24, 343–56.

————. 1863. "Mémoires sur la propagation des chiffres indiens." *Journal Asiatique*, 6ᵉ série 1:27–79, 234–90, 442–529.

Wolf, A. 1952. *A History of Science, Technology, and Philosophy in the Eighteenth Century*. Second edition revised by D. McKie. London: Allen & Unwin.

Woollett, E. L. 1980. "Physics and Modern Warfare: The Awkward Silence." *American Journal of Physics* 43:104–11.

Wright, Gary A. 1978. "Social Differentiation in the Early Natufian." In Redman et al. (eds.) 1978: 201–23.

Wright, Henry T., and Gregory A. Johnson. 1975. "Population, Exchange, and Early State Formation in Southwestern Iran." *American Anthropologist* 77:267–89.

Yates, Frances. 1964. *Giordano Bruno and the Hermetic Tradition*. London: Routledge & Kegan Paul.

————. 1972. *The Rosicrucian Enlightenment*. London: Routledge & Kegan Paul.

————. 1979. *The Occult Philosophy in the Elisabethan Age*. London: Routledge & Kegan Paul.

Youschkevitch, Alexander A. "Markov." *DSB* IX, 124–30.

Youschkevitch, Adolf P. "Abū'l-Wafā'." *DSB* I, 39–43.

———. "Euler." *DSB* IV, 467–84.

Youschkevitch, Adolf P., and B. A. Rosenfeld. "Al-Khayyāmī." *DSB* VII, 323–34.

Zambelli, Paola (ed.). 1977. Albertus Magnus, *Speculum astronomiae.* (Quaderni di storia e critica della scienza, 10). Pisa: Domus Galileana.

Zilsel, Edgar. 1942. "The Sociological Roots of Science." *American Journal of Sociology* 47:544–62.

Zinner, Ernst. 1968. *Leben und Wirken des Joh. Müller von Königsberg genannt Regiomontanus.* 2., vom Verfasser verb. und erw. Aufl. (Milliaria, 12). Osnabrück: Zeller.

Zoubov, V. P. 1959. "Walter Catton, Gerard d'Odon et Nicolas Bonnet." *Physis* 1:261–78.

———. 1961. "Jean Buridan et les concepts du point au quatorzième siècle." *Mediaeval and Renaissance Studies* 5:43–95.

———. 1968. "Autour des *Quaestiones super Geometriam Euclidis* de Nicole Oresme." *Mediaeval and Renaissance Studies* 6:150–72.

Zuckerman, Solly. 1982. *Nuclear Illusion and Reality.* London: Collins.

# NAME INDEX

NB: References to notes indicate the page where the note begins, not necessarily the page where the relevant part of the note is to be found when the note runs over two consecutive pages.

# SUBJECT INDEX

NB: References to notes indicate the page where the note begins, not necessarily the page where the relevant part of the note is to be found when the note runs over two consecutive pages.

For brevity, the terms "Middle Ages" and "medieval" are used with exclusive reference to Latin Europe. Similarly, "Islam" and "Islamic" refer to the Islamic core culture of the seventh to the fifteenth century C.E.

mathematics *continued*
See also methods versus problems,
dominance by
mathematics and technology as "open
systems", 251
mathematics, general competence
among medieval scholars
fourteenth century, 143
thirteenth century, 131, 132, 139
twelfth century, 129, 132
mathematics, medieval
absence of autonomy, 20, 196
as Liberal Arts, 20
contributing to synthesis, 20
disconnected from military concerns, 231
new disciplines, 20
mathematics, Mesopotamia
and bureaucratic rationality, 77
connection of emergence to bureaucracy, 74
emergence in the Late Uruk period, 70–74
See also division problems, Fara
period; mathematics, unapplied;
metrological tables, Mesopotamia; metrology, Mesopotamia;
Old Babylonian mathematics;
rationality, mathematical versus
technical, Mesopotamia; Seleucid mathematics; tables, mathematical, Mesopotamia
mathematics of astronomy
fifteenth-century, 152, 339n. 17
Islam, 116–118
its teaching in Islam, 117
mathematics, Renaissance
and civic utility, 163
First philosophy to Humanist mathematicians, 149, 162, 207–210,
330n. 94, 331n. 103
in itself a representative of ancient
splendor, 148, 154, 157, 163, 208
Not part of the early Northern
Humanist picture of antiquity,
215, 216

symbol of higher truths according
to Cusanus, 151
transformed utility in the late Renaissance, 168
mathematics teaching, social obligations, 22, 276
mathematics, unapplied
absence from Ur III, 5, 79
disappearance after the Old Babylonian period, 84
emergence in the Fara period, 76
Old Babylonian, 7, 26, 31,
80–82
weak development in Egypt, 86
See also "pure" knowledge in
subscientific mathematics
Mathematics Research Center—United
States Army, 346n. 23, 349n. 42
*mathematikoí:* 10, 11, 284n. 36
mathematization of philosophy, fourteenth century, 21, 141, 142–144
*Measurement of the Circle*
(Archimedes), 319n. 98
Mechanical Arts, 108
in *Margaritha philosophica:* 158
medicine, Islam, 313n. 52
"melting-pot effect" in early Islamic
culture, 103, 104
*Meno* (Plato), 12
Merton College, 141
Merton School
its mathematics absent from Regiomontanus's list, 153
its project different from that of
Galileo and Newton, 141
"Merton thesis", 121
*Metalogicon* (John of Salisbury) 117
*Metaphysica* (Aristotle), 24, 92, 210,
288n. 50
methods versus problems, dominance
by, 7, 24, 26, 28–30, 83
*Metrica* (Hero), 234, 344n. 4
metrological tables, 78, 80
metrology, Mesopotamia
Early Dynastic, 75
protoliterate, 70, 71